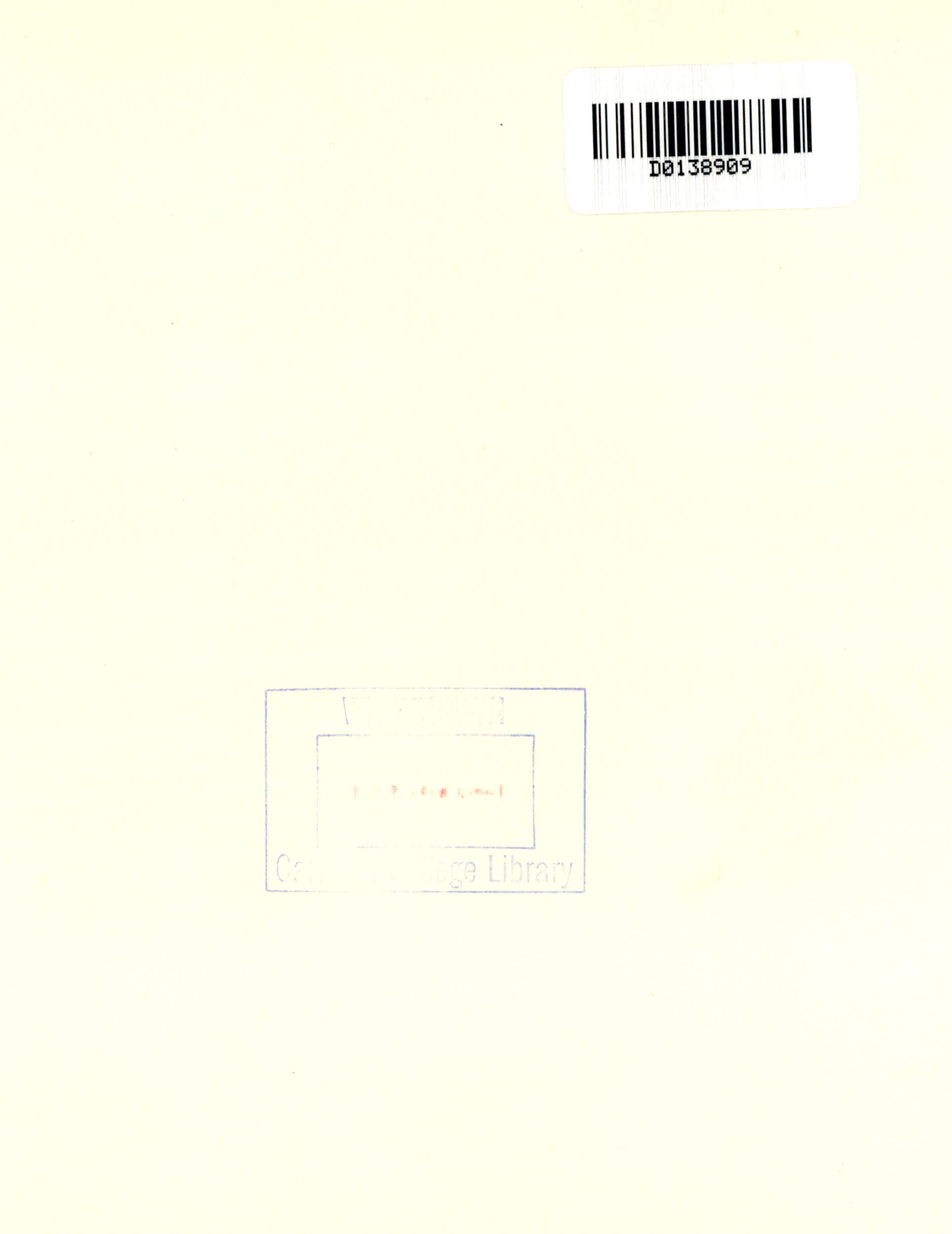
D0138909

# FOUNDATIONS OF STRUCTURAL GEOLOGY

# FOUNDATIONS OF STRUCTURAL GEOLOGY

Third edition

*R.G. Park* *PhD, FGS, CGeol*

*Professor of Tectonic Geology*
*University of Keele, UK*

**CHAPMAN & HALL**
London · Weinheim · New York · Tokyo · Melbourne · Madras

**Published by Chapman & Hall, 2–6 Boundary Row, London SE1 8HN**

Chapman & Hall, 2–6 Boundary Row, London SE1 8HN, UK

Chapman & Hall GmbH, Pappelallee 3, 69469 Weinheim, Germany

Chapman & Hall USA, 115 Fifth Avenue, New York, NY 10003, USA

Chapman & Hall Japan, ITP-Japan, Kyowa Building, 3F, 2-2-1 Hirakawacho, Chiyoda-ku, Tokyo 102, Japan

Chapman & Hall Australia, 102 Dodds Street, South Melbourne, Victoria 3205, Australia

Chapman & Hall India, R. Seshadri, 32 Second Main Road, CIT East, Madras 600 035, India

First edition 1983
Reprinted 1983, 1986
Second edition 1989
Reprinted 1993, 1994, 1995
Third Edition 1997



Typeset in 10/12 Palatino by AFS Image Setters, Glasgow
Printed in Great Britain by The Alden Press, Osney Mead, Oxford

ISBN 0 412 64400 2



A catalogue record for this book is available from the British Library

Library of Congress Catalog Card Number: 97-65877

∞ Printed on acid-free text paper, manufactured in accordance with ANSI/NISO Z39.48-1992 and ANSI/NISO Z39.48-1984 (Permanence of Paper).

# CONTENTS

# PREFACE

In the Preface to the first edition of this book, published in 1983, I explained my reasons for writing the book as follows.

'There are already a number of excellent books covering the various aspects of Structural Geology. Among these are works by Hobbs, Means and Williams, Jaeger and Cook, Price, Ramsay, and Turner and Weiss, all of which I have used extensively in preparing this book and have listed therein as further reading. However, these textbooks are rather advanced for many students commencing the study of geology, and for many years I have been aware of the lack of a suitable elementary book which I could recommend to beginners. My purpose in writing this book, therefore, was to supplement existing textbooks by providing an introduction to the subject which will convey enough information over the whole field of structural geology to stimulate the reader's interest and encourage further study of more advanced textbooks and scientific papers.'

In the intervening 14 years since these words were written, many other textbooks on Structural Geology have been published, and the student is now well served by a variety of excellent books, several of which are referred to in this text. Nevertheless, the demand for a short, inexpensive and reasonably comprehensive elementary textbook has continued to be just as great, which is my justification for producing this third edition.

In this revision I have undertaken a thorough review of all the material, making a large number of corrections and additions to the text that have become necessary as a result of new ideas or approaches over the past eight years, or to correct mistakes uncorrected in the second edition. I have also made numerous corrections and improvements to the illustrations, many of which have been replaced or redrawn, and a number of new ones have been added. The format has been changed to improve the visual attractiveness of the book. Important terms and concepts have been set in **bold** where first defined, and the appropriate page number has been set in **bold** in the index, in order to make it easier for students to find definitions.

In addition, I have taken the opportunity to make some changes to the organization of the book by modifying the somewhat artificial division recognized in the earlier editions between morphology/classification and deformation mechanisms. For example, the purely descriptive or factual aspects of fault and fold structure in the earlier chapters have now been combined with a simple treatment of mechanisms, leaving the more geometrically complex treatment until after the relevant sections on stress and strain. The balance between the more 'traditional' subjects of strain geometry and folding on the one hand and faulting on the other has also been changed to reflect changing preoccupations in recent years, and some subjects are introduced for the first time, e.g. inversion and orogen collapse.

Several chapters have been extensively modified; in particular, chapter 12, on gravity-controlled structures, by emphasizing modern work on salt tectonics; chapter 15, on geological structure and plate tectonics, by expanding the treatment of modern tectonic regimes to show more clearly how the various types of geological structure fit into their plate tectonic context; and a new chapter, 16, has been added on structural interpretation in ancient orogenic belts, by making more detailed reference to the Caledonian orogenic belt of the British Isles, and by completely revising the section on Precambrian orogeny.

It is proposed to issue a companion volume in which the basic geometrical concepts of Structural Geology will be further explained, and which will include a series of simple maps and exercises designed to enable the reader to understand the use of strike lines and stratum contours, and to solve simple geometric problems involving folds, faults, unconformities, igneous intrusions and strain analysis. Particular emphasis will be placed on (1) interpreting structure from geological maps, (2) restoring and balancing sections, and (3) the use of stereographic projection.

In making these changes, I have incorporated many helpful suggestions from colleagues and reviewers, and wish to thank all of them for their help in improving the book. I would also like to reiterate my indebtedness to Paula Haselock, Nick Kusznir and Rob Strachan (all at that time at Keele), and to two anonymous reviewers who read the draft of the first edition and made many useful suggestions for its improvement. I am especially grateful to Bob Standley, then of the City of London Polytechnic, for his meticulous checking of the original manuscript and for a host of valuable suggestions. Many of the original diagrams were drawn by Paula Haselock, whose willing and cheerful help made the task of writing the book much easier.

Finally, I wish to make it clear that I have reluctantly ignored several pieces of good advice in relation to all three editions, usually because of my overriding desire to make the book as short as possible, and that any remaining deficiencies are entirely my own responsibility.

RGP

# INTRODUCTION

## MEANING AND SCOPE OF STRUCTURAL GEOLOGY

It is easier to give examples of geological structures than to define them. The word 'structure' means 'that which is built or constructed'. Structural geologists use the word to signify something that has been produced by deformation; that is, by the action of forces on and within the Earth's crust. Structures consist of a geometric arrangement – of planes, lines, surfaces, rock bodies, etc. The form and orientation of this arrangement reflect the interaction between the deforming forces and the pre-existing rock body.

## GEOLOGICAL STRUCTURES AND DEFORMATION

Because the geometry of structures is so important, a large body of descriptive nomenclature and classification has grown up, which is essential to master if we wish to describe and understand structures. The arrangement of this book reflects my belief that there is little point in discussing such matters as stress, strain and processes of deformation before learning what it is that we wish to explain by such processes. Thus I have discussed the more descriptive aspects of structural geology (morphology) first, then proceeded to introduce the rather complex concepts of stress and strain, after which the more theoretical aspects of deformation mechanisms can be dealt with.

Deformation is the process that changes the shape or form of a rock body – in other words, the process responsible for the formation of geological structures. To understand deformation, it is necessary to understand stress and strain, which deal with the manner in which material reacts to a set of forces. We must also discuss the behaviour of materials, since the way that a rock deforms is dependent on the physical properties of different rocks and on how these change with changes in temperature and pressure, and with time. We can then apply the principles of deformation to the formation of specific types of structures such as faults and folds.

## GEOTECTONICS

In the second part of the book I have attempted to show how structures and deformation may be related to large-scale Earth processes. The subject of geotectonics essentially covers large-scale structural geology – that is, the study of large Earth structures such as mountain belts and continental margins. The value of the plate tectonic theory lies in its ability to explain many types of hitherto unrelated geological phenomena in terms of a unifying theory of crustal movements and processes. It is essential for the structural geologist to see individual structures or deformed areas in their context and to try to relate them to some large-scale pattern, even if the attempt subsequently proves to have been a failure. Only in this way will our understanding of the Earth progress.

## SEDIMENTARY STRUCTURES

Structures produced as a result of processes associated with sedimentation are not of great concern to most structural geologists, who are more interested in the deformation of solid rocks. Such structures are not covered in this book and are adequately dealt with in other textbooks. However, there are areas of overlap between sedimentary and structural geology that should be mentioned here. One important problem for the field geologist is distinguishing between sedimentary and deformational structures. This problem is particularly acute in highly deformed metamorphic terrains where the origin of early and poorly preserved structures is often unclear. Fold-type structures produced by soft-sediment slumping and other processes are open to misinterpretation. When fold-type structures are confined to a single layer (particularly if they are truncated by the layer above) they are likely to be of sedimentary origin and must be treated with caution.

Many sedimentary structures are, of course, of indirect interest to the structural geologist since they reflect tectonic control. Thus features indicating slumping or sliding of soft or unconsolidated sediments are often a direct result of tectonic processes such as earthquakes, fault movements, etc. Certain sedimentary structures are also of value to the structural geologist as indicators of the younging direction of the strata. Cross-bedding, graded bedding and other 'way-up' structures have been used extensively in highly deformed terrains to elucidate the large-scale structure. It is therefore important for students of Structural Geology to acquire at least a basic knowledge of sedimentology.

## MAP INTERPRETATION

The interpretation of geological structures from maps and aerial photographs is another important topic which is essential to the three-dimensional appreciation of structural geometry and will be covered in a companion volume to this book. There are also a number of excellent existing text-books on geological maps and air photo interpretation that the reader may consult. Certain basic stratigraphic and geometric concepts relating to map interpretation of structure are outlined in Chapter 1.

## STEREOGRAPHIC PROJECTION

This is an essential geometric tool in structural geology, and is briefly summarized in the Appendix.

## WARNING TO STUDENTS!

It is easy for a student to be misled into regarding what is printed in a textbook as unquestioned and immutable truth. However, in geology, perhaps more than in other sciences, today's 'facts' may become tomorrow's discarded theories. Much of the material of this book is based on the opinion of established experts based on sound evidence. Some of it, however, is disputed. Be sceptical!

Many textbooks attribute all statements to their original author by reference to the relevant published work, thus establishing an evolving body of 'evidence' built up by numerous individual scientists, sometimes with opposing ideas. This is of course the correct scientific procedure. However, I have chosen in this short book not to follow this procedure because I feel that large numbers of references break up the smooth flow of the text and make for less easy comprehension. Instead, I have listed selected references for further reading at the end of each chapter in the hope that the reader will be encouraged to sample some of the original contributions to the subject and to proceed to more advanced texts.

PART ONE

# GEOLOGICAL STRUCTURES AND DEFORMATION

# BASIC CONCEPTS 1

Before commencing the study of geological structures, it is important to acquire a basic understanding of the geometry of undeformed sedimentary sequences, the geometry of inclined planes and lines, and how the three-dimensional geometry of a deformed area may be portrayed and reconstructed by means of maps and cross-sections. The following account is only a brief summary of the more important aspects, and students of Structural Geology are advised that practice in map interpretation is an essential aid to understanding the subject.

## 1.1 STRATIGRAPHIC TERMS AND CONCEPTS

### BEDDING OR STRATIFICATION: DEFINITIONS AND GEOMETRY

A **bed** is a layer of rock deposited at the surface of the Earth. It is bounded above and below by distinct surfaces (**bedding planes**) which usually mark a break in the continuity of sedimentation caused by a cessation of sedimentation, or a period of erosion, or a change in the type or source of the sediment. Beds are normally sedimentary, but may also consist of volcanogenic material. A thickness in the range from centimetres to metres is usually implied. 'Bed' is synonymous with **stratum**, but the latter term is almost invariably used in the plural (e.g. 'Silurian strata'). Beds may be relatively homogeneous in composition and internal structure, and represent more or less continuous deposition. However, the term is also used for a sedimentary unit composed of numerous thin distinct layers. The term **bedded** means composed of beds: thus 'bedded rocks', 'thin-bedded', 'cross-bedded' etc.; **bedding** is used as a collective noun for the beds in a particular outcrop or area: thus, 'the bedding dips to the west'. The term is also used to describe various characteristics of the beds, such as 'cross-bedding' and 'graded bedding'.

The simplest type of bedding geometry consists of a set of parallel planes, representing a group of beds, or a **formation**, of uniform thickness. However, in practice, beds and formations vary laterally in thickness, in which case the geometry of the formation must be described by two non-parallel bounding surfaces. Thickness variation in such a formation may be described by a set of **isopachytes** (see below).

### UNCONFORMITIES AND ALLIED STRUCTURES

Breaks in the stratigraphic record, representing intervals of geological time not marked by

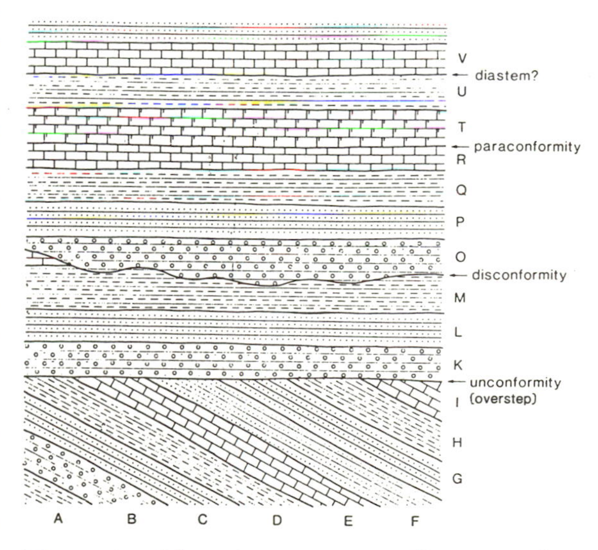


**Figure 1.1** Schematic cross-section illustrating the various types of stratigraphic break. (From Roberts, 1982.)

sediment deposition, are known variously as diastems, non-sequences, paraconformities, disconformities and unconformities (Figure 1.1). **Diastems** represent pauses in sedimentation, marked by abrupt changes in sediment type, producing surfaces of discontinuity (bedding planes) but no other evidence of a time gap. **Non-sequences** (or **paraconformities**) are similar to diastems but exhibit faunal or other evidence of a time gap. **Disconformities** are marked by evidence of erosion during the sedimentary break, but the bedding below the erosion surface is parallel to that above, i.e. there has been no deformation of the lower series of beds prior to erosion.

**Unconformities** are distinguished from other stratigraphic breaks by angular discordance between the older beds below the unconformity surface and the younger beds above. Hence an unconformity represents the following sequence of events:

1. deposition of lower strata;
2. tilting or other deformation of lower strata;
3. erosion;
4. deposition of upper strata.

The structure produced by the discordance of younger upon older strata is termed **overstep**, and the basal beds of the younger series are said to

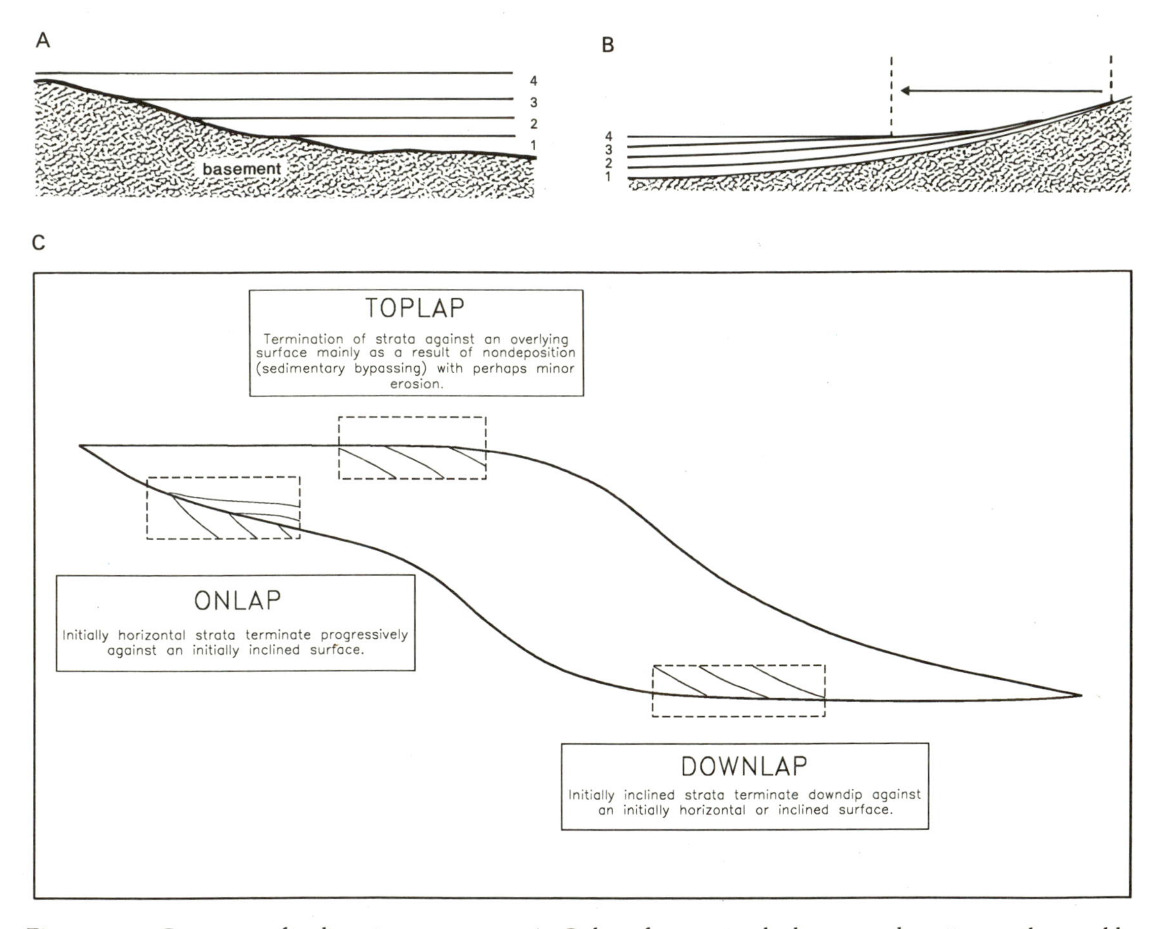


**Figure 1.2** Geometry of sedimentary sequences. A. Onlap of successive beds 1–4, each resting partly on older basement, illustrating transgression. B. Offlap of successive beds 1–4, associated with regression. C. Relationship between onlap, toplap and downlap in a transgressive sequence.

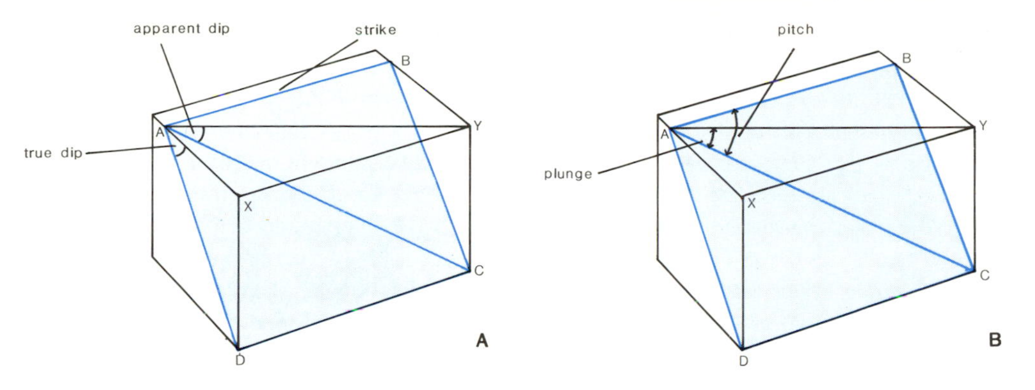


**Figure 1.3** Inclined planes and lines. A. Strike and dip of an inclined plane. The true dip of the plane ABCD is the angle XAD; the angle YAC is an apparent dip. B. Plunge and pitch of an inclined line. The plunge of the line AC is the angle YAC; the pitch is the angle BAC.

'overstep' the various strata of the older series truncated by the erosion surface (Figure 1.1). A **non-conformity** is a type of unconformity where younger strata rest on an erosion surface cut across non-bedded igneous rocks.

GEOMETRY OF SEDIMENTARY SEQUENCES

**Onlap** is the term used to describe a structure formed where successive wedge-shaped beds extend further than the margin of the underlying bed, such that they lie partly on older basement (Figure 1.2A). Such a structure is typical of sedimentary sequences in expanding basins, where the shorelines migrate towards the centre of the landmass, thus decreasing its surface area. This process is called **transgression**. The term **overlap** is synonymous with onlap. **Offlap** is the structure formed where successive wedge-shaped beds do not extend to the margin of the underlying bed but terminate within it (Figure 1.2B). Such a structure is typical of sedimentary sequences in contracting basins, where the shorelines migrate towards the centre of the basin, thus enlarging the land surface. This process is termed **regression** Other terms used to describe the geometry of sedimentary sequences are **toplap** and **downlap** The relationship between onlap, toplap and downlap in a transgressive sequence is illustrated in Figure 1.2C.

## 1.2 GEOMETRY OF INCLINED PLANES AND LINES

The attitude of inclined planes, such as bedding, foliation, faults, etc., is conventionally described in terms of the 'strike' and 'dip' of the plane (Figure 1.3A). The **strike** is the unique direction of a horizontal straight line on the inclined plane and is recorded as a compass bearing (azimuth). The **dip** is the inclination or tilt of a planar surface, e.g. bedding or foliation, measured from the horizontal. The **true dip** of a plane is measured in a vertical plane perpendicular to the strike, and is the maximum angle from the horizontal that can be measured for a given plane. Lines in any other orientation on the plane are at a smaller inclination to the horizontal; such angles are termed **apparent dips**. The apparent dip is thus the angle of inclination of a given plane with the horizontal, measured in a plane that is not orthogonal to the strike. The angle of apparent dip measured in a series of vertical planes varies from zero (parallel to the strike) to a maximum in the direction of true dip. If the angle of apparent dip in two different directions is measured, the true dip can be calculated using a stereogram (see Appendix).

The direction of dip (i.e. the direction in which the plane dips downwards from the surface) is measured either directly as a compass bearing (azimuth) or in relation to the strike direction, which is 90° from the dip direction. Thus a bed may be said to dip at 30° SE, if the strike direction is specified, or at 30° to 110° if it is not. The conventional representation of strike and dip on a geological map is by a line parallel to the strike, with a short tick indicating the dip direction (Figure 1.5A). On older maps, this symbol may be replaced by an arrow parallel to the dip direction with the amount of dip in degrees placed alongside.

The orientation of a linear structure (e.g. a fold axis) is measured in terms of **plunge** or **pitch**. The **plunge** is the angle between the line and the horizontal in the vertical plane. The plunge is given as an angle and a bearing (azimuth), which is the direction of plunge, thus, 30° to 045° or 30° NE. The **pitch** is the orientation of a line, measured as an angle from the horizontal in a specified non-vertical plane. A measurement of pitch must give the strike and dip of the plane of measurement, plus the angle of pitch and the strike direction from which the pitch angle is measured (since there are two possible directions in a given plane for the same pitch angle) (Figure 1.3B). This method is useful in the field where precise measurements of angles within inclined joint, foliation or bedding planes are more convenient than direct measurement of the plunge. The plunge may be easily derived using a stereogram (see Appendix). The instrument used in the field to measure the inclination (dip) of a planar surface or the plunge of a lineation is termed a **clinometer** and is often combined with a compass in order to measure the orientation of planes or lines with reference to geographic coordinates.

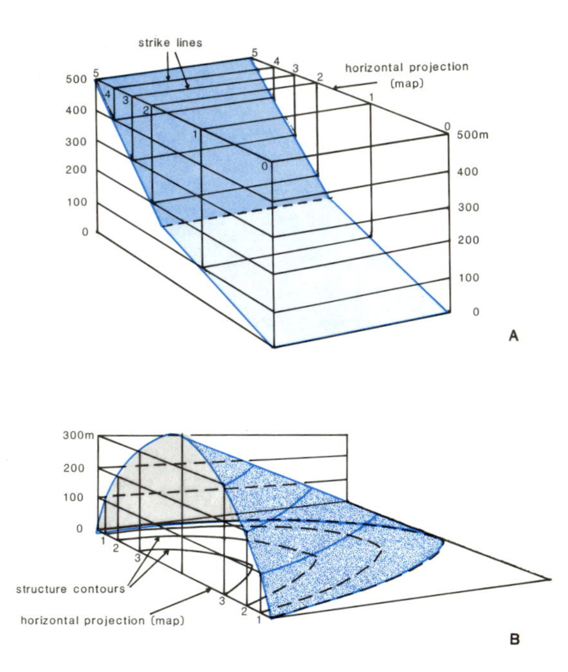


**Figure 1.4** Strike lines and structure (stratum) contours. A. Strike lines at heights of 0 to 500 m on the inclined plane (coloured) project as a set of parallel strike lines (labelled 0–0, 1–1, etc.) on the horizontal plane (i.e the map); the spacing is closer on the steeper part than on the shallower part of the inclined plane. B. An inclined cylinder (plunging fold) intersects a set of horizontal planes at heights of 0–500 m in curved lines termed structure contours or stratum contours, which project on the horizontal plane (map), as shown, to give a map representation of the shape of the structure.

## 1.3 REPRESENTATION OF STRUCTURES ON GEOLOGICAL MAPS

On geological maps, the attitude of planar beds, etc. may be recorded by a set of strike lines drawn parallel to the strike of the plane or set of planes in question. If the dip of the planes is constant, the strike lines are straight, with uniform spacing. An increase in dip produces a decrease in spacing and a decrease in dip produces an increase in spacing (Figure 1.4A). A surface with variable strike is represented by curved strike lines known as **structure contours** (Figure 1.4B; see also Figure 15.9). Structure contours follow a constant height on a geological surface, and a set of such contours, drawn at uniform height intervals, represents the three-dimensional shape of the surface in the same way that topographic contours represent the height variation of the land surface on a topographic map.

A less precise, but often more convenient, method of portraying the structure on a map is to use **form lines**. These are lines drawn on a map to indicate the general direction of the strike of a folded surface (Figure 3.17B–D). A tick on the line indicates the dip. A set of form lines will illustrate the geometry of the folding in a similar way to the outcrop pattern of the strata, for example, but are more precise, and are not affected by topography. They can therefore be used in areas where individual formations have not been mapped. A set of **form line contours** can be drawn in a precise manner such that the spacing is proportional to the dip (see Ragan, 1973). A contoured map constructed by this means will illustrate the shape of a folded surface in the same way as structure contours. A **form surface** is any planar surface that intersects the ground surface as form lines and which may be used for structural mapping.

An **isopachyte** is a line joining points of equal stratigraphic thickness of a formation or group of strata. An isopachyte map is contoured to indicate the three-dimensional shape of a unit of variable thickness. The technique is used, for example, in the study of sedimentary basins and in portraying the geometry of stratigraphic units cut off by unconformities. Isopachyte maps may be prepared from borehole data from which thicknesses are directly obtainable, or by geometric construction using stratum contours for the top and base of the unit, and subtracting the lower from the higher values where they intersect.

The line of intersection of a stratigraphic boundary with a higher stratigraphic boundary such as an unconformity is marked by the zero isopachyte of the rock body between the two boundaries in question. This line is often termed the **feather edge**, e.g. 'the feather edge of the base of the Coal Measures on the base of the Triassic'.

A related term is **subcrop**, which is the 'subsurface outcrop' of a rock unit. A stratigraphic formation may intersect a subsurface plane, e.g. an unconformity or fault, in a subcrop, which represents the area of the plane lying between the lines of intersection (feather edges) of the boundaries of the formation.

## TOPOGRAPHIC EFFECTS

In areas of horizontal or gently-dipping strata, outcrop patterns are controlled mainly by the topographic relief. Younger beds occur at higher topographic levels and older beds at lower levels. Outcrops of younger rocks completely surrounded by older rocks are termed **outliers**, and correspond to hills separated by erosion from other outcrops of the same beds (Figure 1.5). Conversely, an **inlier** is an area of older rocks surrounded by younger rocks, e.g. in a valley cut through younger strata.

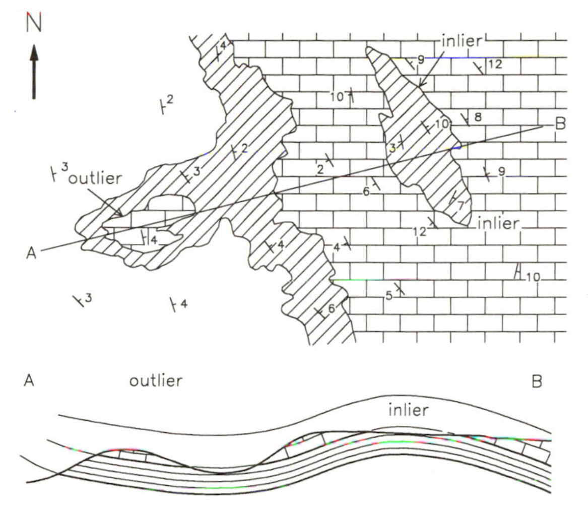


**Figure 1.5** Outlier and inlier produced by the intersection of gently folded strata with topographic relief: the hill produces an outcrop of younger rock surrounded by older rock (outlier), and the valley an outcrop of older rock surrounded by younger rock (inlier). A, map; B, vertical cross-section.

## CROSS-SECTIONS

It is usually necessary to supplement the two-dimensional information on the geological structure of an area provided by the geological map by one or more **cross-sections**, which are diagrammatic representations (normally constructed in the vertical plane) of the geology of an area. Reasonable assumptions must be made about the way in which structures visible at the surface continue downwards, and the surface information may be supplemented by data from boreholes, wells, etc. Cross-sections may be drawn along a particular line or lines on the map, chosen to illustrate the vertical structure most effectively. The combination of map and cross-section should ideally give a good three-dimensional picture of the geological structure of an area (e.g. Figure 1.5B is a vertical cross-section of the map represented in Figure 1.5A). In complex areas, several lines of section may be used to give a better coverage of the structural variation.

Instead of a vertical section, a **down-plunge projection** may be employed; this is a reconstructed profile or cross-section of a fold structure drawn perpendicular to the plunge of the fold axis. This is done to give a more accurate representation of the fold geometry (see section 3.5).

It is important in interpreting the history of an area to be able to visualize the original geometry of a set of rocks before deformation. A geometrical reconstruction, in the form of a map or cross-section, is often employed for this purpose, and is termed a **palinspastic** reconstruction. **Balanced sections** (see section 2.6) are a special type of palinspastic reconstruction much used in the interpretation of complex fold/fault belts.

## FURTHER READING

Maltman, A. (1990) *Geological Maps: an Introduction*, Open University Press, Milton Keynes.

Ragan, D.M. (1973) *Structural Geology: an Introduction to Geometrical Techniques*, 2nd ed, Wiley, New York.

Roberts, J.L. (1982) *Introduction to Geological Maps and Structures*, Pergamon, Oxford.

# FAULTS AND FRACTURES 2

## 2.1 ROCK FRACTURES

A **fracture** is the commonest type of geological structure, and may be seen in any rock exposure. Fractures are cracks across which the cohesion of the material is lost, and may be regarded as planes or surfaces of discontinuity. Where there is a measurable displacement across the fracture plane, that is, where the rock on one side has moved along the fracture relative to the other side, the fracture is termed a 'fault'. A **fault** may thus be defined as a planar fracture across which the rock has been displaced in a direction that is generally parallel to the fracture plane.

Where there is no displacement, or where the displacement is too small to be easily visible, the fracture is termed a **joint**. The distinction between the two is somewhat artificial, and depends on the scale of observation; however, in practice, the great majority of fractures show negligible displacement and are classified as joints.

Fractures are important in a number of ways. Their presence significantly affects the strength of a rock, and they must be carefully studied in civil engineering operations such as those involved in the construction of tunnels and dams. They are also important sites of mineralization, since dilational fractures developed under extension are normally occupied by vein material such as quartz or calcite deposited from aqueous solution in the space created as the fracture opens. Such veins are a valuable source of ore minerals. From a structural point of view, veins are useful in indicating that fractures are dilational, i.e. that the wall rocks have been moved aside to allow the vein material to form (Figure 10.22B).

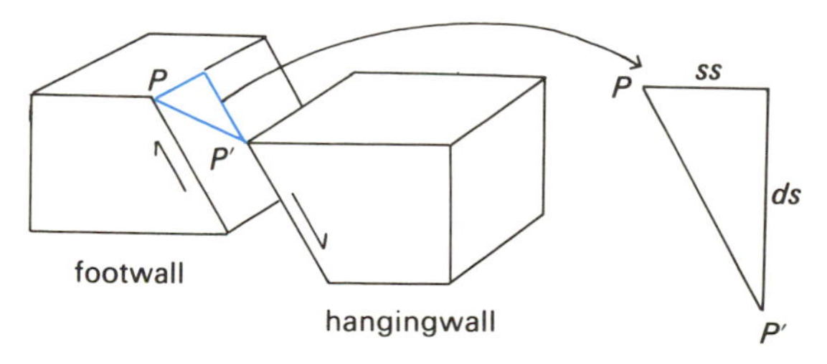


**Figure 2.1** Components of fault displacement: *ss*, strike-slip component; *ds*, dip-slip component; *PP'*, true displacement vector.

## 2.2 FAULT GEOMETRY AND NOMENCLATURE

### GEOMETRY OF DISPLACEMENT

The main elements of the displacement geometry of a fault are shown in Figure 2.1. Where the fault plane is non-vertical, the block above the fault is referred to as the **hangingwall** and the block below the fault as the **footwall**. The inclination of a fault plane may be given as a **dip**, in the same way as bedding (Figure 1.3A), but is sometimes measured as the angle between the fault plane and the vertical, in which case it is termed the **hade**.

The displacement of the fault plane between the two blocks may take any direction within the fault plane. Faults with a displacement parallel to the strike of the fault plane are termed **strike-slip faults** and those with a displacement parallel to the dip of the fault plane are termed **dip-slip faults**. Faults with oblique-slip displacements are regarded as having strike-slip and dip-slip components, as shown in Figure 2.1. Strike-slip faults may also be called **wrench**, **tear** or **transcurrent faults**.

The measurement of the displacement on dip-slip faults is often made with reference to the horizontal and vertical components of the displacement, which

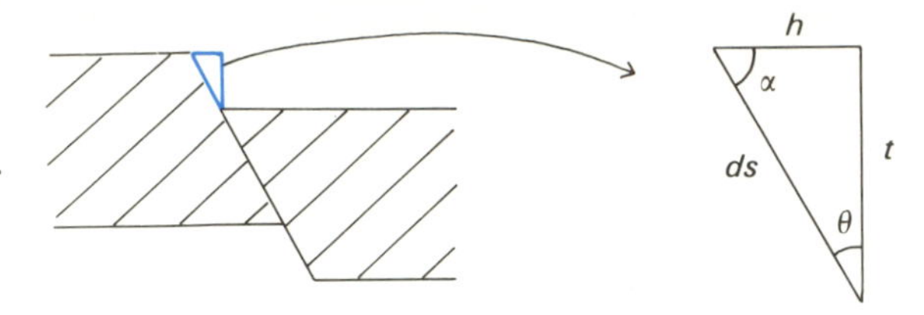


**Figure 2.2** Geometry of dip-slip fault displacement: $h$, heave; $t$, throw; $\alpha$, angle of dip; $\theta$, angle of hade ($= 90° -$ dip angle); $ds$, dip-slip component of displacement.

$$\tan\alpha = \frac{\text{throw}}{\text{heave}} = \frac{t}{h}$$

and

$$\sin\alpha = \frac{\text{throw}}{\text{true displacement}} = \frac{t}{ds}$$

where $\alpha$ is the dip of the fault.

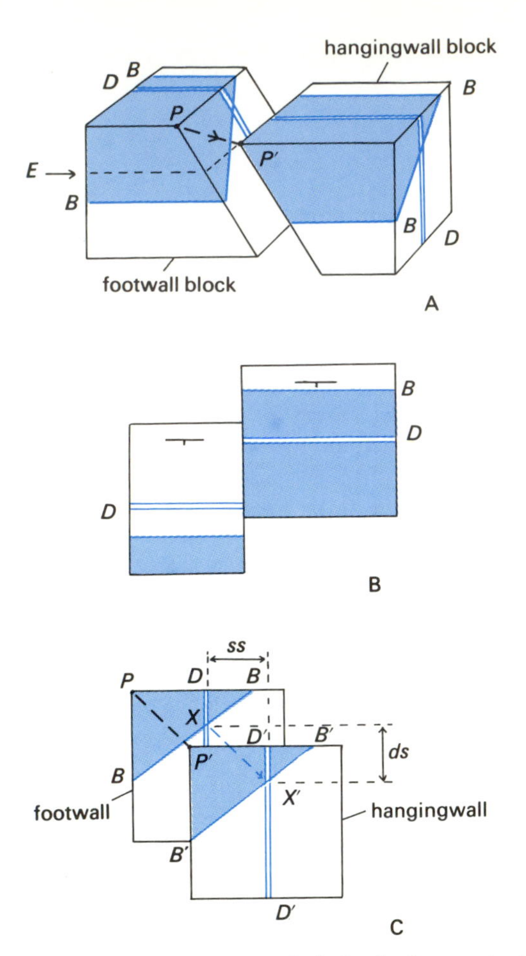


**Figure 2.3** Measurement of fault displacement. A. The fault affects dipping bedding BB and vertical dyke DD. The true displacement vector is $PP'$. B. Map at erosion level E of A shows horizontal displacement of bedding and dyke along fault. Note that the amount of displacement appears to be different. The true horizontal displacement is shown by the vertical dyke DD. C. View of fault plane looking down from the right, showing the trace of bedding BB and dyke DD on the footwall, displaced to positions B′B′ and D′D′ on the hangingwall. The true displacement $PP'$ is given by the movement of intersection X of BB and DD to position X′. The strike-slip component $ss$ is given by the displacement of the vertical dyke DD to D′D′. The dip-slip component $ds$ must be measured using both dyke and bedding displacements.

are termed respectively the **heave** and the **throw** (Figure 2.2). It is the throw, or vertical displacement, that is normally quoted for a dip-slip fault rather than the true displacement. The relationship between these elements is shown in Figure 2.2.

It is important to realize that fault displacements are difficult to measure in practice because it is frequently impossible to match precise points on each side of the fault. If bedding is displaced, we cannot be certain how much of the apparent displacement is due to dip-slip and how much to strike-slip movement (Figure 2.3A, B). The problem is overcome if the direction of movement on the fault plane is indicated by movement striations (see section 2.4) or if there is a measurable offset on a vertical structure, such as a dyke, which can be used to measure the strike-slip component (Figure 2.3C).

SENSE OF DISPLACEMENT

The sense of relative displacement on faults is important and depends upon the orientation of the fault with respect to the direction of compression or extension within the rock (see section 9.2). In the case of dip-slip faults, the displacement is termed **normal** when the hangingwall moves down and **reverse** when the hangingwall moves up, relative to the footwall (Figure 2.4A.).

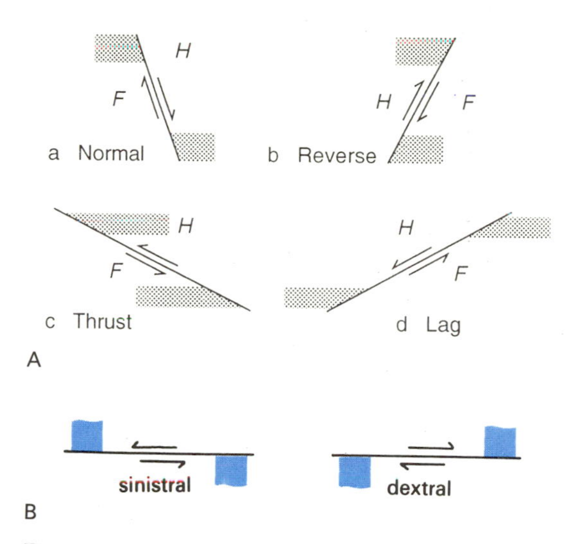


**Figure 2.4** A. Normal and reverse displacements on dip-slip faults (vertical sections). F, footwall; H, hangingwall; a, normal fault; b, reverse fault; c, thrust; d, lag (low-angle normal fault). B. Sinistral and dextral displacements on strike-slip faults (plan view).

An alternative way of expressing the displacement in dip-slip faults is to refer to the direction of throw. The direction of dip of normal faults is towards the downthrown side, whereas in reverse faults the dip is directed towards the upthrown side. The sense of displacement in a reverse fault results in lower (normally older) rocks being placed above higher (normally younger) rocks, whereas the opposite is true in normal faults.

Dip-slip faults dipping at less than 45°, i.e. low-angle faults, are distinguished from high-angle dip-slip faults. If the sense of movement is reverse they are termed **thrusts**, and if the sense of movement is normal they are termed **low-angle normal faults**, or sometimes, **lags**. Thrusts are particularly important in orogenic belts and often have displacements of many tens of kilometres. The Moine thrust, which marks the northwestern margin of the Caledonian orogenic belt in northwestern Scotland (Figure 16.2), has an estimated displacement of about 100 km.

In the case of strike-slip faults, the displacement is termed **sinistral** (or left-lateral) if the opposite block moves to the left, and **dextral** (or right-lateral) if the opposite block moves to the right, as viewed by an observer standing on one side of the fault (Figure 2.4B).

## 2.3 ROCKS PRODUCED BY FAULTING (FAULT ROCKS)

The nomenclature and classification of fault rocks is summarized in Table 2.1.

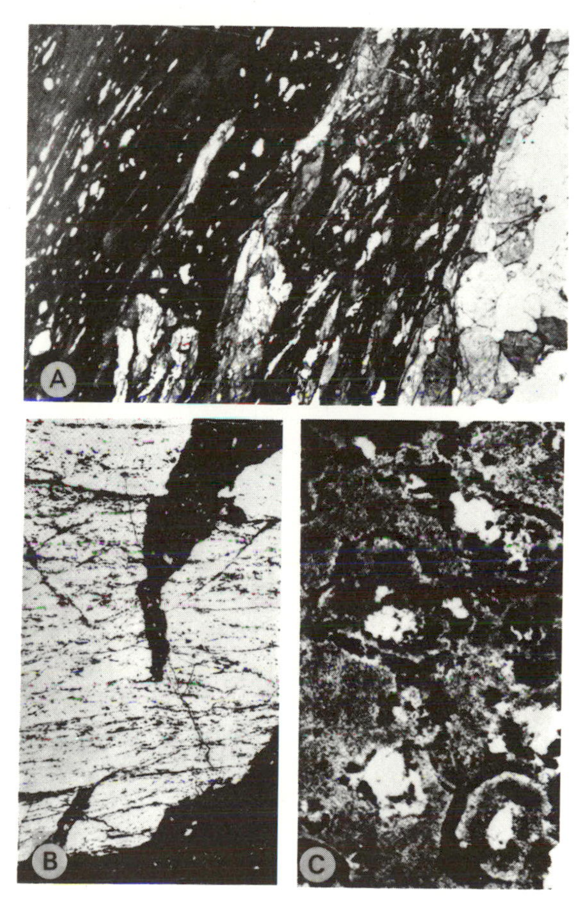


**Figure 2.5** Crush-rocks produced by faulting. A. Gradational change from unaltered rock on the right, through mylonite, to ultramylonite on the left in amphibolite from Ness, Lewis. Plane polars, × 14. (From Sibson, 1977, plate 3.) B. Discordant intrusive veins of dark pseudotachylite cutting foliated gneiss (from Gairloch, NW Scotland). Plane polars, × 5. C. Devitrified spherulitic structure in pseudotachylite from vein cutting Lewisian gneiss at Gairloch, NW Scotland. Plane polars, × 250. (B and C from Park, R.G. (1961) *American Journal of Science*, **259**, 542–50, plate 1.)

**Table 2.1.** Classification of fault rocks. (From Sibson, 1977.)

| | Nature of matrix | Random-fabric | | Foliated | | Proportion of matrix |
|---|---|---|---|---|---|---|
| **Incohesive** | | Fault breccia (visible fragments > 30% of rock mass) | | ? | | |
| | | Fault gouge (visible fragments < 30% or rock mass) | | ? | | |
| **Cohesive** | Glass/devitrified glass | Pseudotachylite | | ? | | |
| | Tectonic reduction in grain size dominates grain growth by recrystallization and neomineralization | Crush breccia | | fragments > 0.5 cm) | | 90–100% |
| | | Fine crush breccia | | (0.1 cm < frags. < 0.5 cm) | | 90–100% |
| | | Crush microbreccia | | (fragments < 0.1 cm) | | 90–100% |
| | | Protocataclasite | Cataclasite Series | Protomylonite | Mylonite series | 50–90% |
| | | Cataclasite | Cataclasite Series | Mylonite (*Phyllonite varieties*) | Mylonite series | 10–50% |
| | | Ultracataclasite | Cataclasite Series | Ultramylonite (*Phyllonite varieties*) | Mylonite series | 0–10% |
| | Grain growth pronounced | ? | | Blastomylonite | | |

FAULT BRECCIA AND GOUGE

Many faults are marked by a zone of broken and crushed rock fragments of varying size. This material is called **fault breccia** where the visible fragments make up an appreciable proportion of the rock. Where the bulk of the rock consists of fine powder, the material is termed **fault gouge**. Since such zones are normally softer and more easily eroded than the unfaulted rock, they give rise to the marked topographic depressions often associated with fault outcrops.

COHESIVE CRUSH ROCKS

Fault breccia and gouge are essentially loose-textured fault rocks found near the surface. At greater depth, various kinds of cohesive crush rocks are found where the rock is lithified and the increased pressure has, in many cases, caused partial recrystallization of the rock texture. Such rocks are termed **crush breccias** where visible fragments dominate the rocks, and **cataclasites** where the fine-grained matrix makes up an appreciable proportion of the rock.

### MYLONITES

Structural geologists generally attempt to distinguish rocks formed under 'brittle' conditions by breaking and crushing of the material (**cataclasis**) from those formed under 'ductile' conditions by continuous recrystallization or flow (see section 7.3 for definitions of brittle and ductile). Finer-grained rocks produced by the latter process are hard and 'flinty', with a platey or streaky texture, and are termed **mylonites** (Figure 2.5A). Where recrystallization is dominant, the rocks are termed **blastomylonites**.

### ULTRAMYLONITE AND PSEUDO-TACHYLITE

Extreme crushing produces a rock composed of broken fragments in a dark, often black, matrix of ultramicroscopic grains. Such material is termed **ultramylonite** (Figure 2.5A). Frictional heating caused by rapid relative movement along the fault may be sufficient to melt some of this material, forming a glassy substance, often containing spherulites, termed **pseudotachylite**, which forms veins intruding into the adjacent fractured rock (Figure 2.5B, C). Since the glassy material is usually devitrified, and contains a high proportion of unmelted fragments, it is often difficult to distinguish from ultramylonite except under high magnification. Pseudotachylite is apparently formed only at depth in the crust under moderate load pressure and a relatively rapid deformation rate. Thus a fault exhibiting soft gouge at the surface might develop pseudotachylite at intermediate depths, and at deeper levels might be replaced by a mylonite.

## 2.4 FEATURES ASSOCIATED WITH FAULT PLANES

### SLICKENSIDES

Fault planes frequently show shiny or striated surfaces caused by the rubbing or polishing action of the opposite face as it moved across. Such features are termed **slickensides**. The grooves or striations indicating the direction of relative movement of the fault are called **slickenside striations** or **slickenlines** (Figure 2.6A). **Slickenfibres** are linear structures which are formed by the growth of fibrous minerals such as quartz or calcite and have their long axes parallel to the direction of motion (Figure 2.6B). These may be used to determine the sense of movement along a fault plane, as shown in Figure 2.6C.

### FLEXURES ASSOCIATED WITH FAULTS

Layered rocks adjoining a fault commonly exhibit open folds or flexures which appear to be related

**Figure 2.6** A. Hand specimen showing slickenslide striations on a fault surface. Scale bar, 6 cm.

B

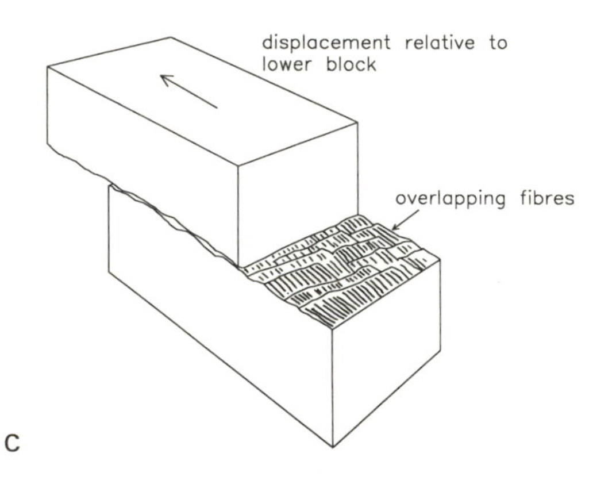


C

**Figure 2.6** cont. B. Calcite growth fibre steps on a fault surface in limestone, Wildhorn nappe, Switzerland. (From Ramsay and Huber, 1983. *The Techniques of Modern Structural Geology*, **1**, Academic Press, London, fig. 13.33.) The observed fault wall has moved upwards relative to the opposite wall, now removed. C. Diagram showing how growth fibre steps form as a result of fault movement.

to the fault movement. The term **normal drag** is used to describe a fold that bends in the direction of fault movement, and **reverse drag** to describe a fold that bends in the opposite sense. Some normal drag folds have originated by the dragging action of the opposing fault block, while others

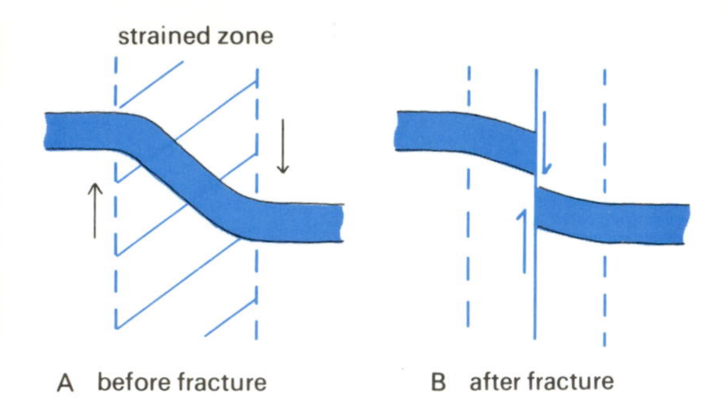


**Figure 2.7** Association of faults and flexures. Normal drag: ductile bending in a strained zone (A) may precede fault movement (B), causing flexure in the rocks adjacent to the fault.

result from an initial ductile strain that precedes fracturing (Figure 2.7). Reverse drag folds in the hangingwall of normal dip-slip faults are referred to as **rollovers** and are important in extensional faulting (see section 2.7).

## 2.5 FAULT ASSOCIATIONS

Faults are usually found in groups of the same type. Thus a major fault with a large displacement may be accompanied by a set of smaller parallel faults with the same sense of displacement; these are known as **synthetic faults**. Faults which dip in the opposite direction to the main set are termed **antithetic**.

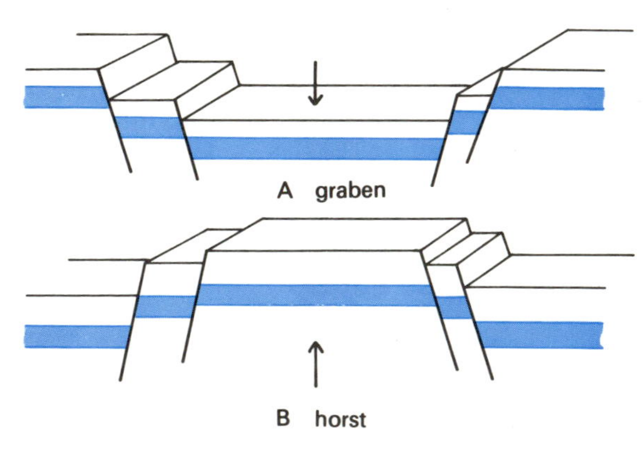


**Figure 2.8** Graben and horst structure.

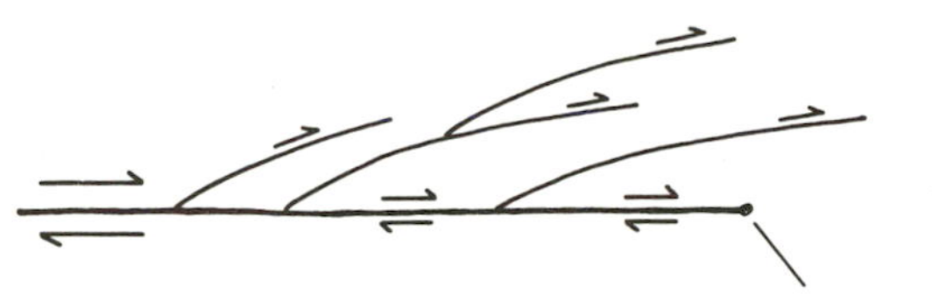


**Figure 2.9** Splay faults at the termination of a main fault. The displacement on the main fault may be spread over a large area around the termination of the main fault by the cumulative effect of displacements on smaller branching faults.

Major uplifted or depressed blocks may be bounded by sets of faults of the same type but opposite sense of movement. A depressed block bounded by normal faults is termed a **graben** and an elevated block a **horst** (Figure 2.8). Major graben features which extend for long distances are called **rifts**. Two well-known examples are the great African rift system (Figure 15.1) and the Rhine rift, both of which consist of a series of connected graben extending for hundreds of kilometres.

Fault sets need not consist of parallel faults. Quite complex arrangements of branching faults may accompany a major fault (Figure 15.8) and very often the termination of a fault is marked by branching **splay faults** which spread the displacement over a large area (Figure 2.9).

Faults may terminate against other faults of different types; for example, two thrusts may be linked by a strike-slip fault so that the displacement is transferred from one to the other. Such a fault is known as a **transfer fault** (Figure 2.10).

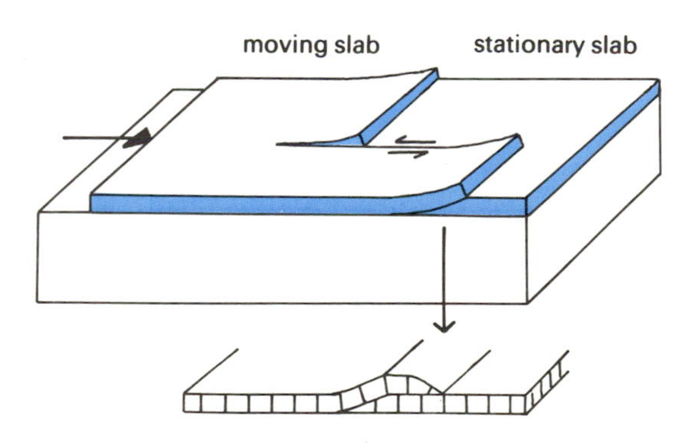


**Figure 2.10** Transference of displacement from thrust to strike-slip fault.

FAULT SYSTEMS

Sets of faults of the same age can generally be ascribed to some regional tectonic control which, for example, might exert a uniform compression or extension across the region. We shall see in chapter 15 that constructive plate boundaries are associated with divergent tectonic regimes, destructive boundaries with convergent regimes, and transform faults with strike-slip regimes. Each of these regimes is characterized by a specific fault system; thus convergent regimes are characterized by thrust systems, and divergent regimes by extensional fault systems. Although faults of different types may be found in association with those of the main fault system, it is the latter that are dominant and from which the nature of the regional tectonic control may be determined. Three systems will now be considered: thrust systems, extensional fault systems and strike-slip fault systems.

## 2.6 THRUST SYSTEMS

The detailed structure of many thrust belts is extremely complex (see for example section 16.1 and Figures 16.2 and 16.3). However, much of the complexity may be explained by careful reconstruction following certain simple basic rules, e.g. by constructing balanced sections as described below.

THRUST GEOMETRY

Where thrusts affect a set of bedded rocks that are near-horizontal in attitude, the thrusts generally follow a **staircase** path made up of alternating **ramps** and **flats** (Figure 2.11A). The flats are where the thrust sheet slides along a relatively weak bedding plane (often called a **detachment** or **décollement** plane) and the ramps are sections where the thrust cuts upwards through the stratigraphic sequence at an angle of typically around 30° to the horizontal. In uninverted strata, thrust displacements of this type always place older strata upon younger strata.

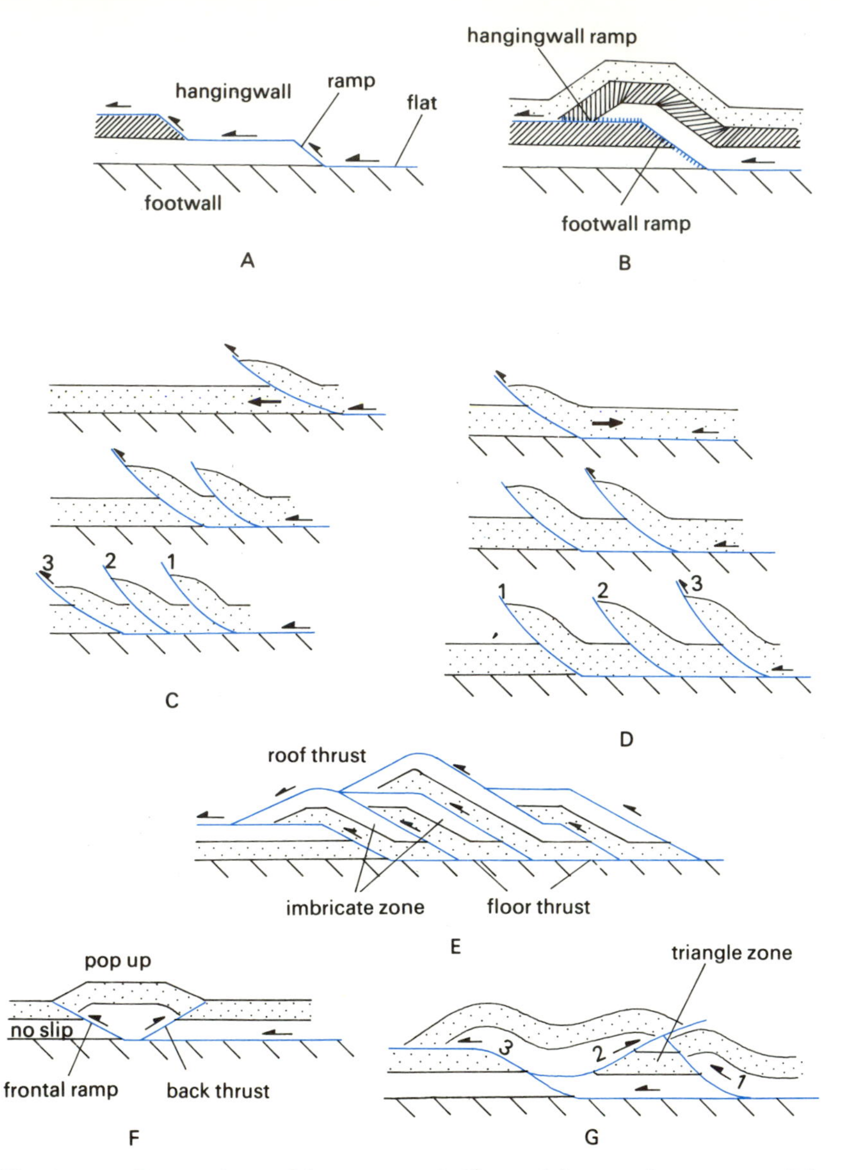


**Figure 2.11** Structures and terminology of thrust zones. A. Shape of thrust surface: ramps and flats. B. Hangingwall geometry: a fold in the hangingwall must result from a ramp. C. Piggyback thrust sequence (new thrusts develop in the footwall). D. Overstep thrust sequence (new thrusts develop in the hangingwall). E. Structure of a duplex: imbricate thrust slices are contained between a floor thrust and a roof thrust. F. Pop-up structure formed by backthrusting (see text). G. Triangle zone formed by backthrusting (see text). (After Butler, R.W.H. (1983) *Journal of Structural Geology*, **4**, 239–45.)

The ramp that is cut in the hangingwall is oblique to the bedding in the hangingwall sheet. Therefore, as this sheet moves up the footwall ramp and along a flat, the bedding will be folded as shown in Figure 2.11B. Folds formed in this way are a geometric consequence of the thrust movement.

Ramps do not necessarily strike perpendicular to the direction of transport, but are also found oblique or even parallel to the transport direction, in which case there will be a strike-slip component of movement along them.

## SEQUENCE OF THRUSTING

Thrusts may develop in sequence either forwards or backwards from the first thrust (Figure 2.11C, D). Where the later thrusts develop in the footwall of the original thrusts, the earlier hangingwalls are carried forwards by the later hangingwalls to form a **piggyback sequence** (Figure 2.11C). Conversely, if the thrusts migrate backwards so that later thrusts develop in the hangingwalls of the earlier thrusts, an **overstep sequence** develops so that the higher thrusts will be the later thrusts (Figure 2.11D).

Piggyback sequences are considered to be the normal mode of propagation in thrust systems, but many thrust belts exhibit local out-of-sequence thrusts that originate at or below the level of the sole thrust and cut up through the imbricate stack behind the thrust front, often using pre-existing fault planes. Such thrusts may locally violate the normal rule that thrusts place older strata upon younger strata.

## FORMATION OF A DUPLEX

Thrust sequences often result in the stacking up of many thrust sheets, making up an **imbricate zone**. In sequences of piggyback type, an imbricate zone may be bounded at the top by the original thrust surface, forming a **roof thrust**, and at the base by the currently active thrust surface, forming a **floor thrust** (Figure 2.11E). The whole thrust package is termed a **duplex**, and the individual imbricate sheets within the duplex are called **horses**. A typical duplex therefore consists of a roof thrust and a floor thrust enclosing a stacked-up pile of horses. The currently active thrust surface which lies at the base of the thrust sequence, and which extends beyond it to the edge of the displaced zone, is termed the **sole thrust**.

## BACK THRUSTS AND EXTENSIONAL FAULTING

Thrusts with a displacement direction opposite to that of the main thrust movement (i.e. antithetic thrusts) are occasionally found in thrust belts, and are known as **back thrusts**. These may be explained by the additional layer-parallel compression induced by gravity as the sheet climbs up the ramp. The uplifted hangingwall block thus formed is termed a **pop-up** (Figure 2.11F). If the backthrust truncates an earlier thrust, a **triangle zone** is formed (Figure 2.11G).

Another cause of complication in thrust belts is the occurrence of low-angle normal faulting due to extension at or near the thrust front. This has been attributed to gravitational sliding induced by the over-thickened thrust stack.

## BALANCED SECTIONS

Section 'balancing' has been developed as a method of unravelling complex thrust and extensional fault zones by restoring them to their original lengths, in order to measure the fault displacement and to reconstruct the sequence of movements responsible for their often complex geometry (Figure 2.12).

The section must be taken perpendicular to the 'orogenic strike', (i.e. the main fold trend). Individual horizons in the deformed section are restored to their original lengths and the section is said to be 'balanced' if the restored lengths of all the measured horizons are equal, and equivalent to the original length. Deformed sections are normally balanced down to a basal detachment plane or sliding horizon below which the rocks are assumed to be unaffected by the shortening. Two common explanations for sections that do not balance are (1) inter-layer slip (very common in thrust tectonics) and (2) different amounts of layer-parallel shortening (see Figure 10.7) in different beds (Figure 2.12C, D). There is no unique solution to a section-balancing problem and usually some assumptions have to be made to achieve a solution. A common assumption is that restored thrust dips should be at 30°–35°. A three-dimensional study of fault displacements can be made if sufficient data are available and, except in the simplest types of fault belts, movements in the third dimension should be considered.

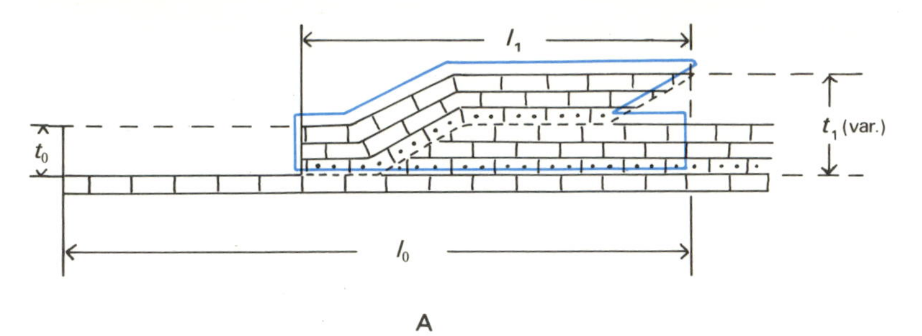

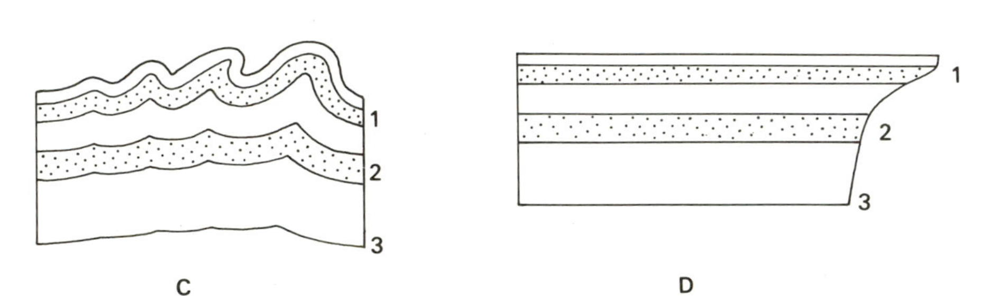


**Figure 2.12** Construction of balanced sections. A. A section is restored by putting each layer back to its original position. The section 'balances' if all the restored layers are of equal length. Sections may not balance for several reasons. In B, the lower layers have been strongly distorted by the shear resistance at the ramp, so that the displacement on the left side is 3 units compared with 5+ units on the right side. Such distortion is often achieved by varying amounts of inter-layer slip. In C, the restored length of layers 1, 2 and 3 are different (D) because the lower layers have deformed more by shortening and thickening than by folding. (Based on Dahlstrom, C.D.A. (1969) *Canadian Journal of Earth Sciences*, **6**, 743–7.)

## 2.7 EXTENSIONAL FAULT SYSTEMS

It is a comparatively common feature for specific regions of the crust to exhibit sets of related faults on which the individual displacements produce a net extension in the system as a whole. Such fault systems are characteristic of the extensional, or divergent, tectonic regimes associated with (1) oceanic and continental rift zones and (2) regions of continental back-arc extension on the upper plate of subduction zones, such as the Basin and Range Province of the western USA.

Classical views on faulting, and on the formation of graben and rifts, visualized extension being accommodated by dip-slip movements on steep normal faults (Figure 2.8) or by the filling of extensional fissures by magma. However, the normal fault/graben model can only provide a relatively small amount of extension, and in more recent times the importance of low-angle normal faults has been emphasized as a method of achieving the

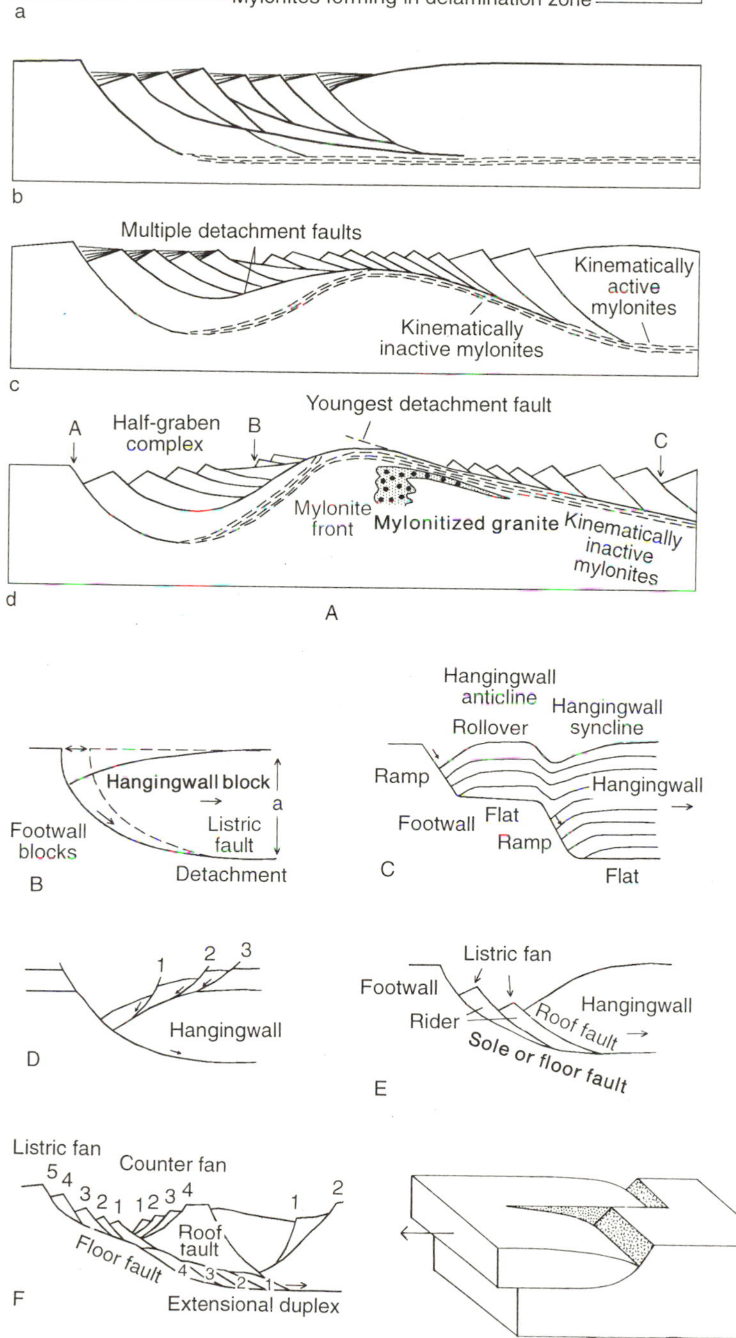


**Figure 2.13** Extensional fault systems. A. Evolutionary model illustrating progressive extension accommodated by block rotation above a low-angle extensional fault. (After Lister, G.S. and Davis, G.A. (1989). The origin of metamorphic core complexes, *Journal of Structural Geology*, **11**, 65–94.) B. Listric fault with hangingwall rollover anticline. Areas A and B are equal. The adjustment in hangingwall shape implies internal deformation. C. Flat/ramp geometry of fault produces geometrically necessary folding in hangingwall. D. Antithetic faults in hangingwall. E. Synthetic faults in footwall forming listric fan. F. Extensional duplex with listric fan and antithetic counter fan. (B–F after Gibbs, A.D. (1984) *Journal of the Geological Society of London*, **141**, 609–20.) G. Two listric faults linked by a transfer fault; all three detach on the same sole fault.

much larger extensions that have been estimated in certain extensional provinces, such as the Basin and Range Province in the western USA (Figure 15.4).

Three principal geometric elements have been identified in fault systems associated with large extensions: (1) low-angle normal faults acting as detachment planes (sole faults), (2) rotated fault blocks, and (3) curved or listric fault surfaces (Figure 2.13A). The widespread occurrence of such geometries has been confirmed by seismic profiles obtained in the petroleum exploration of marine basins.

## LISTRIC FAULTS AND LARGE EXTENSIONS

A **listric fault** is a curved normal fault which achieves a rotation in the hangingwall as a geometric consequence of the displacement, and may be accompanied by an accommodation fold, known as a **rollover anticline** (Figure 2.13B). The detachment or sole fault in extensional systems has the same function as that in thrust fault systems, and may possess a similar ramp-flat geometry (Figure 2.13C) necessitating complex accommodation structures in the hangingwall. The accommodation may take place by the formation of a set of antithetic faults, which have the effect of extending and thinning the hangingwall (Figure 2.13D). As extension proceeds, the sole fault may migrate into the footwall, producing a set of synthetic listric faults known as a **listric fan** (Figure 2.13E), in an analogous way to the propagation of thrusts into the foreland in piggyback thrust sequences. The fault blocks formed in this process are termed **riders**. The migration of the sole fault into the footwall may create an **extensional duplex** bounded by a roof fault. Figure 2.13F shows the complex type of geometry that may be produced by a combination of hangingwall and footwall collapse resulting from progressive extension.

The rotated blocks formed by these processes form **half-graben** at the surface and these may become filled with sediment (see Figure 2.13A). Upper-crustal extension on fault systems of this type may be transferred to mid-crustal levels and below by low-angle mylonite zones

The above listric geometries are based on observations and interpretations of shallow seismic profiles of marine sedimentary sequences. A problem with the listric fault model, however, is that large, currently active extensional faults observed on land are steep and planar down to about 10 km. It is necessary therefore to speculate as to how the listric curvature and the low-angle detachment are produced. The answer is thought to lie in the more ductile properties of the middle and lower parts of the crust, where movement zones are typically low-angle. The effect of progressive extension is thought to deform and rotate earlier-formed planar faults into a listric geometry as shown in Figure 2.13A.

## TRANSFER FAULTS IN EXTENSIONAL SYSTEMS

Extensional fault systems frequently contain steep transfer faults with strike-slip displacements. Such faults are integral to the system and transfer displacement from one dip-slip plane to another. They are smaller-scale counterparts of oceanic transform faults (see section 14.3). Figure 2.13G shows how two dip-slip faults may be linked by a transfer fault, each detaching on the same sole fault. Transfer faults may separate imbricate fault-fold packages that are geometrically distinct and uncorrelatable across the fault.

**Figure 2.14** Strike-slip fault systems. A. Local compressional and extensional structures produced by fault terminations and fault overlaps in strike-slip faulting. B. Formation of raised and depressed wedge-shaped blocks by local transpression and transtension on a branching strike-slip system. (A and B after Reading, 1980.) C. Positive and negative flower structures produced by convergence and divergence respectively in strike-slip motion. Dot and cross symbols within circles indicate out-of-page and into-page components of motion respectively. D. Diagrams illustrating the formation of strike-slip duplex structure in transpression (a) and transtension (b). Note that the structures are analogous morphologically to compressional and extensional dip-slip fault duplexes respectively. (C and D from Park, R.G. (1988) *Geological Structures and Moving Plates*, Blackie, Glasgow and London.)

## 2.8 STRIKE-SLIP FAULT SYSTEMS

Sets of related faults on which the individual movements produce a net strike-slip displacement in the system as a whole are characteristic of tectonic regimes associated with major continental transform fault zones (see section 15.5), such as the San Andreas fault zone (Figure 15.8). An

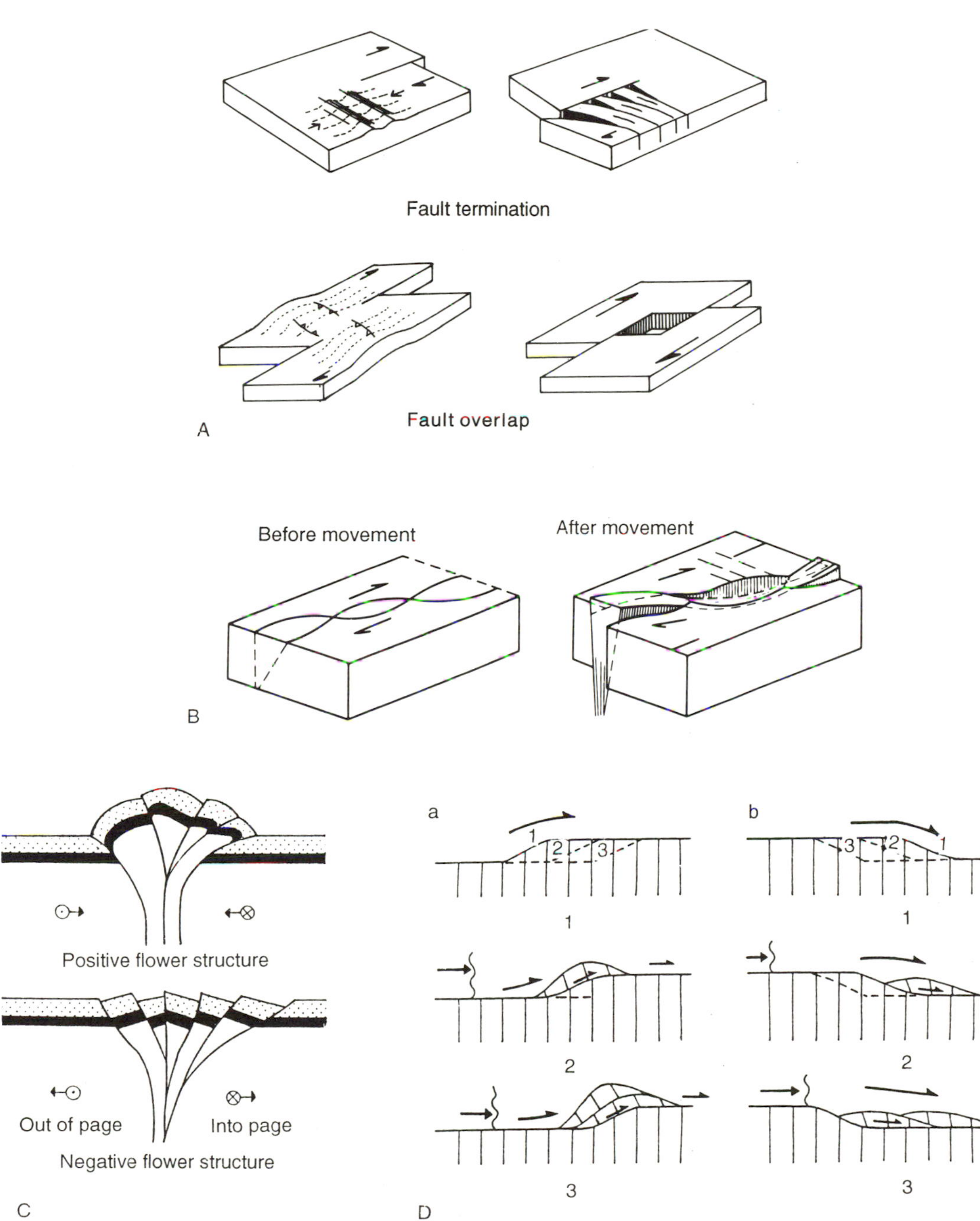

important feature of continental strike-slip regimes is that in addition to strike-slip faults, they also contain examples of compressional and extensional structures. Such zones are also characterized by differential vertical movements, which create rapid alternations between uplifted blocks and depressed basins.

Major strike-slip zones, such as the San Andreas fault zone, are of the order of 100–200 km across, within which quite complex tectonic effects take place. Strike-slip faults, which typically form a branching network, are the most important type of structure, but important secondary geometrical effects result from the irregularity of the fault geometry. For example, localized strains are produced around fault terminations that are either compressional or extensional, depending on the sense of movement (Figure 2.14A). Fault overlaps thus create local zones of compression or extension (Figure 2.14B); compressional zones may be characterized by reverse faults or by folds, and extensional zones by normal faults.

**Pull-apart basins** formed in local extensional zones as a result of this process are a diagnostic feature of strike-slip regimes.

#### TRANSTENSION AND TRANSPRESSION

Local changes of direction in strike-slip faults create local zones where the fault trend is oblique to the movement direction, causing either compression or extension across the fault. Where strike-slip movement is combined with extension, the process is termed **transtension**; where strike-slip and compression are combined, the result is **transpression**. In complex curving fault networks, where branching faults have opposed dips, overall strike-slip displacement leads to oblique slip movement on individual fault segments. The resulting convergence and divergence of wedge-shaped blocks gives rise to the formation of alternate raised and depressed zones (Figure 2.14C). Combinations of folds and faults produced in local zones of transpression or transtension are termed **flower structures**. These are positive for uplifted blocks and negative for depressed blocks (Figure 2.14D).

#### STRIKE-SLIP DUPLEXES AND DISPLACED TERRANES

Strike-slip duplexes (Figure 2.14E), which are analogous geometrically to thrust and extensional duplexes, may be formed at bends in a strike-slip fault as a result of the progressive migration of the active fault into one wall. Transtensional duplexes are morphologically equivalent to extensional duplexes, and transpressional duplexes to thrust duplexes respectively. Large pieces of crust may become detached and isolated from one block, transferred to the opposing block, and transported far from their sites of origin, becoming 'displaced terranes' (see section 15.5).

## 2.9 INVERSION

The term **inversion** is used to describe the process of regional reversal of tectonic movement from subsidence to uplift, say, or from extension to compression. Typically, inversion involves the reactivation of faults with the opposite sense of movement, such that normal faults experience a change to reverse movement, or thrusts become low-angle reverse faults. In **positive inversion**, a region changes from subsidence to uplift, and in **negative inversion**, from uplift to subsidence. Such movements form an integral part of the tectonic evolution of orogenic belts.

The recognition of inversion in individual structures and groups of structures has become very important in hydrocarbon exploration, since where inversion is involved the interpretation of structures at depth may be completely different to that with unmodified extensional or compressional structures. Inversion may create a set of structures that display variation in net sense of movement; thus compressional re-activation of normal faults may give rise to a mixture of apparently normal and apparently reverse faults.

Because the crust is stronger in compression than in extension, and because extensional structures are consequently more widespread, when regional compression takes place it will typically be easier to re-activate old extensional planes of weakness, given that they are suitably oriented, than to create

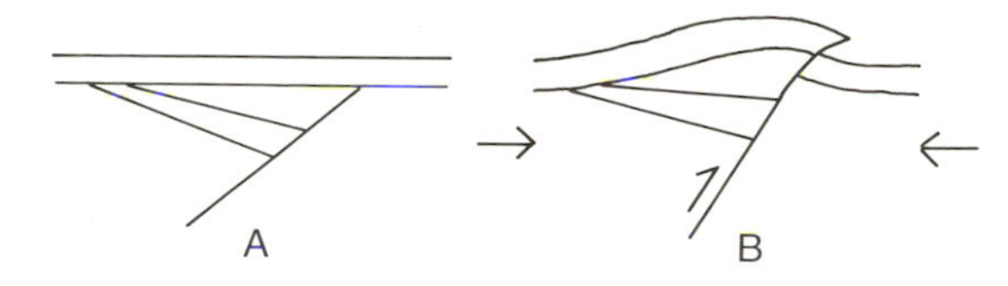


**Figure 2.15** Inversion: extensional half-graben re-activated in compression. Note that compressional folds coexist with net normal displacement.

new faults. Thus, for example, the Alpine compression of northwest Europe has created widespread Tertiary folds and reverse faults by inversion of Mesozoic extensional normal faults.

Figure 2.15 shows the basic geometry of an extensional half-graben subjected to compression during inversion. The strata above the half-graben are shortened and folded, although the fault may still appear to be normal.

## 2.10 JOINTS

Joints may occur in sets of parallel, regularly spaced fractures, and several sets may occur in the same

**Figure 2.16** Joint sets in thin-bedded Devonian sandstones, Lligwy Bay, Anglesey.

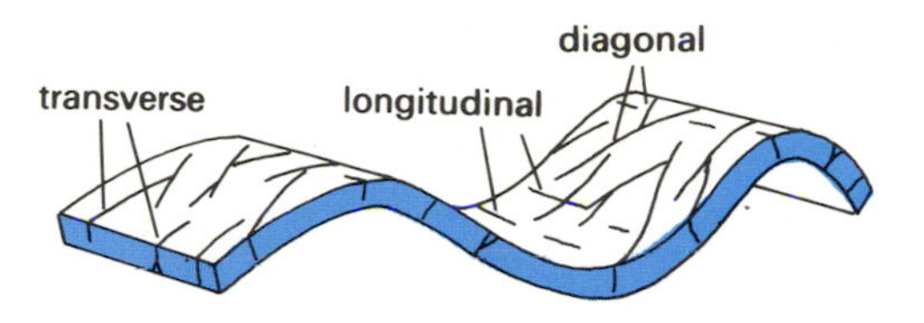


**Figure 2.17** Relationship of joint sets to major folds. Regular joint sets perpendicular to bedding may be divided into longitudinal (parallel to fold axes), transverse (perpendicular to fold axes) and diagonal (oblique to fold axes).

rock, giving a conspicuous blocky appearance to the outcrop (Figure 2.16). More commonly, however, joints are much less regular and systematic. Where a recognizable joint set exists, it can usually be related in some way both to a regional tectonic control and to the geometry of the rock body containing the joints. For example, joint sets are frequently found both perpendicular and parallel to the bedding in layered rocks. The perpendicular joints may form two or more intersecting sets which bear a simple relationship to the regional fold geometry (Figure 2.17). Under favourable circumstances, it is possible to relate regular joint sets that occur regionally in various different rock types to a regional compression or extension in the same way as folds.

### PRESSURE-RELEASE (UNLOADING) JOINTS

Many joints are due to the release of 'stored' pressure. The weight of a great thickness of overlying strata causes deeply buried rock to be compressed. However, once the overlying rock has been eroded, this load pressure is reduced and the rock expands by the development of tensional joints which are often parallel to bedding surfaces in sedimentary strata, or to the temporary erosion surface in massive igneous rocks, where they are termed **sheet joints**.

### COOLING JOINTS

Another common cause of joint formation is the contraction that takes place in a cooling igneous body. Tabular igneous bodies, i.e. dykes and sills,

**Figure 2.18** Polygonal cooling joints forming columnar structure in basalt, Giant's Causeway, Co. Antrim. (From Holmes, A. (1978) *Principles of Physical Geology*, Nelson, figure 5.7 (photograph J. Allan Cash).)

frequently exhibit polygonal columnar jointing perpendicular to the cooling surfaces. The spectacular columnar jointing of the Giant's Causeway in Antrim (Figure 2.18) is a well-known example of this type of structure.

## FURTHER READING

Coward, M.P. (1994) Inversion Tectonics, in *Continental deformation* (ed. P.L. Hancock, Pergamon, Oxford, pp. 289–304.

Coward, M.P., Dewey, J.F. and Hancock, P.L. (eds) (1987) *Continental Extensional Tectonics, Special Publication of the Geological Society of London*, **28**.

Hobbs, B.E., Means, W.D. and Williams, P.F. (1986) *An Outline of Structural Geology*, 2nd edn, Wiley, New York. [An excellent general textbook.]

McClay, K.R. and Price, N.J. (eds) (1981) *Thrust and Nappe Tectonics, Special Publication of the Geological Society of London*, **9**.

Reading, H.G. (1980) Characteristics and recognition of strike-slip fault systems. *Special Publication of the International Association of Sedimentologists*, **4**, 7–26.

Sibson, R.H. (1977) Fault rocks and fault mechanisms. *Journal of the Geological Society of London*, **133**, 191–213. [Gives a useful classification of fault rocks and discusses the relationship between mylonite, pseudotachylite and other fault rocks.]

Twiss, R.J. and Moores, E.M. (1992) *Structural Geology*, Freeman, New York. [Chapters 3–7 give considerable additional information about fractures, faults and joints, with numerous illustrations.]

# FOLDS 3

While fractures are the commonest expression of rock deformation, there can be no doubt that folds are the most spectacular. When we see a bed of rock that was originally flat and planar bent into a huge arch many hundreds of metres in height, we are forced to recognize in a very direct way the existence of the tremendous forces that act upon the Earth's crust. The sight of a large-scale fold in a mountainside, as can be seen in the Alps and other mountain belts, is much more satisfying as an expression of deformation than reconstructions based on mapping, although the latter may be equally convincing.

### 3.1 MEANING AND SIGNIFICANCE OF FOLDS

A **fold** is a structure produced when an originally planar surface becomes bent or curved as a result of deformation. We have seen that fractures result from brittle deformation that causes the rock to break completely along discrete planes. Folds, however, are an expression of a different type of deformation which produces gradual and more continuous changes in a rock layer, both in its attitude and internally, as the rock accommodates to changes in shape. Such a deformation is more pervasive than that responsible for faulting and is typically more ductile. The difference between brittle and ductile deformation and the way rocks accommodate physically to deformation are discussed in Chapter 7.

### 3.2 BASIC FOLD GEOMETRY AND NOMENCLATURE

FOLD HINGE AND LIMBS

If we consider a single folded surface (Figure 3.1A, B), the main elements of the geometry of the fold shape are, firstly, the **hinge** (or **closure**), which is the zone of maximum curvature of the surface, and secondly, the **limbs**, which are the areas between the hinges. A single fold comprises a hinge and two limbs which enclose the hinge. In a series of folds, each limb is common to two adjacent folds. Depending on the actual shape of the fold, that is, on the way in which a surface changes curvature, the hinge may be very sharply defined and the limbs relatively straight (Figure 3.1B), or the curvature may be more constant around the fold if it approaches a cylindrical shape (Figure 3.1C). If it is possible to define a line along which the maximum curvature of the fold takes place, this line is called the **hinge line**. In a truly **cylindrical** fold, where the fold surface corresponds to part of the surface of a cylinder, neither hinge nor limb can be defined, but such examples are uncommon. The term 'cylindrical' used in this sense should not be confused with 'cylindroidal' which is defined in section 3.7.

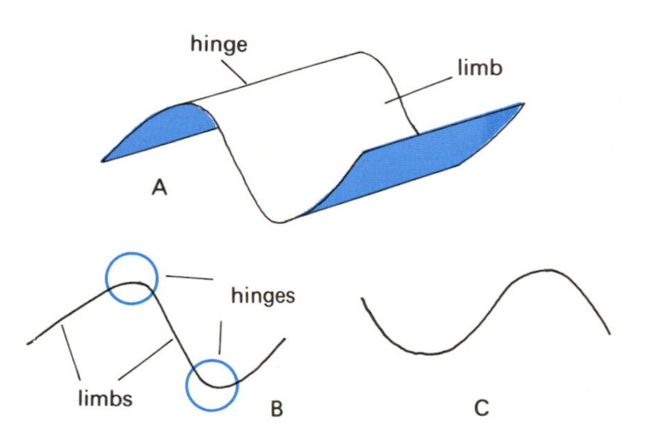


**Figure 3.1** Hinge and limbs of a fold.

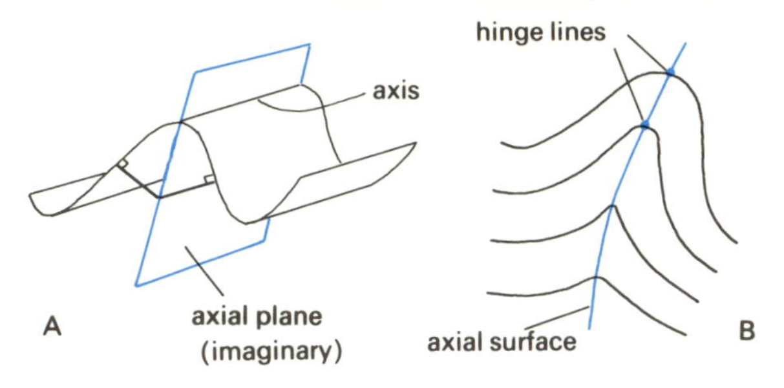


**Figure 3.2** Axis, axial plane and axial surface of a fold.

FOLD AXIS AND AXIAL PLANE

In describing the attitude of a fold, it is useful to refer not to the attitudes of both limbs but to the imaginary plane that is equidistant from each limb and which bisects the angle between them. This plane is termed the **axial plane** and cuts the hinge zone of the fold along a line termed the **fold axis** (Figure 3.2A). Provided that the limbs of a fold are sufficiently well defined, the determination of the orientation of the axial plane and axis will give a precise description of the orientation of the fold. It must be remembered, however, that these terms refer only to a single fold surface.

FOLD AXIAL SURFACE

When we wish to describe a fold consisting of a number of folded layers, the axial planes of the different folded surfaces may not correspond. In such a case, it is usually more convenient to refer to the **fold axial surface**, which is defined as a single surface passing through the hinge lines of each successive fold surface (Figure 3.2B). Such a surface is, in general, not planar, and may not correspond with the axial planes of the individual layers in the fold. The hinge lines are not always easy to locate by eye if the hinge zone is broad, and the concept is only appropriate for relatively tight folds with well-defined hinges.

INTER-LIMB ANGLE

The smaller angle made by the limbs of a fold is termed the **inter-limb angle** or **fold angle**. The tightness or openness of a fold as expressed by this angle is a useful method of classifying folds (Figure 3.7A) and reflects the amount of deformation.

AMPLITUDE AND WAVELENGTH

Amplitude and wavelength are convenient measures of the size of a fold (Figure 3.3). The **wavelength** of a fold is the distance between the hinges on each side of the fold. If these are not visible, the half-wavelength may be measured instead by taking the distance between the two **inflexion points** on each side of the fold hinge. The inflexion points are where the sense of curvature changes from one fold into the next (Figure 3.3A). The **amplitude,** or 'height', of a fold may be measured by taking half the perpendicular distance from the hinge to the line joining the two hinges on each side, or alternatively the perpendicular distance from the hinge to the line joining the two nearest inflexion points.

### 3.3 FOLD ORIENTATION

The orientation of a fold may be obtained in the field by measuring the attitudes of both limbs, and if the limbs are relatively straight and regular, this information is sufficient for a complete description of the fold attitude. However, it is more convenient,

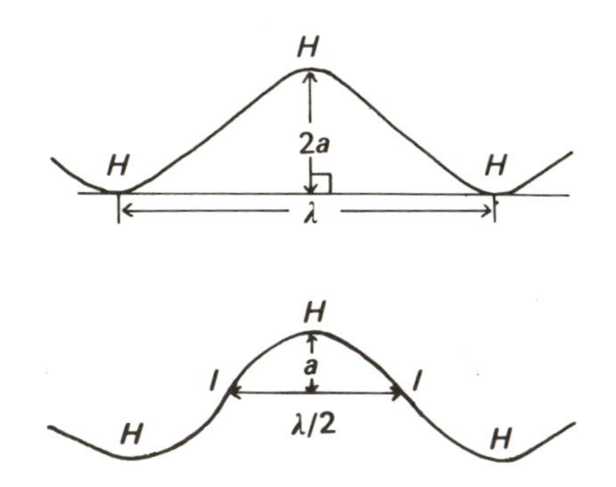


**Figure 3.3** Amplitude and wavelength of a fold. $\lambda$, wavelength; $a$, amplitude; H, hinge; I, inflexion point (see text).

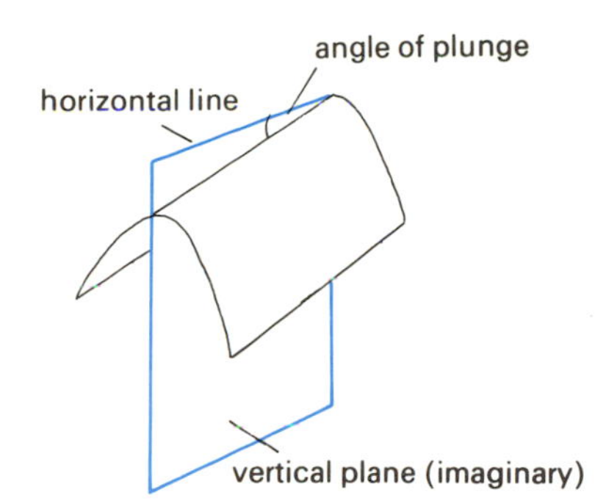


**Figure 3.4** Plunge of a fold. The angle of plunge is measured from the horizontal, in a vertical plane.

and conveys more useful information, to describe and record folds in terms of their axial planes or surfaces, and axes. Thus on a map, the position of a fold is shown by drawing the outcrop of the fold axial surface as a line. This line is termed the **fold axial trace**. The dip and strike of the axial plane or axial surface may be plotted in the same way as for a bedding plane.

The attitude of the fold axis is measured as the angle between the axis and the horizontal. This angle, which must be measured in a vertical plane (like the dip angle) is termed the **fold plunge** (Figure 3.4), so a complete description of the orientation of a fold axis is given as an amount and direction of plunge; for example, a fold may be described as having a plunge of 30° towards (on a bearing of) 105°.

## 3.4 CLASSIFICATION OF FOLDS

The classification of folds is based on four main features or properties: direction of closing, attitude

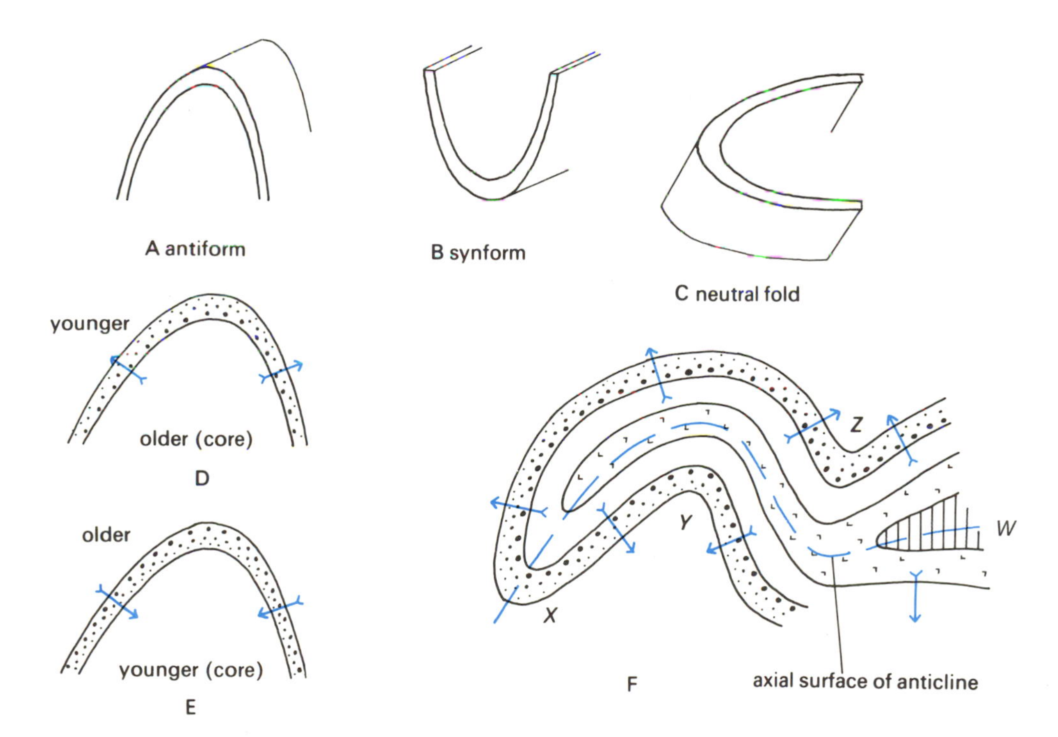


**Figure 3.5** Closing and facing directions of a fold. A, antiform – an upward-closing fold. B, synform – a downward-closing fold. C, neutral fold – a sideways-closing fold. D, anticline – older rocks in core; fold faces upwards. E, syncline – younger rocks in core; fold faces downwards and is an antiformal syncline. F, refolded recumbent anticline, W; fold at X is a synformal anticline, lower part of fold at Y is an antiformal syncline, and Z is a synformal syncline.

of axial surface, size of inter-limb angle and shape of profile.

## CLOSING AND FACING DIRECTION

Folds that close upwards, that is where the limbs dip away from the hinge, are termed **antiforms** (Figure 3.5A) and those that close downwards, where the limbs dip towards the hinge, are termed **synforms** (Figure 3.5B). Folds that close sideways are termed **neutral folds** (Figure 3.5C). Under normal conditions where the bedding becomes younger upwards, an antiform will contain older rocks in its core; it is then given the more familiar name 'anticline'. Thus the term **anticline** strictly applies only to a fold with older rocks in its core (Figure 3.5D). Conversely, a **syncline** is a fold that contains younger rocks in its core (Figure 3.5E).

In areas of more complex folding, where strata are commonly inverted, it is possible to find downward-closing anticlines or upward-closing synclines (Figure 3.5F). In such cases it is convenient to speak of the **facing** direction of a fold, which is defined as the direction along the axial surface in which the strata become younger. Thus in Figure 3.5F, fold X is a synformal anticline and faces downwards, since the strata are inverted. The lower part of fold Y is an antiformal syncline, and also faces downwards. Fold Z is an upward-facing synform, i.e. a synformal syncline. The main fold at W, where it has a subhorizontal axial surface, is a neutral fold, facing towards the left.

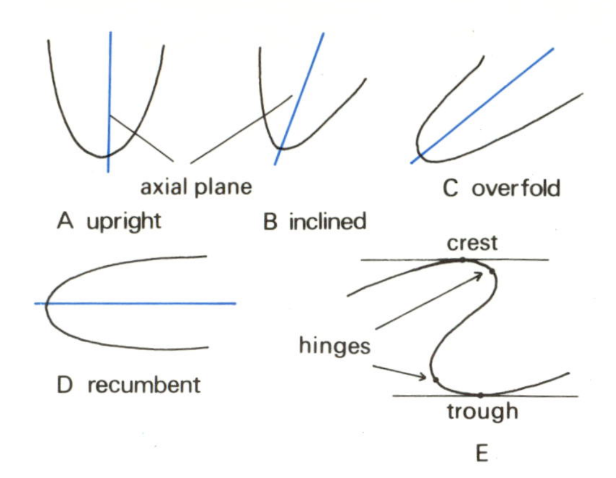


**Figure 3.6** Attitude of the fold axial plane. A, upright. B, inclined. C, overfold. D, recumbent. E, distinction between crest and trough lines and hinge lines in an inclined fold.

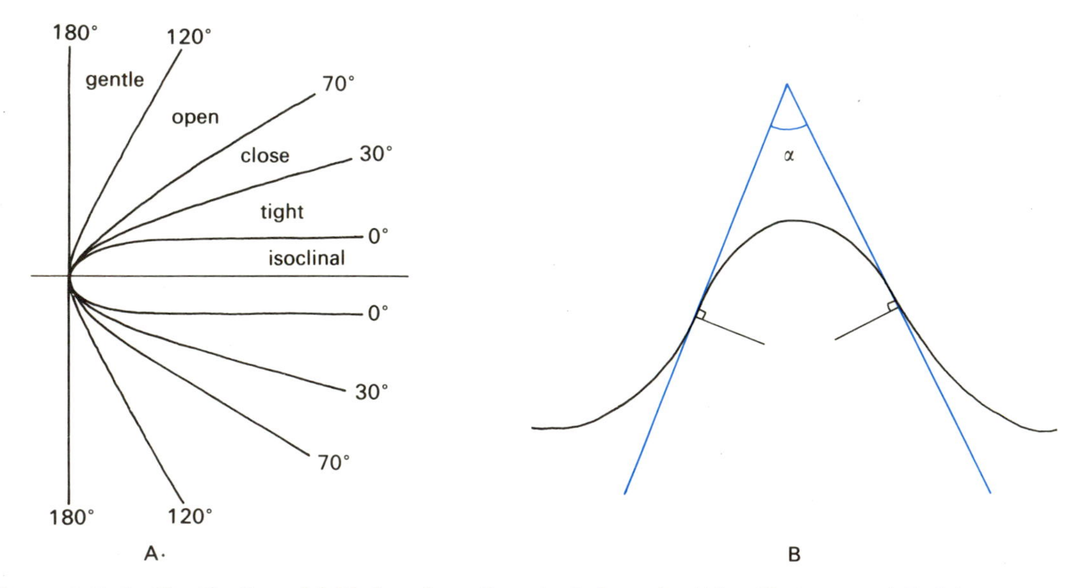


**Figure 3.7** A. Classification of folds based on the inter-limb angle. (After Fleuty, 1964.) B. Measurement of inter-limb angle $\alpha$ in folds with rounded profile. Tangents are drawn to the fold surface at the points of inflexion.

ATTITUDE OF AXIAL PLANE

Folds may be divided into three groups based on the dip of the axial plane, or surface. Folds with steep to vertical axial planes are termed **upright**, those with moderately dipping axial planes are termed **inclined** and those with subhorizontal axial planes are termed **recumbent** (Figure 3.6A–D). The division between these classes is not rigidly defined. Inclined folds where one limb is inverted are often termed **overfolds** (Figure 3.6C). In inclined folds, the highest and lowest points on the fold surface do not in general correspond with the hinges and it is sometimes useful to use the terms **crest** and **trough** respectively for these positions (Figure 3.6E). The crest is of interest in petroleum exploration, since it may form a trap for oil or gas, which may be contained in a permeable layer sealed by an impermeable layer above.

INTER-LIMB ANGLE

The size of the inter-limb angle measures the degree of tightness of a fold and reflects the amount of compression experienced by the folded strata. Figure 3.7A shows a classification scheme that subdivides folds into five classes: **gentle** (180°–120°), **open** (120°–70°), **close** (70–30°), **tight** (30°–0°) and **isoclinal** (0°). These limits are intended only as a general guide; precision is better achieved, if required, by stating the angle. Gentle folds are often termed **flexures**. Problems arise in measuring the fold angle where the limbs are not straight. In such

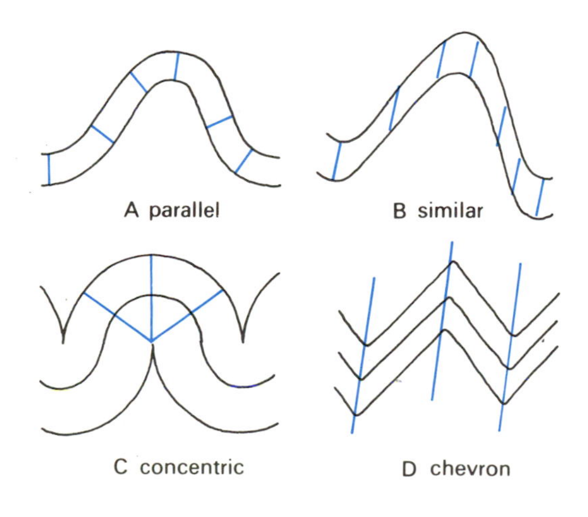


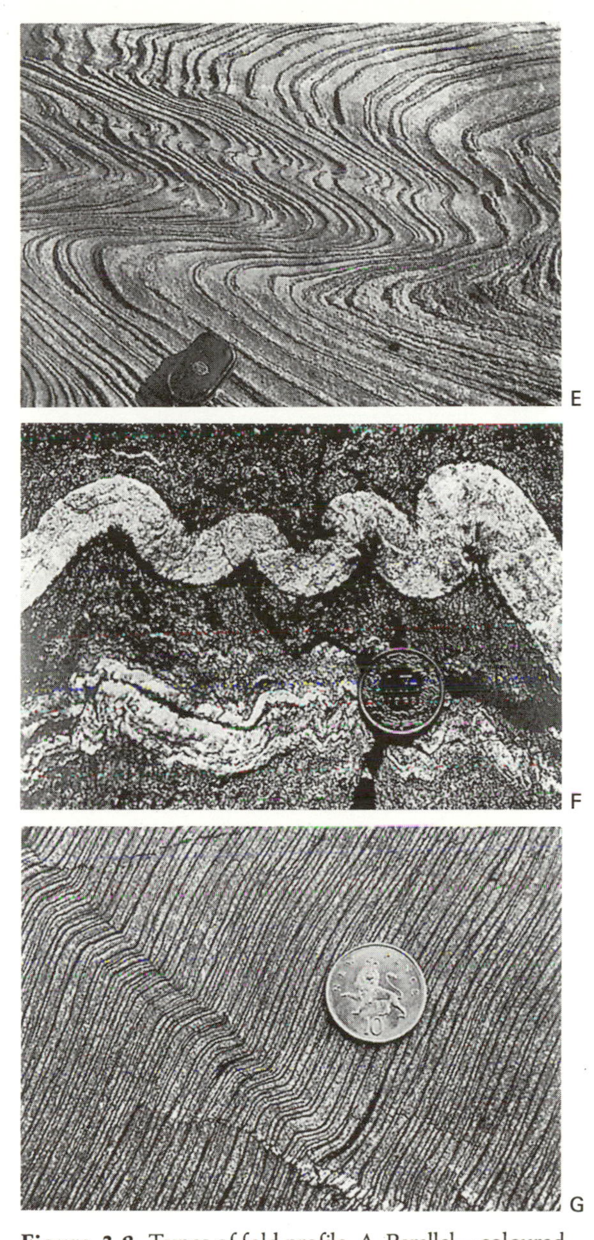


**Figure 3.8** Types of fold profile. A. Parallel – coloured lines show constant layer thickness measured perpendicular to fold surface. B. Similar – coloured lines show constant layer thickness measured parallel to axial surface. C. Concentric – coloured lines are radii of a circle. D. Chevron – coloured lines are kink planes (axial surface traces) separating straight fold limbs. E. Folds of generally similar style in marble, Sokumfiell, Norway. F. Folds of generally parallel style in aplite layer (above coin) in gneiss, Cristallina, Switzerland. G. Kink band in laminated siltstone, Bigsbury, south Devon. (E–G from Ramsay and Huber, 1987.)

a case, tangents can be drawn to the curve at the point of inflexion (Figure 3.7B). The inter-limb angle is the smaller angle made by the two tangents.

PROFILE

The **fold profile** is the shape of a folded layer observed in the plane perpendicular to the fold axis. Folds exhibit considerable variation in profile, and since this variation partly reflects differences in the mechanism of formation (see Chapter 10), a precise description of this profile is often very important.

The main categories of profile are shown in Figure 3.8. The simplest is the **parallel fold**, where the fold surfaces bounding the folded layer are parallel; in such a fold, the thickness of the folded layer measured perpendicular to the fold surface (i.e. the **orthogonal thickness**) is constant (Figure 3.8A, E). A special case of parallel fold is termed a **concentric fold**. This is where adjacent fold surfaces are arcs of a circle with a common centre, known as the **centre of curvature** of the fold (Figure 3.8C).

Parallel folds affect only a limited thickness of layers as a consequence of their geometry, so that upright parallel folds die out both upwards and downwards. This property is especially obvious in the case of concentric folds (Figure 3.8C), where the extent of the fold is limited by the centres of curvature. Beyond the centres of curvature, the shortening would be accommodated by faulting or by a different type of folding.

Another type of fold is the **similar fold**, in which the orthogonal thickness of the folded layer changes in a systematic way such that the fold maintains a constant thickness measured parallel to the axial surface (Figure 3.8B, F). In a true similar fold, the shape of adjacent curves should correspond precisely, and this property enables a similar fold to maintain its shape indefinitely along the axial surface through successive layers.

Folds that possess planar limbs and sharp angular hinges are known as **chevron folds** (Figure 3.8D); the term **accordion fold** is used synonymously. Such folds exhibit the curious property of being apparently both similar and parallel – in the sense that many of the individual folded layers may be parallel whereas the fold as a whole is usually similar. Where such a fold is markedly asymmetric, the superimposed short limbs give the appearance of bands running across the rock. Such bands, which are effectively contained within adjacent axial surfaces, are known as **kink bands** (Figure 3.8G).

These fold types represent end-members of continuous series of natural folds. Thus concentric to chevron represents one series, from extreme rounded to extreme angular, and parallel to similar represents another. Folds may be described in terms of how closely they approach these ideal types. For more exact work in folds, it is necessary to record the changes in curvature in more detail, as described in the following section.

## 3.5 GEOMETRY OF THE FOLD PROFILE

A useful way of portraying the geometry of the fold profile, and thereby of comparing it with one of the ideal fold models, is to plot the change in thickness of the folded layer as it varies away from the hinge. At any particular position on the fold surface, the tangent to the curve is drawn at an angle $\alpha$ with a line perpendicular to the axial surface (Figure 3.9). The thickness of the layer at that angle, $t_\alpha$, is then measured by taking the perpendicular distance between the two parallel tangents making the angle $\alpha$ with the inner and outer arcs respectively. The value of $\alpha$ reaches a

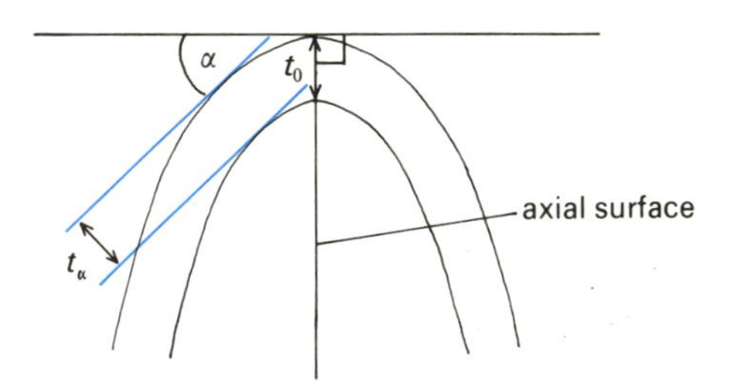


**Figure 3.9** Method of measuring variation in layer thickness $t_\alpha$ in a fold. Measurement is made at the point where the tangent to the outer curve makes an angle $\alpha$ with the perpendicular to the axial surface. (After Ramsay, J.G (1967) *Folding and Fracturing of Rocks*, McGraw-Hill, New York, figure 7.18.)

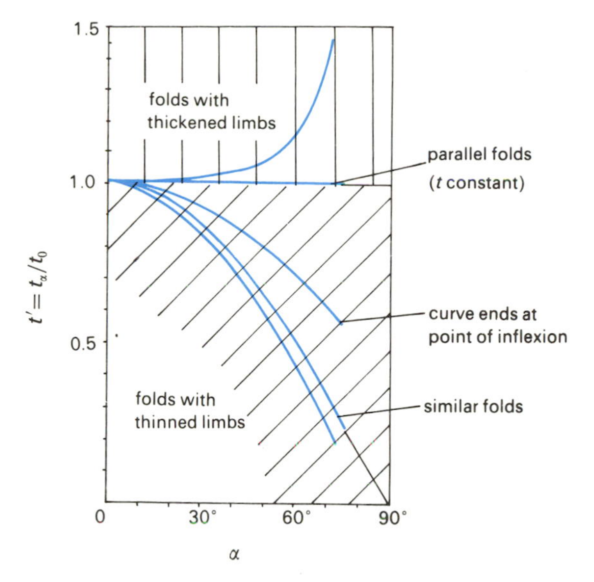


**Figure 3.10** Graph showing curves of $t'$ against $\alpha$ for various types of fold. $t' = t_\alpha/t_0$ where $t_0$ is the layer thickness at the hinge. (After Ramsay, J.G (1967) *Folding and Fracturing of Rocks*, McGraw-Hill, New York, figure 7.25.)

maximum at the point of inflexion, where the sense of curvature changes into the next fold. A plot is then made of the variation of $t'$ with $\alpha$, where $t' = t_\alpha/t_0$ and $t_0$ is the thickness at the hinge. The curves of $t'$ against $\alpha$ are characteristically different for the different fold models (Figure 3.10).

## DIP ISOGON METHOD

An alternative method of portraying the fold profile is to draw a set of lines joining points of equal limb dip (i.e. equal values of $\alpha$) in successive layers through the fold profile. These lines of equal dip are called **dip isogons**. The method is illustrated in Figure 3.11. The dip of the isogons reflects differences in curvature between the outer and inner arcs of the folded layer. Thus if the isogons converge towards the core of the fold, the curvature of the outer arc is less than that of the inner arc. Conversely if the isogons diverge towards the core of the fold, the curvature of the outer arc is greater than that of the inner arc. If the isogons are parallel, the curvature of the two arcs is equal, and the fold is therefore similar. Isogons that are perpendicular to the folded layer indicate a parallel fold. These rules lead to a convenient fold classification.

## FOLD CLASSIFICATION BASED ON DIP ISOGONS

The fold classification is shown in Figure 3.12. There are three classes: class 1 – folds with convergent isogons; class 2 – folds with parallel isogons (similar folds); and class 3 – folds with divergent isogons.

Class 1 is subdivided into three subclasses according to the degree of convergence. Thus class 1A folds are strongly convergent, the isogons making an angle greater than $\alpha$ with the axial surface. Class 1B folds are parallel folds where the isogons are perpendicular to the fold surface and make an angle $\alpha$ with the axial surface. Class 1C folds are weakly convergent, the isogons making an angle less than $\alpha$ with the axial surface.

These classes can be shown on the $t'/\alpha$ plot (Figure 3.10). There is a large field of folds with thinned limbs, which is divided by the curve representing class 2 similar folds, where the orthogonal thickness changes according to the relationship $t' = \cos\alpha$. It is rather important to measure such a

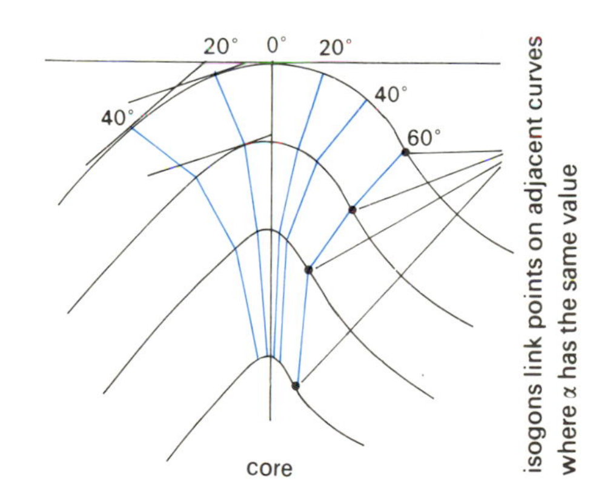


**Figure 3.11** Construction of dip isogons. The isogons join points on successive fold surfaces with the same inclination $\alpha$. In this example, the isogons converge towards the fold core. (After Ramsay, J.G (1967) *Folding and Fracturing of Rocks*, McGraw-Hill, New York, figure 7.18.)

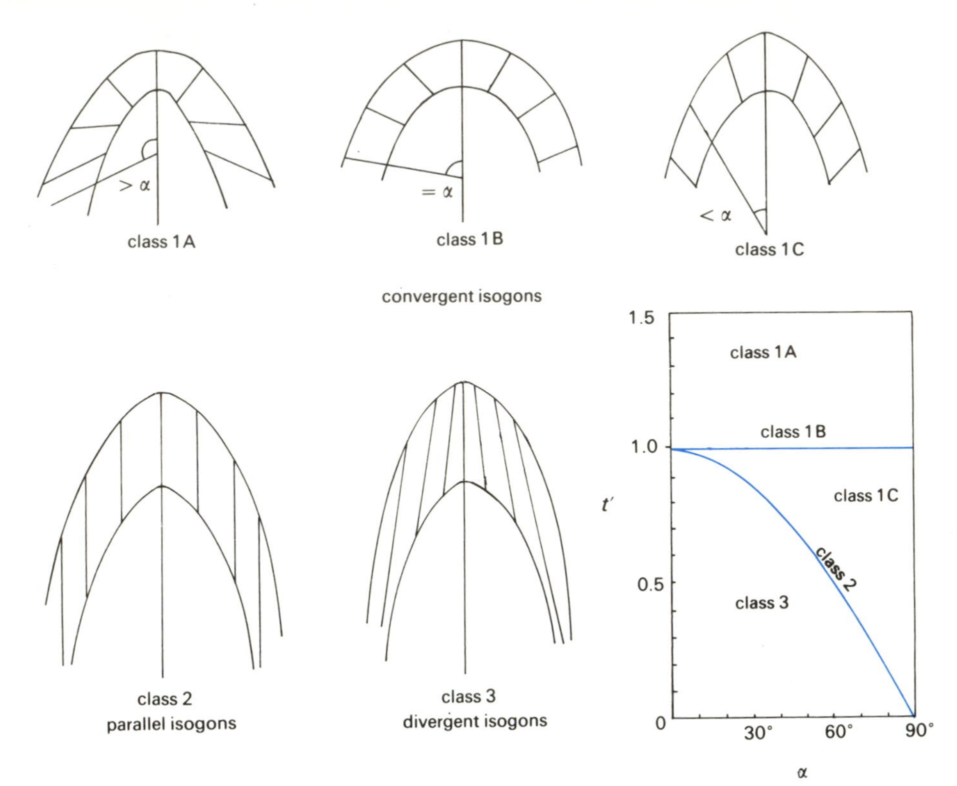


**Figure 3.12** Fold classification based on dip isogons – see text. (After Ramsay, J.G (1967) *Folding and Fracturing of Rocks*, McGraw-Hill, New York, figures 7.24 and 7.25.)

fold precisely, since many similar-looking folds actually belong to class 1C and may have formed by a different mechanism from the class 2 folds (see section 10.2).

## 3.6 DESCRIPTION OF FOLD SYSTEMS

The way in which folds are arranged in systems, and how they are related to each other, is just as important in understanding their process of formation as the shape of the individual folds. We shall examine several important features of fold systems: their symmetry, the existence of different scales of folding in the same layer, and the variation in profile shape between different layers.

### FOLD SYMMETRY

A set of folds is regarded as **symmetric** if the limbs are of equal length, and **asymmetric** if the limbs are of unequal length (Figure 3.13). Folds showing systematic differences in thickness between two sets of limbs are also asymmetric. A **monocline** is a special type of asymmetric fold where one limb is very short in relation to the adjoining limbs. The term monocline is normally used for large-scale folds produced by a local steepening in dip.

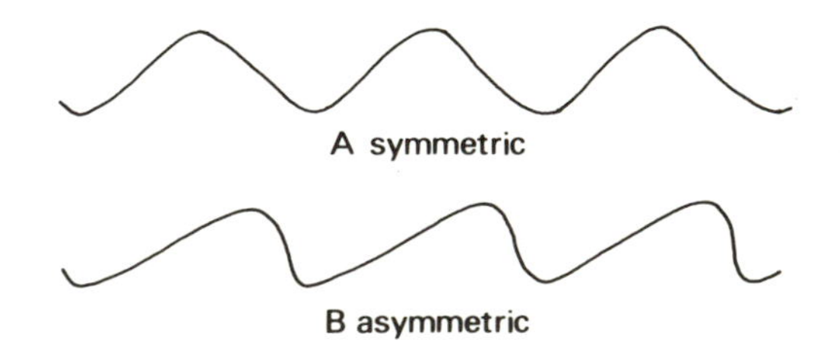


**Figure 3.13** Symmetric and asymmetric folds.

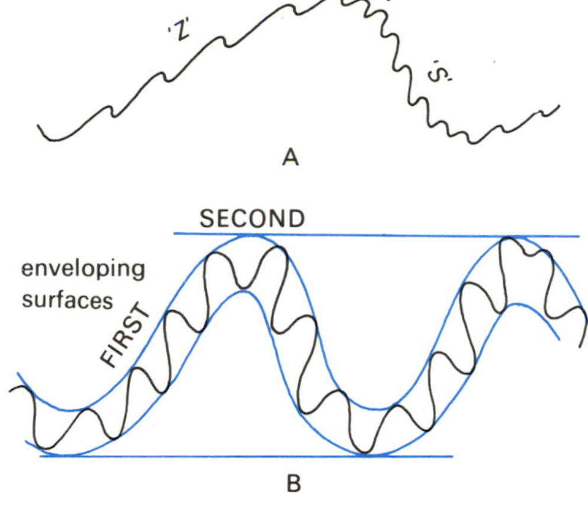


**Figure 3.14** Parasitic folds and the use of the enveloping surface. A. Parasitic folds have an asymmetric 'Z' profile on the left-hand limb of the anticline, a symmetric 'M' profile on the crest and an asymmetric 'S' profile on the right-hand limb. B. A complex profile may be simplified by drawing one or more enveloping surfaces joining the parasitic fold hinges.

## FOLD VERGENCE

A set of asymmetric folds is said to **verge** in the direction indicated by the shorter limbs of the antiforms of the set. Thus a set of folds may be described as indicating southwesterly **vergence**, or 'verging southwest'. The concept of vergence is applied more usually to overturned or highly asymmetric folds, often associated with thrusts, which exhibit a sense of overturning or thrusting in a particular direction (for an example see Figure 16.3).

## PARASITIC FOLDS

Very often a set of folds with a small wavelength is found superimposed upon folds of larger wavelength. The smaller folds, which occur on the limbs or hinge of the larger folds, are known as **parasitic folds** (Figure 3.14A). In many cases, there is a systematic relationship between the symmetry of the parasitic folds and their position on the larger folds. Thus when both small-scale and large-scale folds are generated together, the sense of asymmetry will generally change from Z-shaped on the left-hand limbs of anticlines, through M-shaped over the hinge area, to S-shaped on the right-hand limbs, when viewed in profile.

The existence of folds of different orders of magnitude can produce a very complicated shape in the folded surface. Locally measured bedding dips on such a surface will give very little idea, in general, of the overall shape of the main structure. However, this shape can be considerably simplified by drawing an **enveloping surface**, which is a surface drawn through the hinge lines of all the folds (Figure 3.14B). If the enveloping surface is itself folded, a second enveloping surface may be drawn to produce further simplification.

## HARMONIC AND DISHARMONIC SYSTEMS

Where sets of folds in adjacent layers correspond with each other in wavelength, symmetry and general shape, the system is called **harmonic**. However, in many cases, the wavelength and shape of folds in adjacent layers are quite different; this is often due to differences in physical properties or thickness (see section 10.2). Such fold systems are known as **disharmonic** (Figure 3.15; see also Figures 10.5 and 10.6).

## CONJUGATE AND POLYCLINAL SYSTEMS

A pair of asymmetric folds with opposite senses of asymmetry such that the axial surfaces dip towards each other are termed **conjugate folds**

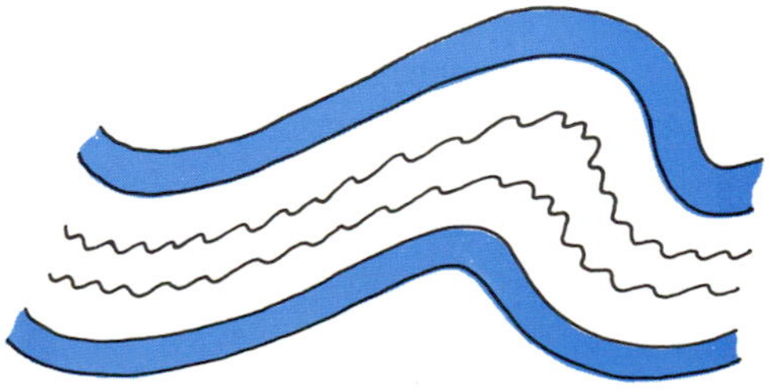

**Figure 3.15** Disharmonic folds. The wavelength of the inner, thinner layers is much shorter than that of the outer layers.

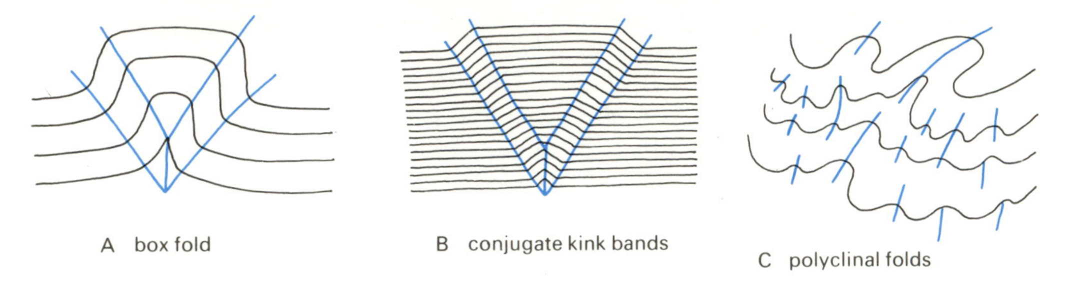


**Figure 3.16** Conjugate and polyclinal folds. A. Box fold – a symmetric fold with four hinges. B. Conjugate kink bands producing an asymmetric structure. C. Polyclinal folds with variable axial surfaces.

(Figure 3.16B). A common type of conjugate fold is a **box fold** (Figure 3.16A), where the fold angles are approximately 90°, forming an almost rectangular structure. A **polyclinal fold** system is a more complex structure formed when the axial surfaces of adjacent folds have differing orientations (Figure 3.16C).

## 3.7 FOLDS IN THREE DIMENSIONS

In the preceding sections, we have been discussing a geometrical classification of folds based essentially on the two-dimensional fold profile, ignoring the third dimension. Folds that maintain a constant profile are termed **cylindroidal folds**. Such folds may be generated by a line moving parallel to itself so that the fold surface produced contains a set of parallel lines. However **non-cylindroidal folds,** which vary in profile shape, are very common, and we shall discuss several special types.

### PERICLINES, DOMES AND BASINS

A **pericline** is a fold whose amplitude decreases regularly to zero in both directions, so the fold has precise limits in space (Figure 3.17A, B). The term pericline is usually applied only to large-scale folds. Periclines may be either anticlinal (antiformal) or synclinal (synformal). Anticlinal periclines are sometimes termed **brachyanticlines**, and synclinal periclines, **brachysynclines**.

A **dome** is a special type of antiformal pericline where the dip is radial, that is, in plan view the structure is close to circular. The synformal counterpart of a dome is called a **basin** (Figure 3.17C, D).

### CULMINATIONS AND DEPRESSIONS

In a surface affected by non-cylindroidal folds, whether periclinal or more complex, the fold axes are generally curved and vary in height. Points of maximum elevation along curved fold axes are termed **culminations**, and points of minimum elevation are termed **depressions**. In certain cases, culminations on antiformal fold axes will be domes and depressions on synformal fold axes, basins.

### INTERFERENCE PATTERNS AND SUPERIMPOSED FOLDS

Many complex fold systems are the result of **interference** between two or more fold sets of simpler geometry. If a layer which already possesses a set of folds is refolded by a second set of folds, a complicated three-dimensional shape is produced (Figure 3.18A). The second set of folds is said to be **superimposed** on the first, and the resulting geometry is termed an **interference structure**. Such structures are easily recognized by the outcrop patterns they produce. The type of outcrop pattern depends on the geometry of the two fold sets and on their relationship to each other.

Three characteristic types of interference structure are shown in Figure 3.18B: (1) closed **dome and basin** shapes, (2) **crescent and mushroom** or 'stirrup' shapes, and (3) hooked **double zigzag** shapes. In the latter two cases the order of superimposition is clear from the fact that the earlier

**Figure 3.17** Periclines, domes and basins. A. Oblique air photograph of part of a large anticlinal pericline in limestone, Asmari Mountain, Iran. (Aerofilms Educational Series, sheet 22, courtesy Aerofilms Ltd.) B. Plan view of synclinal (S) and anticlinal (A) periclines showing typical 'canoe'-shaped outcrops. C, D. Plan views of a basin and a dome. Arrowheads indicate dip direction.

folds and their axial traces are folded around the later fold axes. An example of a large-scale map interference pattern from the Loch Monar area of Scotland is shown in Figure 3.18C, D. The geometry of superimposed folds is discussed in more detail in section 10.3.

## 3.8 FOLDING MECHANISMS AND FOLD GEOMETRY

Several different mechanisms or methods of forming folds have been suggested. The most important distinction between these methods lies

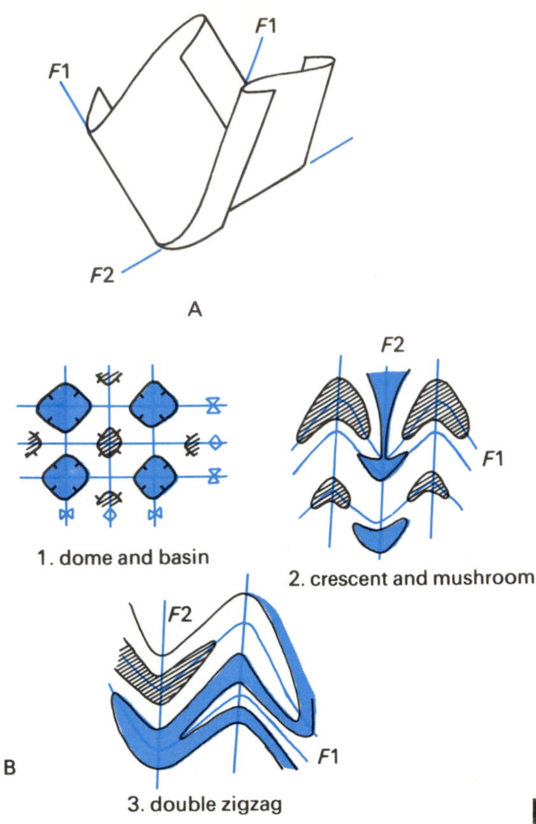


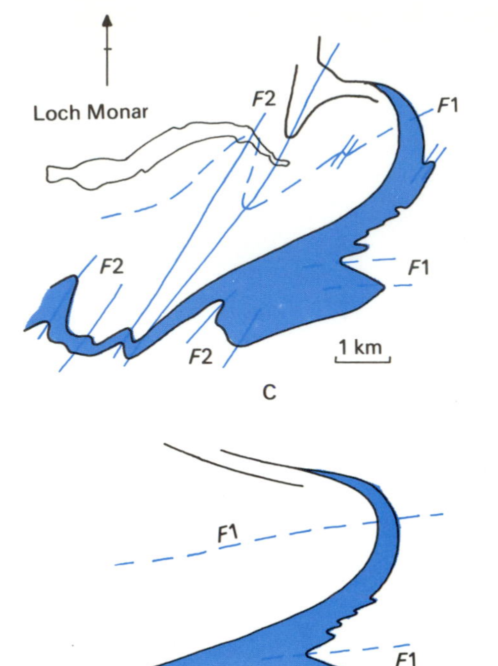

**Figure 3.18** Superimposed folds. A. Refolding of an already folded layer, with folds F1, by a second fold F2, producing a complex three-dimensional shape. B. Three main types of outcrop pattern (interference structure) produced by superimposition. Fold axial traces shown in colour. C. An example of an interference structure from Loch Monar, northwest Scotland. D. The Loch Monar fold before F2 refolding. (C and D after Ramsay, J.G (1967) *Folding and Fracturing of Rocks*, McGraw-Hill, New York, figure 10.20.) E. Interference structure caused by superimposition of steeply inclined open to tight folds on earlier isoclinal folds in Precambrian banded gneiss, Härmanö, Orust, southwest Sweden.

in whether a layer responds actively to a compressive stress applied parallel to its length in forming the fold, or whether it responds passively to a change in shape or in position brought about by movements taking place outside the layer and obliquely to it. The first of these methods may be illustrated by the **buckling** of a thin sheet under lateral pressure, and the second by the gravitational **bending** of a layer draped over a depressed basement block (Figure 3.19). Both are unrealistic models in the sense that they ignore the effect of the material enclosing the layer, which in practice

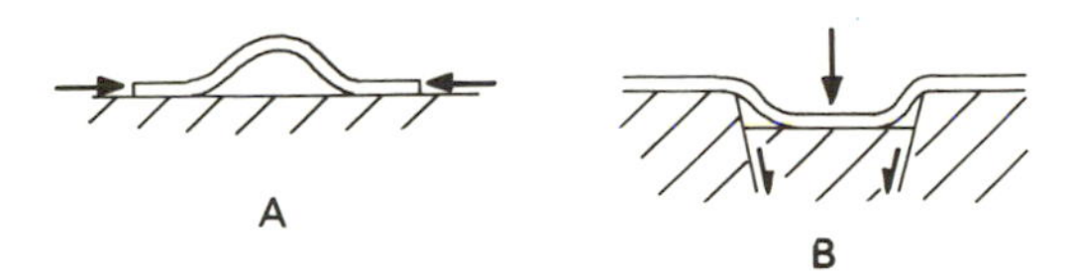


**Figure 3.19** Active buckling (A) and passive bending (B).

has a very important role in determining or modifying the fold geometry.

BUCKLING

Ideally, in a fold produced by buckling under lateral compression of a single layer, the layer maintains its thickness throughout so that a parallel or concentric fold is produced (Figure 3.8A). The deformation produced within the layer is dictated by extension around the outer arc and compression in the inner arc, separated by a neutral surface near the centre of the layer (Figure 10.1A). The geometry of natural buckle folds is much more complex, however, and is discussed in section 10.2.

BENDING

The most common types of fold produced by bending are the accommodation structures formed in thrust and extensional systems, such as hangingwall anticlines and rollovers (see sections 2.6 and 2.7). In such folds the role of gravity is dominant, and the layer is deformed passively by accommodating to changes in shape of the surface over which it is moving. In practice, of course, the layers are also subject to layer-parallel compressive or extensional forces, which trigger other mechanisms such as buckling or kinking. Moreover, the response of a layer to passive bending requires an additional mechanism to produce the change in shape.

OTHER MECHANISMS

Folds produced by buckling, for example, may be modified by other processes. These will be discussed in detail in Chapter 10. Some processes involve internal deformation within the folded layers, which modifies the profile geometry from a parallel towards a similar shape (Figure 3.8A, B). The process of **flexural slip** involves slip between successive layers deformed by buckling or some other mechanism (Figure 10.1B). This type of folding characterizes the deformation of relatively strong layers separated by planes or thin zones of weakness. **Kinking** forms folds of the kink band or chevron type which typically have straight limbs and sharp hinges (Figure 3.8D). Their geometry is controlled by the rotation of sets of layers which remain planar between the kink planes, whereas rapid changes of orientation occur along the kink planes. The limbs of the fold deform by flexural slip. Kinking often produces folds of overall similar profile, although the individual layers exhibit different geometries.

## 3.9 RELATIONSHIP BETWEEN FAULTS, FOLDS AND SHEAR ZONES

At the beginning of this chapter it was pointed out that whereas faults are the product of rather sudden brittle failure, folds were typically formed by slow continuous changes. However, the two processes are not completely separate, as will be seen when we discuss the physical behaviour of rocks under deformation in Chapter 7. For example, under certain conditions, folding may lead to fracture as deformation progresses. Conversely, the process of fault displacement itself may lead to folding, as shown by the development of geometrically necessary folds in the hangingwall of thrusts and extensional faults (Figures 2.11 and 2.13). Moreover, layers of stronger material interbedded with weaker material may exhibit fracturing, while the weaker material shows only folding (Figure 3.20).

Rocks become more ductile at deeper levels in the crust, and a brittle fault at the surface may pass downwards into a structure where the displacement between the two fault blocks is taken up by a type of ductile structure called a 'shear zone' (see below), which embodies characteristics of both folds and faults.

**Figure 3.20** Faults developed during folding, Lady anticline, Saundersfoot, Pembrokeshire. Note thrusts developed in competent layers to accommodate to the tight chevron shape.

SHEAR ZONES

A **shear zone** is a zone of ductile or brittle–ductile deformation between two blocks that have moved relative to each other (Figure 3.21). There are no discrete fracture planes in an ideally ductile shear zone, but in practice there is a complete gradation between a fault zone and a shear zone, with intermediate stages being represented by brittle–ductile shear zones containing faults (Figure 3.22). The structure and mode of formation of shear zones are discussed in more detail in section 10.6.

SLIDES

The term **slide** was originally used for a fault that developed during folding. The type examples of such structures in the Scottish Highlands are low-angle thrusts or extensional faults developed in association with large-scale recumbent folds, and often themselves folded by subsequent deformation. In practice it is very difficult to distinguish between faults that precede the folding and those that develop contemporaneously. Some of the famous slides of the Scottish Highlands (the Sgurr Beag slide for example) are actually ductile shear zones developed under metamorphic conditions and associated with wide belts

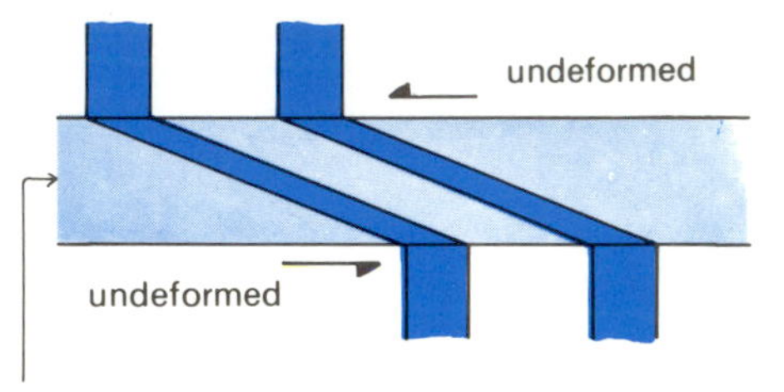

**Figure 3.21** A. Simplified geometry of a shear zone. (Note that in most real shear zones, the marker layers would be curved through the zone rather than straight). B. Shear zone in gneiss, Tshaba River, Zimbabwe.

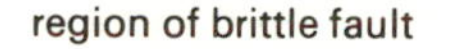


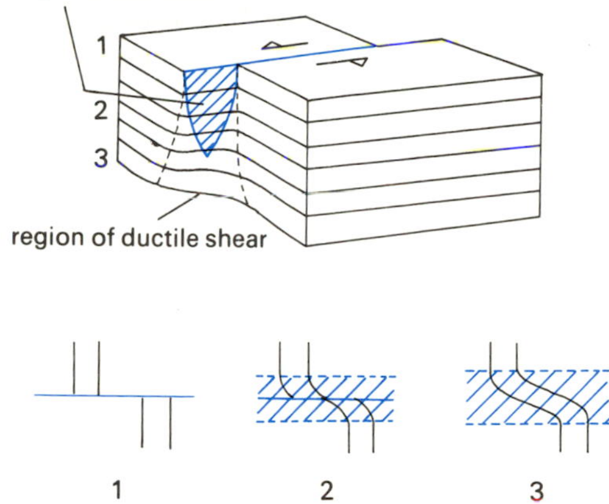


**Figure 3.22** Transition from brittle fault to ductile shear at depth.

of intense deformation. Structural geologists generally now use the term 'slide' to describe structures of the latter type, where the slide corresponds either to a shear zone or to one or more faults within and genetically related to a shear zone. The term is not widely used outside the British Isles.

## FURTHER READING

Fleuty, M.J. (1964) The description of folds. *Proceedings of the Geological Association*, **75**, 461–89.

Hobbs, B.E., Means, W.D. and Williams, P.F. (1986) *An Outline of Structural Geology*, 2nd edn, Wiley, New York.

Price, N.J. and Cosgrove, J.W. (1990) *Analysis of Geological Structures*, Cambridge University Press. [See chapter 10.]

Ramsay, J.G. and Huber, M.I. (1987) *The Techniques of Modern Structural Geology, Vol. 2: Folds and Fractures*, Academic Press, New York. [See chapter 2 for a comprehensive treatment of fold morphology.]

See also further reading for Chapter 10.

# FOLIATION, LINEATION AND FABRIC 4

Structures found in rocks from the deeper levels of the crust are characteristically different from those formed at higher levels. The difference is due mainly to the effect of the increased temperature and pressure in these regions, which increases the ductility of the rocks, and to strong compression leading to intense and repeated folding.

Three important generalizations may be made concerning the structures at deeper crustal levels.

1. Folding, rather than faulting, is the typical mode of deformation.
2. Sets of new planar surfaces (cleavage, schistosity, etc.) are commonly developed.
3. Pervasive recrystallization under compression results in the internal rearrangement of the rock texture producing a new 'fabric', or structural texture.

## 4.1 FOLIATION

A **foliation** is a set of new planar surfaces produced in a rock as a result of deformation. Foliation is a general term covering several different kinds of structure produced in different ways. Examples of common types of foliation are shown in Figure 4.1.

'Slaty cleavage', 'schistosity' and 'gneissosity' are all examples of foliations. Many rocks exhibit several generations of foliation, which are distinguished chronologically using the system S1, S2, S3, etc. Earlier foliations are deformed and cut by later foliations, enabling the structural history of the rocks to be established. The bedding is usually the first recognizable planar surface, and may be designated SS or S0. Sometimes, however, the bedding is obliterated by deformation and the first visible planar surfaces are tectonic; in other cases the origin of the first set of surfaces may be indeterminate. In many areas there is a foliation parallel to bedding, often termed a **bedding foliation**. This may arise either as a result of the load pressure of overlying strata, or through deformation associated with folding, in which case the foliation will locally cut the bedding in fold hinges (see Figure 4.6B).

### TYPES OF FOLIATION

The nomenclature of the various types of foliation is rather confusing. This reflects the fact that for many years the origin of such structures as slaty cleavage and gneissose banding, for example, was not fully understood. The term **cleavage** refers to the fissility of a rock, which allows it to be split along a set of foliation planes, and embraces structures of various origins, formed at low metamorphic grade. It is replaced at higher grades by 'schistosity' and 'gneissosity' (see below). The four main types of cleavage are: (1) **slaty cleavage**, which is the very pervasive cleavage found in typical roofing slates, owing to the parallel alignment of elongate minerals or grains; (2) **fracture cleavage**, a set of closely spaced fractures; (3) **crenulation cleavage**, a banded structure produced by microfolding; and (4) **solution cleavage**, in which differential solution and deposition of the more soluble minerals in the rock has produced a compositional banding. These cleavage types are end-members between which continuous variation can occur, and any particular example of cleavage may incorporate elements of two or more of these end-members. This may be illustrated by the cleavage tetrahedron (Figure 4.2).

Foliations may either be **penetrative**, i.e. pervasive, affecting all parts of the rock, or they may be **non-penetrative**, i.e. occur at intervals,

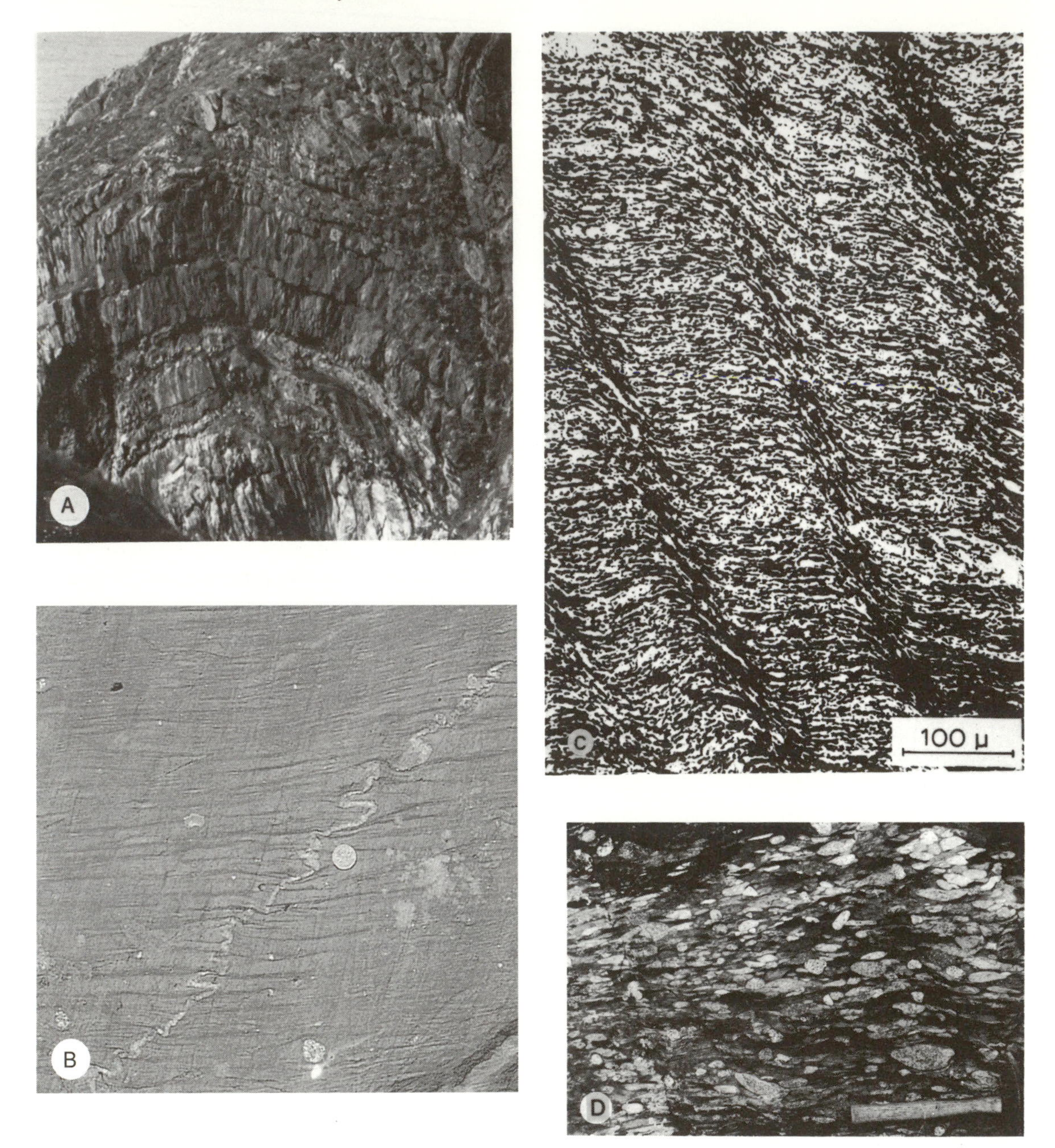


**Figure 4.1** Examples of rock cleavage. A. Axial-plane cleavage in folded sandstones of the South Stack series, Rhoscolyn, Anglesey. B. Pressure-solution cleavage in Huronian tillite, Whitefish Falls, Ontario, Canada. Note concentrations of dark material forming 'pressure shadows' around pebbles. C. Photomicrograph showing the development of crenulation cleavage in slate. The cleavage 'planes' correspond to zones of superimposed short limbs of asymmetric crenulations which are enriched in micas, etc. relative to quartz, thus forming darker layers. (From photograph by W.D. Means (1977) in *Atlas of Rock Cleavage* (eds B.M. Bayly, G.J. Borradaile and C.McA. Powell), University of Tasmania, Hobart.) D. Shape fabric foliation produced by the alignment of flattened and elongated pebbles in a deformed conglomerate. Stevenson Lake, Manitoba. (Photograph by I.F. Ermanovics.)

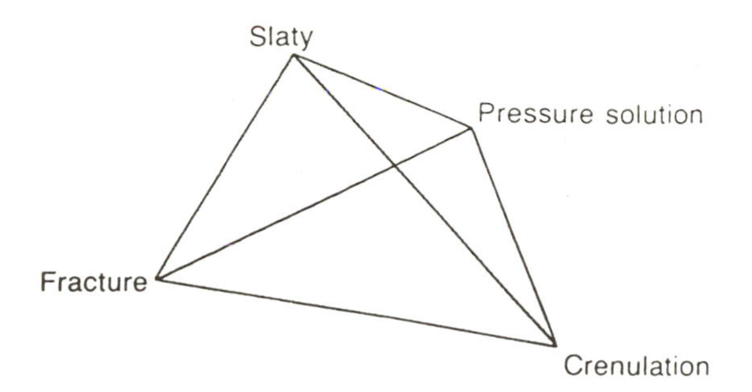


**Figure 4.2** The cleavage tetrahedron (see text).

separated by unaffected rock. Non-penetrative cleavage is often termed **spaced cleavage**. Whether or not a foliation is regarded as penetrative depends on the scale of observation; penetrative foliation at outcrop or hand-specimen scale may appear as discrete planes in thin section.

**Schistosity** is a foliation produced by parallel alignment of tabular or elongate minerals, such as micas or hornblende, in rocks that have undergone more intense metamorphic recrystallization. **Gneissose banding** or **gneissosity** is another type of foliation, and is produced by compositional layering, similar to bedding but of metamorphic or deformational origin. These various types of foliation will now be described in more detail.

## SLATY CLEAVAGE

This type of cleavage is best shown in fine-grained rocks such as mudstones that have been deformed under very low-grade metamorphism. Consequently the nature of the internal changes in the rock that have produced this penetrative fissility is not usually obvious at outcrop or in hand specimen. Under the microscope, however, the nature of the cleavage becomes much clearer. The cleavage planes are then seen to be due partly to the parallel orientation of sheet-like minerals such as muscovite and clays, and partly to the parallel arrangement of tabular or lensoid aggregates of particles.

The origin of this structure becomes clear when we examine slates containing deformed objects of known initial shape, such as fossils, or reduction spots. The plane of the cleavage is seen to correspond to the plane of flattening of such deformed objects, which leads us to conclude that the parallel orientation of grain aggregates and tabular minerals is due to intense compression of the rock in a direction perpendicular to the plane of the cleavage. This compression has resulted in the rotation of previously formed minerals and has also controlled the growth of new minerals, causing them to be aligned in the direction of the cleavage. Slaty cleavage appears to form only when a suitable rock has been compressed by about 30% of its initial length.

## FRACTURE CLEAVAGE

As the name suggests, a fracture cleavage consists of parallel, closely spaced fractures. Fracture cleavage is usually easy to distinguish from slaty cleavage because it consists of discrete planes separated by slabs of uncleaved rock, called **microlithons**. Displacement of the rock across these planes may often be visible in thin section, showing that the planes are microfaults. This type of cleavage is formed under brittle conditions at low temperatures and is typical of deformed relatively strong rocks such as sandstones.

Fracture cleavage may accompany slaty cleavage, either in the same rock or in adjoining layers, and in some cases displacements may take place on previously formed slaty cleavage planes. This has caused some confusion in the past over the origin of slaty cleavage.

Fractures may be compressional in origin (shear fractures) or extensional, in which case they will often be filled by quartz or calcite (see Figure 6.12). Certain examples of extensional fracture cleavage are thought to be due to **hydraulic fracturing** caused by the pressure of water forced out of sediments by the load effect of the strata above.

## CRENULATION CLEAVAGE

Crenulation cleavage is caused, as the name suggests, by small-scale folding (crenulation) of very thin layers or laminations within a rock. If the axial surfaces of such crenulations are closely-spaced and parallel, they produce a marked foliation

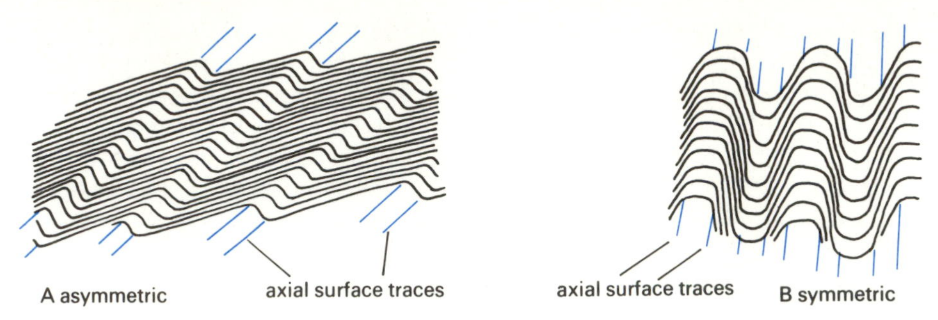


**Figure 4.3** Crenulation cleavage. A, asymmetric; B, symmetric. The cleavage is parallel to the axial surface traces (colour) of the crenulations.

(Figure 4.3). This foliation is often enhanced by selective recrystallization, leading to a concentration of certain constituents in layers (see *Solution cleavage*, below). Thus micas, for example, may become concentrated in one set of limbs of asymmetric crenulations or in both sets of limbs of symmetric crenulations, as a result of the migration of quartz or calcite into limbs or hinges (Figure 4.1C). The combination of compositional banding and parallel orientation of platy minerals provides planar weaknesses that can impart a strong fissility to the rock.

Crenulation cleavage is very commonly associated with the deformation of rocks that already possess a strong cleavage or schistosity as a result of an earlier deformation. The early foliation planes provide a well-laminated structure which assists the crenulation process.

Previously undeformed rocks may also possess a crenulation cleavage in suitable lithologies – generally very finely laminated shales. If the crenulations are on a sufficiently small scale, the resulting cleavage becomes indistinguishable from other types of slaty cleavage in hand specimen.

## SOLUTION CLEAVAGE

A new compositional banding can arise through the migration of certain constituents of a rock during deformation, and, for example, often accompanies the formation of a crenulation cleavage (see above). This phenomenon, which is common in low-grade or unmetamorphosed rocks, is caused by the process of **pressure solution** and the resulting cleavage is termed **solution cleavage** (Figure 4.4A).

The solution and accompanying deposition appear to be part of a diffusion process which takes place by means of a grain-boundary fluid phase (see section 4.4). Solution occurs on grain or layer

**Figure 4.4** Compositional layering produced by deformation. $A_1$. Solution cleavage in synform affecting Devonian slates, Tor Cross, Devon. The light-coloured silt layers are cut by a penetrative cleavage defined by seams enriched in dark material, presumably owing to the solution and redeposition of the light material elsewhere. $A_2$. Close-up of $A_1$. B. Photomicrograph of A showing dark layers where material is thought to have been dissolved. Bedding is clearly shown dipping steeply to the right (scale bar 2 mm). (A and B from photograph by H.R. Burger (1977) in *Atlas of Rock Cleavage* (eds B.M. Bayly, G.J. Borradaile and C.McA. Powell), University of Tasmania, Hobart.) C. Gneissose banding produced by deformation. The three photographs show three stages in the deformation of a coarse-gained leucogabbro from the Archaean of West Greenland. In $C_1$ the mafic areas are easily recognizable although generally elongate and recrystallized. In $C_2$ the rock has been transformed into a gneiss with alternating lensoid mafic and felsic components. In $C_3$ further deformation has resulted in a striped rock whose origins could not be determined from this outcrop alone. (From Bridgwater, D., Keto, L., McGregor, V.R. and Myers, J.S. (1976) in *Geology of Greenland* (eds A. Escher and W.S. Watt), Geological Survey of Greenland, Copenhagen, figures 48–50.)

A1

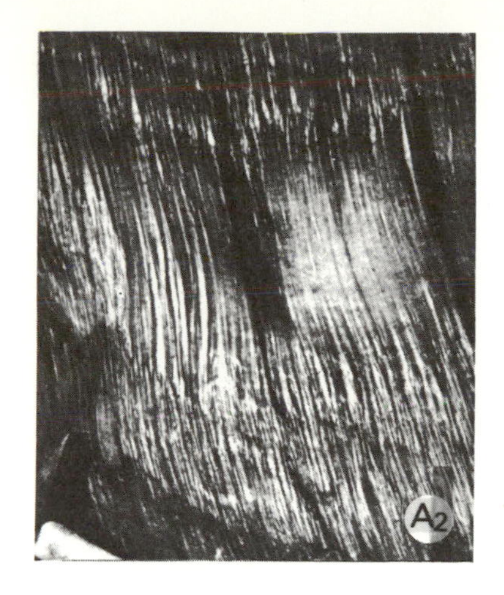
A2

B

C2

C3

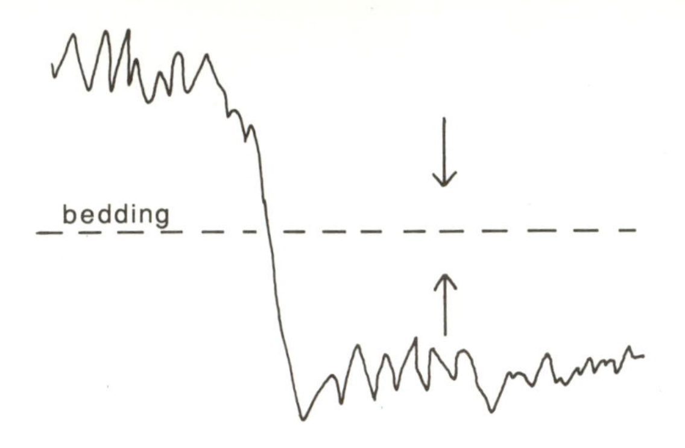


**Figure 4.5** Stylolites in Niagaran limestone. (From Price, N.J. and Cosgrove, J.W. (1990) *Analysis of Geological Structures*, Cambridge University Press, Cambridge, figure 15.3.)

boundaries that are perpendicular to the direction of greatest compression, and deposition takes place on surfaces that are perpendicular to the direction of extension or least compression. In its most common form, this structure consists of alternating lighter and darker bands, where the darker bands are produced by the removal by solution of calcite or quartz (Figure 4.4B). The same process is responsible for the formation of **stylolites**, which are a common feature of limestones (Figure 4.5); these structures form usually along bedding surfaces by pressure solution at localized sites on the surface, resulting in distortion of the originally planar surface. The new surface is typically highly irregular, and is marked by concentrations of insoluble dark material.

SCHISTOSITY

With increasing metamorphic grade, slates are transformed to schists by an increase in the size of the newly formed metamorphic minerals. In slates, the aligned planar minerals that produce the slaty cleavage are invisible to the naked eye, whereas in schists, the individual tabular crystals of mica, hornblende, etc. are large enough to be visible in hand specimen. A foliation marked by the parallel orientation of such tabular minerals in a metamorphic rock with a sufficiently coarse grain size is called a **schistosity**.

A schistosity can be produced directly from a slaty cleavage merely by a coarsening of the grain-size, consequent on an increase in temperature. Crenulation cleavage may also pass into schistosity as a result of grain-size coarsening. Many schistose rocks show a combination of mineral alignment (true schistosity) and a tabular or lensoid arrangement (i.e. a shape alignment), similar to that seen in many slates, produced by compression, but on a larger scale.

GNEISSOSITY

A parallel banding produced by alternating layers of different composition is an important feature of many deformed metamorphic rocks, and can arise in several different ways. We have already discussed how the migration of certain constituents during the formation of a crenulation cleavage can produce such a banding. Preferential redistribution of minerals in a metamorphic rock is termed **metamorphic segregation** or **metamorphic differentiation** and is very important in producing compositional layering in deformed rocks.

Compositional layering is a characteristic feature of most gneisses and is termed **gneissosity** or 'gneissose banding'. Gneisses are coarse-grained metamorphic rocks, typically quartzo-feldspathic in composition. The distinction between schists and gneisses is not clear cut, and individual geologists have their own preferences as to where the dividing line should be drawn. Many metamorphic rocks display both schistosity and gneissosity.

The origin of the compositional layering in gneisses has occasioned much debate in the past and it is now clear that several different processes are involved. Many gneisses with a layered or lensoid structure (**augen gneisses**) have clearly originated as coarse-grained plutonic rocks, often rather variable in composition, with abundant cross-cutting veins. Under intense deformation such rocks become layered, partly as a result of the flattening and elongation of large crystals, and also due to the rotation of veins and other heterogeneities into the plane of flattening (Figure 4.4C). Metamorphic segregation is also important in the production of gneissose banding, either by enhancing a deformational layering or by producing a new layering by pressure solution.

Intensely deformed gneisses of sedimentary origin (paragneisses) and derived from sediments of mixed composition, e.g. greywackes or arkoses, are often very difficult to distinguish from those of igneous origin (orthogneisses). In view of the above, it is clear that the presence of a compositional banding cannot be assumed to indicate a sedimentary origin.

### PLANAR SHAPE FABRIC

The structure produced by a set of parallel, dimensionally oriented objects within a deformed rock is called a **shape fabric,** and a planar shape fabric is a type of foliation. These aligned objects may be grains, or grain aggregates, such as pebbles, ooliths, fossils, etc. (see Figure 4.1D). Planar shape fabric is one of the elements contributing to slaty cleavage. The subject of fabrics is discussed in more detail in section 4.4.

### RELATIONSHIP BETWEEN FOLDS AND FOLIATIONS

The deformation that is responsible for the formation of a foliation will normally also produce folds in a suitably layered rock. Since the two structures are produced by the same compression, they generally bear a simple geometrical relationship to each other. A foliation that corresponds to a plane of flattening is parallel or sub-parallel to the axial surface of a related fold (Figure 4.6A) or forms a fan-shaped structure arranged symmetrically about the axial surface (Figure 4.6C, D). Under intense deformation, tight to isoclinal folds are produced whose limbs may become thinned and modified until they are indistinguishable from the foliation, and only the fold hinges are clearly definable. Such folds are termed **intrafolial folds** (Figure 4.6B). Where the hinges are completely detached from the limbs, the folds are called **rootless intrafolial folds**.

The association between folds and foliations conveys valuable information about the conditions of deformation of a rock and is discussed more fully in Chapter 10.

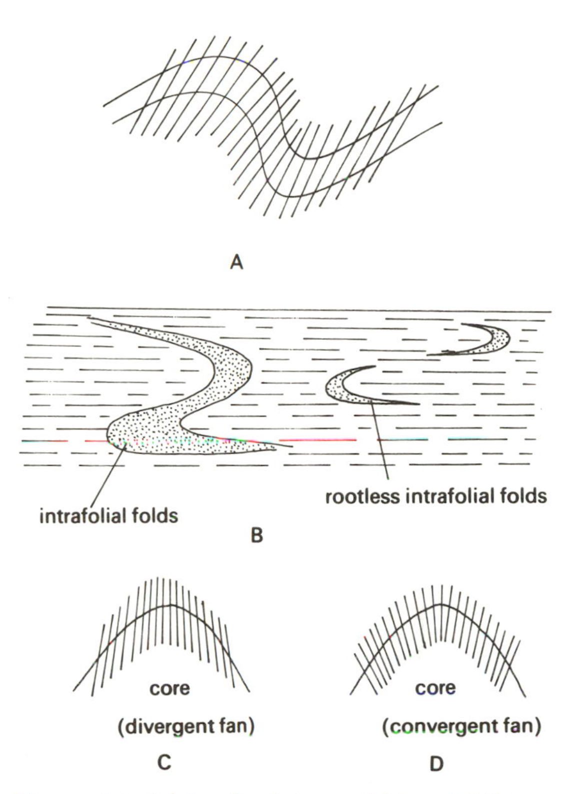


**Figure 4.6** Relationship between folds and foliation. A. A foliation developed during folding is often subparallel to the fold axial surface. B. Intense deformation may cause thinning and rotation of fold limbs into the plane of the foliation, ultimately causing the limb to disappear (see text). Types of folds thus formed are intrafolial and rootless intrafolial folds. C, D. Foliation developed during folding often has a simple fan-shaped arrangement about the axial surface – C, divergent fan; D, convergent fan.

## 4.2 LINEATION

A **lineation** is a set of linear structures produced in a rock as a result of deformation, and is therefore the linear counterpart of a foliation.

There are many types of lineation, and one of the problems facing an inexperienced geologist in the field is to distinguish significant from insignificant lineations. Since any two planes intersect in a line, the more planar surfaces there are in an exposure, the more potential lineations there will be. It is important to identify the nature of the surface(s) on which the lineations occur. Common examples

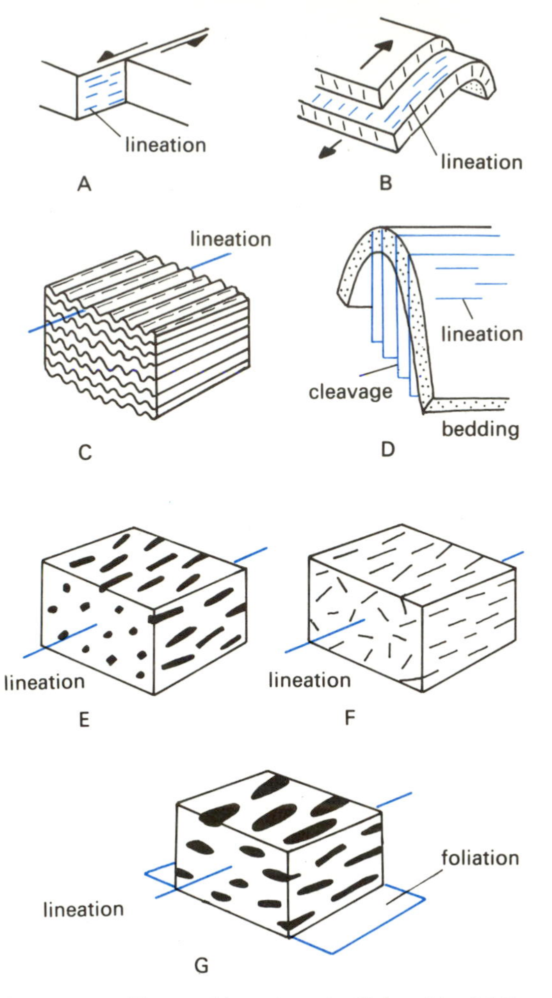


**Figure 4.7** Types of lineation. A. Slickenside striation on a fault. B. Striations on a bedding plane caused by flexural slip. C. Crenulation lineations. D. Intersection lineation between bedding and cleavage. E. Mineral elongation lineation. F. Mineral intersection lineation. G. Fabric that is partly planar and partly linear.

of lineation are slickenside striations, crenulation fold axes, elongate pebbles in a deformed conglomerate, lines of intersection of bedding and cleavage in slate, and alignment of the longer dimension of elongate minerals.

A convenient subdivision of lineations can be made as follows:

1. lineations indicating the direction of movement along a surface (e.g. slickenside striations);
2. axes of parallel crenulations or small-scale folds;
3. dimensional elongation of a set of deformed objects such as pebbles, ooids, megacrysts, etc.;
4. parallel orientation of elongate minerals (mineral lineations);
5. intersections of sets of planes (intersection lineations).

LINEATIONS INDICATING MOVEMENT DIRECTION

Fault surfaces showing slickensides (section 2.4) commonly exhibit grooves or striations indicating the direction of relative movement (Figure 4.7A). A similar type of lineation is often found on bedding surfaces involved in 'flexural slip folding' (see section 10.1), where successive layers have moved over one another as the folds tighten (Figure 4.7B). Such lineations usually make a large angle with the fold axis. Both these types of lineation lie in a particular surface and do not permeate the body of rock in which they are found. They are thus non-penetrative structures.

CRENULATION AXES

Rocks that are finely laminated and affected by intense small-scale folding (crenulation) exhibit a strongly developed linear structure due to the abundant parallel fold hinges that permeate the rock (Figure 4.7C). Slaty or schistose rocks frequently exhibit two or more sets of such crenulations and crenulation cleavages, and individual foliation surfaces may then contain two intersecting sets of crenulation lineations forming a small-scale interference pattern.

INTERSECTIONS OF PLANES

One of the commonest types of lineation is formed when two sets of planar structures intersect. Strongly developed lineations are often found at the intersection of bedding and cleavage (Figure 4.6D) or of two foliations – a schistosity and a crenulation cleavage, for example. Such intersection lineations are very often parallel to local fold axes

A

**Figure 4.8** Examples of linear structure. A. Strongly developed rodding lineation in Lewisian siliceous schists from the Loch Maree Group, Gairloch, northwest Scotland. B. Mullion structure in vertical bedding surface between sandstone and slate – the slate has been mostly removed by erosion – North Eifel, Germany. (From Hobbs, Means and Williams, 1986, figure 6.9.)

and crenulation lineations that belong to the same episode of deformation.

Great care must be taken in the field to determine the origin of a lineation before it is recorded and measured, since the structural significance of different lineations may be quite different. Many insignificant lineations may be observed at an exposure caused, for example, by the intersection of a foliation on a random joint surface, or on the surface of the exposure itself, and these should be ignored. Crenulation lineations and intersection lineations parallel to the local fold axes were often called 'b-lineations', and lineations parallel to the movement direction were called 'a-lineations' in the older literature, but these terms are not recommended.

### DIMENSIONAL ELONGATION

An important type of lineation termed an **elongation lineation** is formed by the parallel alignment of a set of elongate objects within a rock body as a result of deformation. There is a wide variety of objects that may define such an elongation lineation. Obvious examples are pebbles, ooids and spherulites, but individual grains or grain aggregates, or indeed any kind of heterogeneity within the rock, may acquire an elongate shape as a result of deformation and contribute to an elongation lineation. In strongly deformed rocks, such a lineation may pervade the whole rock and consist of many different types of oriented feature.

### RODDING AND MULLION STRUCTURE

Many highly deformed rocks possess a rodded structure whose origin is not immediately apparent (e.g. Figure 4.8A, B). Rods may be formed by the hinges of crenulated quartz veins, or may represent extremely elongate pebbles. In rocks that have been subjected to very high strains, pebbles that are only a few centimetres across may measure one or more metres long, in which case the true nature of the lineation may not be apparent at first sight. A rodded structure with dimensions of tens of centimetres across is often termed **mullion structure**. A common type of mullion is formed at the interface between competent and incompetent layers (sandstone and shale for instance) and results partly from the folding of the interface and partly from the effect of bedding–foliation intersection (Figure 4.8B). Other types of rodding or mullion structure may represent an elongation lineation, and lie parallel to the direction of extension; in cases of boudinage (see section 4.3), the rods or mullions may be perpendicular to the extension direction. Each example should be carefully examined to determine its true nature.

### MINERAL ORIENTATION

Parallel linear orientation of individual minerals is very common in deformed metamorphic rocks. Lineations of this type are usually called **mineral lineations** (Figure 4.6E). The parallel alignment of the crystals forming the lineation may be due to rotation into a favoured attitude as a result of the deformation, or to the effect of recrystallization under pressure, when certain crystallographic orientations are encouraged and others suppressed. Minerals that grow preferentially in a particular direction under stress are said to show **growth anisotropy**.

Mineral lineations may be caused by the alignment of crystals with an elongate habit, e.g. hornblende (Figure 4.7E), but they may also be caused by the alignment of minerals with a platy habit, e.g. micas, if they are arranged in such a way as to form an intersection lineation (Figure 4.7F). In the latter case, they may be arranged randomly in the plane perpendicular to the lineation.

Many mineral lineations are associated with foliations, particularly where the latter are also formed by mineral orientation, or, as is commonly the case, by a combination of dimensional and crystallographic planar orientation (Figure 4.7G). In such cases, the planar and linear elements are both aspects of the same three-dimensional geometry which reflects the way the rock has been deformed (see section 6.3). We may envisage a continuous progression from a purely linear structure with no planar element, reflecting elongation without flattening, through a variety of structures produced by combinations of elongation and flattening, to a purely planar structure with no linear element, which reflects purely flattening.

## 4.3 BOUDINAGE

When relatively strong layers of rock are stretched and become elongated during deformation, they may separate into blocks (Figure 4.9C) or form lensoid or pillow-shaped structures separated by narrow 'necks' (Figure 4.9A, D). Such structures are called **boudins** (or 'pull-apart' structures) and the process of elongation that produced them is called **boudinage**. Where the separation is incomplete and the layers show a narrowing or 'necking', the structures are often termed **pinch-and-swell** structures. Boudins may either be linear and form an elongation lineation (Figure 4.9A) or they may form a two-dimensional **chocolate-tablet structure** if extension has occurred in both directions within the plane of the layer (Figure 4.9B).

Boudins characteristically form in relatively thin layers (usually up to about 1 m in thickness) that are stronger and more competent than the enclosing rock, which therefore tends to stretch in a ductile fashion and flow into the spaces between the boudins. Many boudins, particularly of the 'pull-apart' type, are separated by veins of quartz or calcite or, in high-grade metamorphic rocks, pegmatitic material.

Boudins are particularly useful as indicators of the directions of extension in very intensely deformed rocks. The relationship between boudinage and folding during progressive deformation is discussed in section 6.8.

## 4.4 FABRIC

The **fabric** of a rock body is the geometrical arrangement of all the structural elements within

boudin

A linear boudinage

B chocolate-tablet boudinage

C

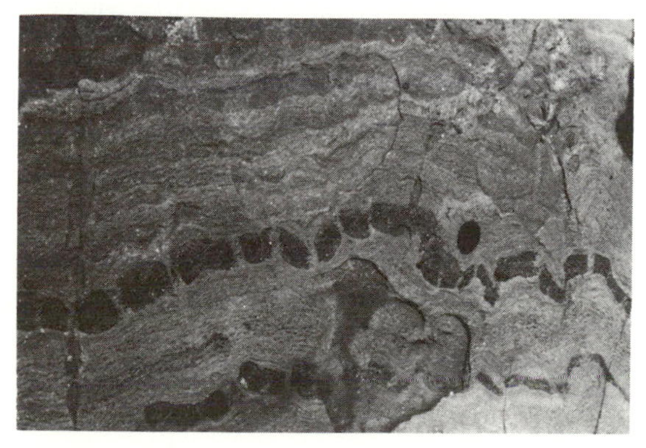

D

**Figure 4.9** Boudinage. A. Linear boudinage. B. 'Chocolate-tablet' boudinage produced by two directions of extension. C. Boudinage (pinch-and-swell) structure in quartzite layer. D. Folded boudinage shown by thin dykes in Precambrian gneisses at Tromøy, southern Norway.

the body. It can be regarded as the structural 'texture' of the rock. Only structural elements observed on a relatively small scale (hand-specimen or microscopic size) are normally considered as fabric, not large-scale structures. Thus fabric study effectively involves structural analysis at grain-size level.

A fabric is made up of a number of fabric elements each consisting of a group of geometric features of the same kind. These elements are usually either planar or linear. Planar fabric elements include the small-scale expressions of a foliation, e.g. the dimensional or crystallographic orientation of grains and grain aggregates. They also include various kinds of planar discontinuities, such as grain boundaries, twin planes and dislocations within crystals. Fabric elements usually produce a planar or linear orientation, but it is possible to have 'random' fabrics, consisting of randomly oriented fabric elements. A fabric composed of dimensionally oriented objects (e.g. grains or grain-aggregates) is known as a **shape fabric**.

The fabric of a rock expresses the geometry of the deformation in the same way as foliations and lineations do, and reflects the way in which the rock has accommodated itself to deformation by changes in the shape, pattern and orientation of the grain network.

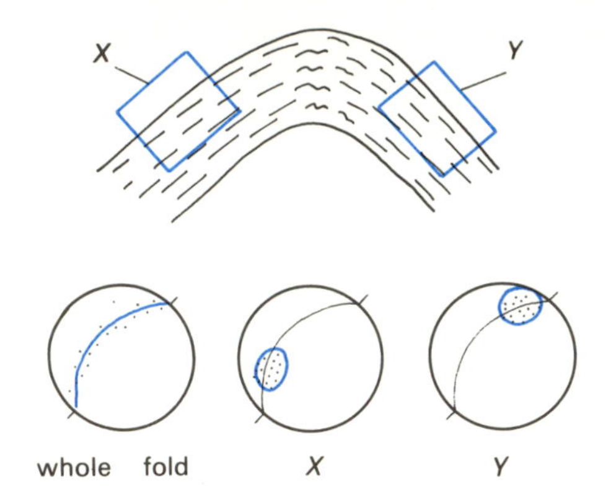


**Figure 4.10** Homogeneous domains (X, Y) in a heterogeneous fabric. Stereograms show poles to planar fabric over the whole fold and in domains X and Y.

## HOMOGENEITY OF FABRICS

The fabric of a rock is said to be **homogeneous** if any two parts, similarly oriented, show identical structure. The term is usually applied to a particular fabric element (a foliation, for example). The scale of observation is important, since the fabric element in question may be homogeneous at hand-specimen scale but heterogeneous at thin-section scale.

**Heterogeneous** fabrics can be divided into homogeneous domains in order to simplify the analysis (Figure 4.10).

## MICROFABRIC ELEMENTS

Detailed study of how deformation affects individual grains, carried out particularly with the electron microscope, has shown that deformation on the submicroscopic scale is very heterogeneous and takes place by displacements along a series of dislocations or discontinuities within grains and on grain boundaries. The processes involved in the production of the crystal microfabric are discussed in more detail in section 7.10 (see also Hobbs, Means and Williams, 1986).

There are a number of different types of discontinuity at individual crystal scale. In addition to the grain boundaries, these discontinuities consist principally of crystal defects, the nature of which is controlled by the molecular structure of the crystal. Planar defects are normally parallel to one of the crystallographic planes in a crystal and allow displacements to take place within the crystal. These displacements enable the crystal to change its shape to accommodate the deformation. Many crystals show planar defects, usually of limited extent, across which there is a very small displacement of the crystal structure. These defects are called **stacking faults** and are important in the propagation of dislocations that may subsequently form more pervasive deformation planes.

**Sub-grain boundaries** are planar defects within grains; they separate regions of slightly different lattice orientation and are visible as a small change in extinction angle (Figure 4.11A, B). Such

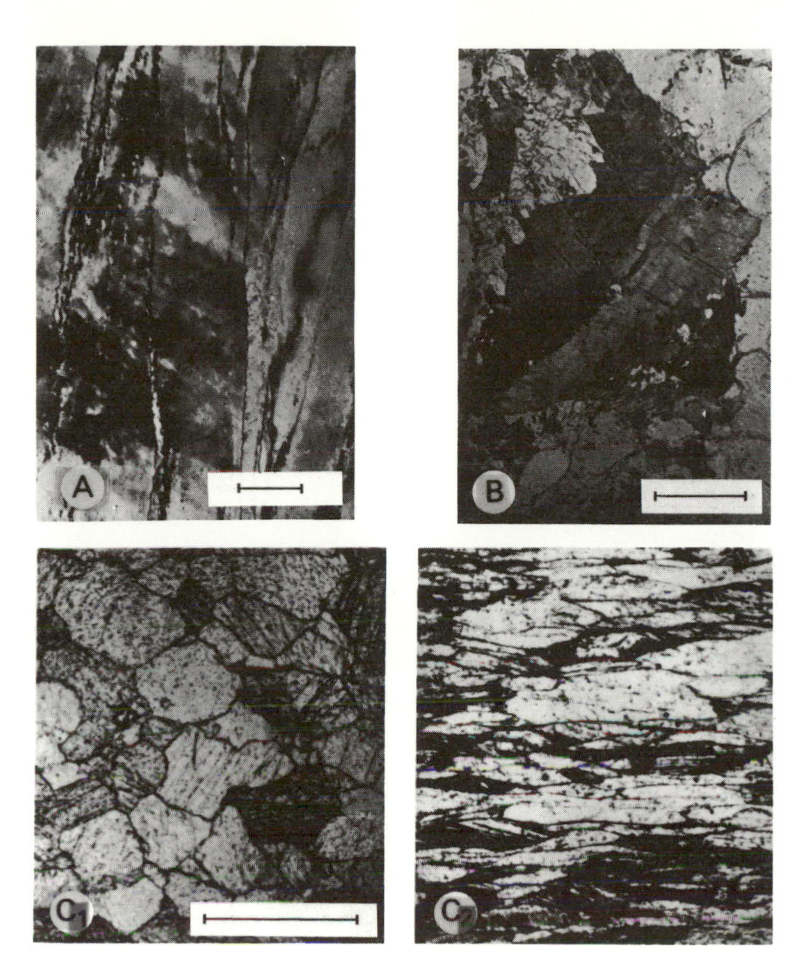


**Figure 4.11** Microfabric at crystal scale. A. Deformation bands in naturally deformed quartz. B. Kink boundaries in naturally deformed biotite. Both A and B from Arunta complex, central Australia; scale mark, 0.3 mm; crossed polars. (From Hobbs, Means, 1986, figure 2.11.) C. Flattened grain texture resulting from intragranular gliding and twinning in experimentally deformed calcite. $C_1$, undeformed Carrera marble; $C_2$, shortened by 50% at 400 °C and 1.5 kbar confining pressure at $10^{-5}$/s strain rate. Plane polars. (Photographs by E.H. Rutter and N.D. Shaw (1977) in *Atlas of Rock Cleavage* (eds B.M. Bayly, G.J. Borradaile and C. McA. Powell), University of Tasmania, Hobart.)

boundaries cause the well-known phenomenon of **undulose extinction** in quartz.

**Deformation bands** are narrow planar zones containing material that has deformed differently from the adjoining parts of the crystal, either by a small change in lattice orientation or by a more complicated series of changes. **Deformation lamellae** are a special type of deformation band that possess a uniform structure but exhibit a different refractive index from the host crystal. Another type of deformation band is caused by **deformation twinning** (Figure 4.11B), which is common in many crystals, especially calcite. This structure is produced by a slight rotation of the crystal lattice, in a homogeneous manner, between the twin planes.

## FURTHER READING

Hobbs, B.E., Means, W.D. and Williams, P.F. (1986) *An Outline of Structural Geology*, 2nd edn, Wiley, New York. [Contains an excellent section on microfabrics.]

Turner, F.J. and Weiss, L.E. (1963) *Structural Analysis of Metamorphic Tectonites*, McGraw-Hill, New York. [This is still the most comprehensive descriptive work on tectonic fabrics, the geometrical analysis of structures and the use of stereographic projection in structural geology.]

The following should only be read after Chapters 5–10 of this book have been tackled:

Cosgrove, J.W. (1976) The formation of crenulation cleavage. *Journal of the Geological Society of London*, **132**, 155–78.

Gray, D.R. and Durney, D.W. (1979) Crenulation cleavage differentiation: implications of solution-deposition processes. *Journal of Structural Geology*, **1**, 73–80.

Platt, J.P. and Vissers, R.L.M. (1980) Extensional structures in anisotropic rocks. *Journal of Structural Geology*, **2**, 397–410. [Discusses boudinage in the context of a general treatment of extensional structures.]

Price, J.N. and Hancock, P.L. (1972) Development of fracture cleavage and kindred structures. *Proceedings of the 24th International Geological Congress*, Section 3, pp. 584–92. [Discusses the hydraulic fracturing mechanism.]

# STRESS 5

In the following six chapters, 5–10, structures are discussed and explained in terms of the processes that governed their formation (i.e. processes of deformation). To do this, we need to start with a theoretical treatment of deformation in terms of the causal forces or pressures that act on a rock body, and of the geometrical changes resulting from these. The behaviour of materials in response to deformation is then discussed, together with the various physical controls, such as temperature and confining pressure, that influence this response. We shall also examine methods of quantitative determination of deformation in rocks. Having discussed the more theoretical aspects of deformation, we shall then be in a position to deal with the deformation mechanisms involved in faulting, folding and the emplacement of igneous intrusions.

The **deformation** of a material is the process whereby physical changes are produced in the material as a result of the action of applied forces. The forces that act on the rocks of the Earth's crust arise in various ways. The most important of these forces are due to gravity and to the relative movements of large rock masses in the crust and upper mantle. Since gravitational force is proportional to mass, the weight of an overlying column of rock constitutes a very significant force on rocks at depth in the crust.

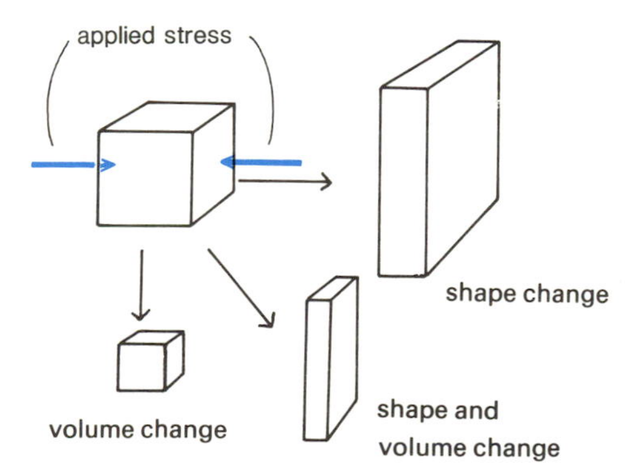


**Figure 5.1** Effects of stress on a cube: a change in shape and/or volume.

The forces acting on a portion of rock produce a set of 'stresses' (see below), and the amount of deformation caused by these stresses is measured by the change in dimensions of the body. This change may consist of a change in shape, or volume, or both shape and volume (Figure 5.1) and constitutes the 'strain' (see Chapter 6).

## 5.1 FORCE AND STRESS

In order to understand the concept of stress, we must first define 'force': a **force** is the product of a mass and its acceleration. Force is a vector quantity, and thus possesses both amount and direction; it can be represented by a line whose length specifies the amount and whose orientation specifies the orientation of the force. The sense of direction may be indicated by an arrow (see Figure 5.2).

### RESOLUTION OF FORCES

Figure 5.2A shows how a force $F$ may be resolved into two components $F1$ and $F2$ at right angles and, conversely (Figure 5.2B), how any two forces $F1$ and $F2$ may be represented by their resultant $F$. By extending this principle, it is clear that any system of forces acting at a point can be represented by a single resultant force.

### DEFINITION OF STRESS

In rock deformation, we usually neglect any overall

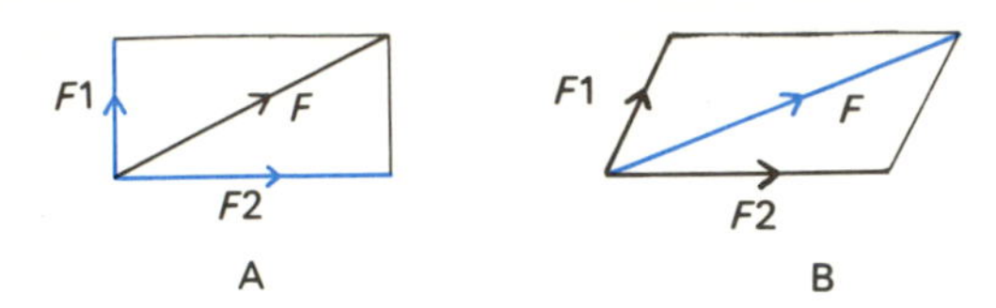


**Figure 5.2** Resolution of forces. A. Force $F$ resolved into two components $F1$ and $F2$. B. Two forces $F1$ and $F2$ represented by resultant $F$ (see text).

acceleration of a body and treat the system of forces as closed, i.e. opposing forces cancel out. This situation is governed by Newton's third law of motion, which states: 'For a body at rest or in uniform motion, to every action there is an equal and opposite reaction'.

We can now define stress: a **stress** is a pair of equal and opposite forces acting on unit area of a body. Thus a stress results from a force acting on a surface (either real or imaginary) surrounding or within a body, and comprises both the force and the reaction of the material on the other side of the surface. The magnitude of the stress depends on the magnitude of the force and on the surface area over which it acts. Thus:

$$\text{Stress} = \text{Force/Area}.$$

The force of gravity can give rise to a stress which is measured by calculating its effect across a surface. Gravity makes an important contribution to the stress field governing the formation of folds and faults.

### UNITS OF MEASUREMENT

The standard SI unit of force is the **newton** (N), which is defined as:

1 newton = 1 kilogram metre per second squared ($1\,\text{kg}\,\text{m}\,\text{s}^{-2}$).

The SI unit for both pressure and stress is the **pascal** (Pa), which is defined as:

1 pascal = 1 newton per square metre ($1\,\text{N}\,\text{m}^{-2}$).

A more commonly used unit is the **bar** or the **kilobar**, where:

1 bar = $10^5$ pascals = 0.1 MPa.

The dimensions of stress are thus (mass) × ($\text{length}^{-1}$) × ($\text{time}^{-2}$).

## 5.2 NORMAL STRESS AND SHEAR STRESS

A force $F$ acting on a body can be resolved into a **normal stress** acting perpendicular to a surface within the body and a **shear stress** acting parallel to the surface (Figure 5.3A). Normal stresses are conventionally given the symbol $\sigma$ (sigma) and shear stresses $\tau$ (tau). In three dimensions, it can be seen that $\tau$ can be resolved into two further components $\tau_1$ and $\tau_2$ at right angles to each other. Thus we have converted the force $F$ into three mutually perpendicular stresses (Figure 5.3B). Note that stresses cannot be resolved in the same way as forces – they have to be converted into forces by multiplying them by the area over which they act.

### APPLICATIONS TO GEOLOGICAL STRUCTURES

The application of normal and shear stresses can be illustrated with reference to two simple geological examples: the stress at a fault plane (Figure 5.4A) and the stress at a bedding plane undergoing flexural slip folding (Figure 5.4B), both resulting from opposed compressive forces $F$. Clearly the sense of fault displacement and bedding-plane slip can be predicted if the direction of the force is known, and vice versa.

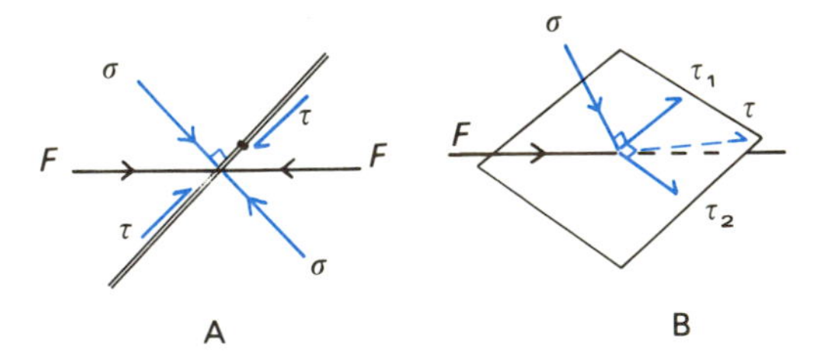


**Figure 5.3** A. Normal stress $\sigma$ perpendicular to the plane and shear stress $\tau$ parallel to the plane produced by opposed forces $F$ acting on a plane (in two dimensions). B. In three dimensions, the shear stress $\tau$ can be further resolved into $\tau_1$ and $\tau_2$ at right angles, giving three stresses, all mutually at right angles, resulting from the forces $F$.

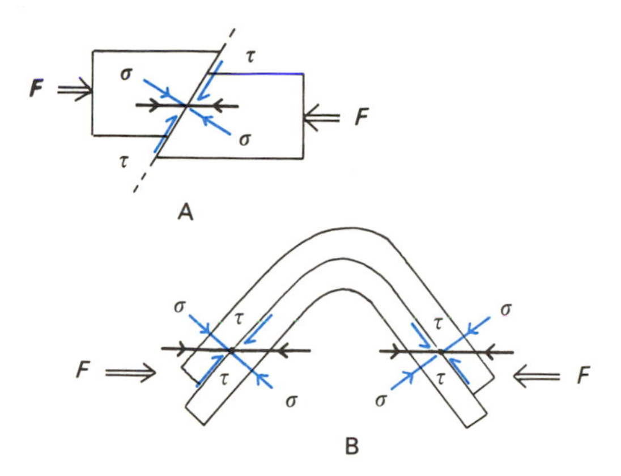


**Figure 5.4** Normal and shear stresses at a fault plane (A) and a bedding plane during flexural slip folding (B) produced by resolving opposed compressive forces *F*.

## 5.3 STRESS AT A 'POINT' – THE STRESS COMPONENTS

In order to consider the state of stress at a point in three-dimensional space, we must imagine the effect of a system of forces on an infinitesimal (vanishingly small) cube (Figure 5.5). The system of forces can be resolved into a single force *F* which acts at the centre of the cube. Since the cube is very small, we can consider the forces acting on each face of the cube to be equal to *F*. If we make the edges of the cube parallel to orthogonal axes *x*, *y* and *z*, the components of stress acting on the cube are as shown in Figure 5.5. There are nine stress components, three on each face (Figure 5.3B).

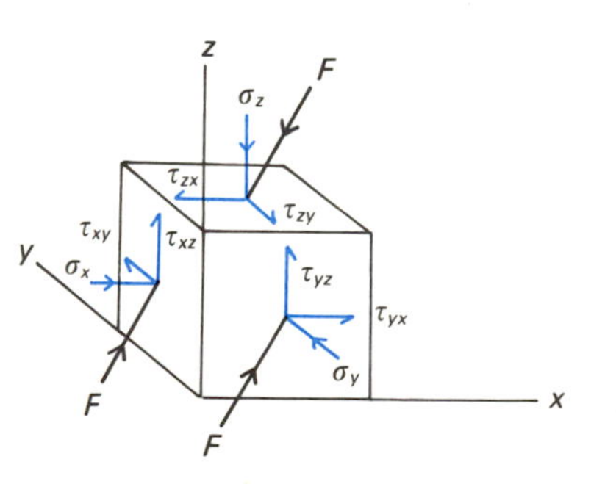


**Figure 5.5** Stress components for an infinitesimal cube acted on by opposed compressive forces *F* (see text).

Because the forces are equal and opposite, the stresses on opposite faces are identical. The nine components are:

$$\begin{matrix} \sigma_x & \tau_{xy} & \tau_{xz} \\ \sigma_y & \tau_{yx} & \tau_{yz} \\ \sigma_z & \tau_{zx} & \tau_{zy} \end{matrix}$$

Since the definition of stress precludes any contribution from an overall rotation of the cube, opposing shear stresses about the axes *x*, *y* and *z* must balance – otherwise the cube would rotate about these axes. Thus:

$$\tau_{xy} = \tau_{yx}, \tau_{xz} = \tau_{zy} \quad \text{and} \quad \tau_{yz} = \tau_{zy}$$

leaving six independent stress components, three normal stresses, $\sigma_x$, $\sigma_y$, $\sigma_z$ and three shear stresses, $\tau_{xy}$, $\tau_{yz}$, $\tau_{zx}$. Therefore, for an arbitrarily chosen set of orthogonal axes *x*, *y* and *z*, six independent quantities are necessary to specify completely the state of stress at a point.

## 5.4 PRINCIPAL STRESSES AND THE STRESS AXIAL CROSS

Rather than use arbitrary axes *x*, *y* and *z*, it is convenient to choose new axes *a*, *b* and *c* such that the shear stresses are zero. That is:

$$\tau_{ab} = \tau_{bc} = \tau_{ca} = 0$$

The three mutually perpendicular planes on which the shear stress is zero are called **principal stress planes**, and the normal stresses across them are called the **principal stress axes**. These are given the conventional notation $\sigma_1$, $\sigma_2$ and $\sigma_3$ (where $\sigma_1 > \sigma_2 > \sigma_3$) or greatest, intermediate and least principal stresses, respectively. A state of stress is specified completely by giving both the direction and the size of each of the three principal stresses.

### THE STRESS AXIAL CROSS

The three mutually perpendicular stress axes are often termed the **stress axial cross** in which the lengths of the axes may be drawn proportional to the magnitudes of the principal stresses (Figure 5.6).

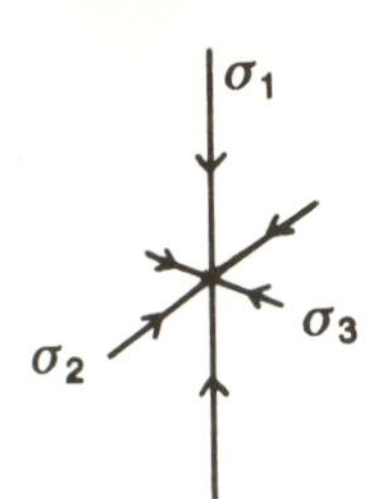


**Figure 5.6** The stress axial cross (principal stress axes $\sigma_1 > \sigma_2 > \sigma_3$; see text).

## 5.5 STRESSES ACTING ON A GIVEN PLANE

If the principal stresses are known, the stresses acting on any plane with known orientation can be calculated. The problem is easier to visualize in two dimensions.

Consider the stresses acting on a plane AB whose normal makes an angle $\theta$ (theta) with $\sigma_1$ in a two-dimensional stress field with principal stresses $\sigma_1$ and $\sigma_2$ (Figure 5.7). Since we cannot resolve stresses $\sigma_1$ and $\sigma_2$ we must convert these stresses to forces. Let the line AB represent unit length (one side of a square of unit area in three dimensions). Then OA $= \sin\theta$ and OB $= \cos\theta$. The forces acting along OA and OB are thus $\sigma_1 \cos\theta$ and $\sigma_2 \sin\theta$ respectively (from force = stress × area). Resolving these forces perpendicular and parallel to the plane AB, the normal stress $\sigma$ and shear stress $\tau$ are as follows:

$$\sigma = \sigma_1 \cos^2\theta + \sigma_2 \sin^2\theta \qquad (5.1)$$

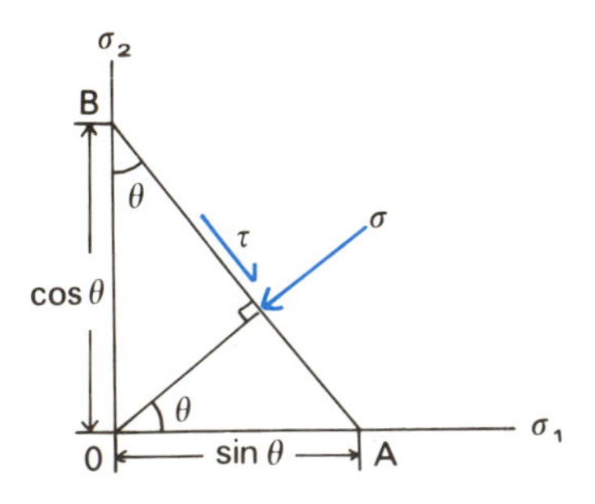


**Figure 5.7** Normal and shear stresses on a plane inclined to the principal stress axes (in two dimensions). See text for explanation.

and

$$\tau = \sigma_1 \sin^2\theta - \sigma_2 \cos^2\theta \qquad (5.2)$$

Since

$$\cos^2\theta = \tfrac{1}{2}(1 + \cos 2\theta)$$

and

$$\sin^2\theta = \tfrac{1}{2}(1 - \cos 2\theta)$$

we can rewrite these equations as follows:

$$\sigma = \tfrac{1}{2}(\sigma_1 + \sigma_2) + \tfrac{1}{2}(\sigma_1 - \sigma_2)\cos 2\theta \qquad (5.3)$$

and

$$\tau = \tfrac{1}{2}(\sigma_1 - \sigma_2)\sin 2\theta \qquad (5.4)$$

### MAXIMUM SHEAR STRESS

The value of $\tau$ in the last equation is a maximum when $2\theta = 90^\circ$ and $\sin 2\theta = 1$. Thus the planes of maximum shear stress make an angle of $45^\circ$ with $\sigma_1$ and $\sigma_2$ regardless of the values of $\sigma_1$ and $\sigma_2$.

In these positions,

$$\tau = \tfrac{1}{2}(\sigma_1 - \sigma_2) \qquad (5.5)$$

### STRESS IN THREE DIMENSIONS

The geometry in three dimensions can be derived from the above by considering a plane of unit area making angles of $\theta_1$, $\theta_2$ and $\theta_3$ with the three principal stress axes $\sigma_1$, $\sigma_2$ and $\sigma_3$. The normal stress on this plane is:

$$\sigma = \sigma_1 \cos^2\theta_1 + \sigma_2 \cos^2\theta_2 + \sigma_3 \cos^2\theta_3 \qquad (5.6)$$

and the shear stress is given by:

$$\begin{aligned}\tau^2 = &(\sigma_1 - \sigma_2)^2 \cos^2\theta_1 \cos^2\theta_2 \\ &+ (\sigma_2 - \sigma_3)^2 \cos^2\theta_2 \cos^2\theta_3 \\ &+ (\sigma_3 - \sigma_1)^2 \cos^2\theta_3 \cos^2\theta_1 \qquad (5.7)\end{aligned}$$

The reader should refer to a more advanced textbook (e.g. Ramsay, 1967) for the derivations of the last equations.

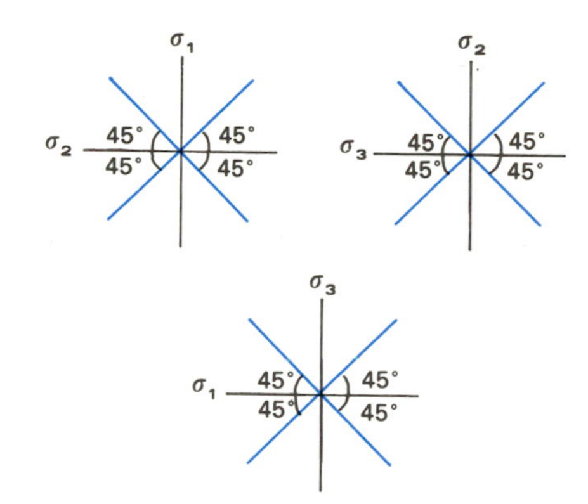


**Figure 5.8** Planes of maximum shear stress (colour) make angles of 45° with the principal stress axes. There are three sets of planes intersecting in $\sigma_1$, $\sigma_2$ and $\sigma_3$.

PLANES OF MAXIMUM SHEAR STRESS

From equation 5.7, it can be shown that there are three sets of planes of maximum shear stress, each plane making an angle of 45° with one pair of principal stresses and intersecting the third (Figure 5.8).

DIRECTION OF MAXIMUM SHEAR STRESS IN A PLANE

If a plane makes an angle with all three principal stress axes (Figure 5.9), the direction of maximum shear stress in the plane will depend on the relative magnitudes of $\sigma_1$, $\sigma_2$ and $\sigma_3$, and on the angles that the plane makes with the three stress axes (see Ramsay, 1967, for further details).

## 5.6 HYDROSTATIC AND DEVIATORIC STRESSES

Where the principal stresses are equal, the state of stress is said to be **hydrostatic**, i.e. it corresponds to the stress state of a fluid. It may be seen from equation (5.5) that the shear stress $\tau$ is zero in this situation. Hydrostatic stress will cause volume changes but not shape changes in a material.

In a system with unequal principal stresses $\sigma_1$, $\sigma_2$ and $\sigma_3$, it is convenient to recognize a mean

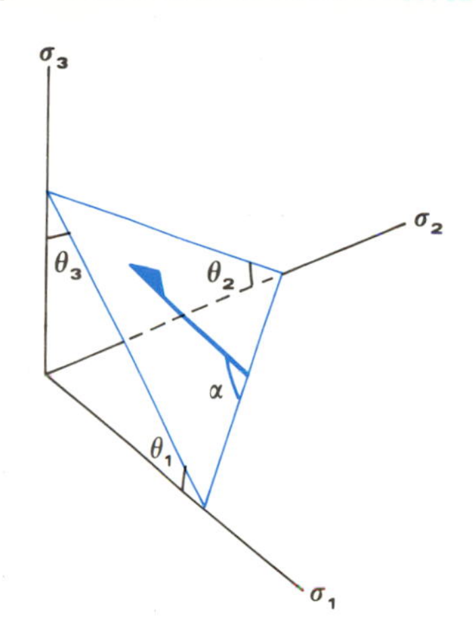


**Figure 5.9** Shear stress on a plane inclined at angles $\theta_1$, $\theta_2$ and $\theta_3$ to the principal stress axes. The shear direction is oblique to all three axes and makes an angle $\alpha$ with the strike of the plane.

stress $P$, which represents the hydrostatic stress component of the stress field. Thus:

$$P = (\sigma_1 + \sigma_2 + \sigma_3)/3 \qquad (5.8)$$

The remaining part of the stress system is referred to as the **deviatoric stress** component, which consists of three deviatoric stresses $\sigma_1 - P$, $\sigma_2 - P$ and $\sigma_3 - P$. These deviatoric stresses measure the departure of the stress system from symmetry and control the extent of shape change or distortion in a body, whereas the hydrostatic stress component controls the change in volume (Figure 5.10).

In rocks at depth, stresses that are hydrostatic and due solely to the weight of overlying rock are termed **lithostatic**. The vertical component of lithostatic stress has the value $\rho g z$, where $\rho$ is the density of the overlying rock, $g$ the value of gravity and $z$ the depth. Note that the lithostatic stress (or pressure) will not in general correspond to the mean stress, $P$, since $P$ depends also on the values of the horizontal stresses.

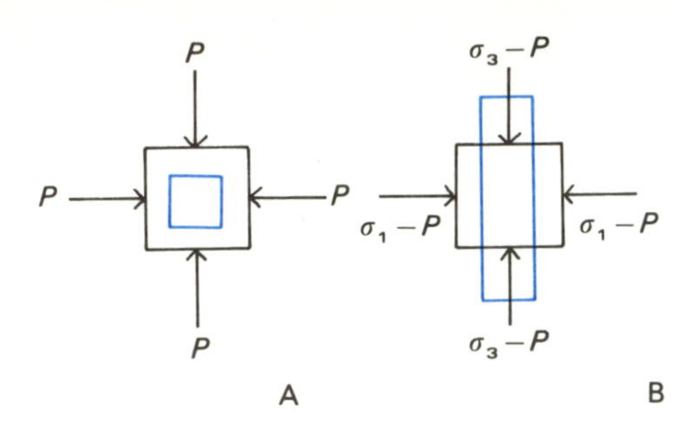


**Figure 5.10** Effects of hydrostatic and deviatoric stresses (shown in two dimensions). A. Hydrostatic stress $P$ causes a volume change. B. Deviatoric stresses $\sigma_1 - P$ and $\sigma_3 - P$ cause a shape change.

## 5.7 STRESS FIELDS AND STRESS TRAJECTORIES

Until now we have been considering stress at a 'point', but normally stresses will vary throughout a rock body, forming what is known as a **stress field**. Stress variation can be portrayed and analysed using **stress trajectories**, which are lines showing continuous variation in principal stress orientation from one point to another through a body. Two-dimensional stress trajectories of $\sigma_1$, and $\sigma_2$ are shown in Figure 5.11. Individual trajectories may be curved or bent, but obviously the principal stresses must remain at right angles to each other at each point in the curves. Examples of the use of stress trajectories in analysing fault and dyke patterns are given in subsequent chapters (Figures 9.10 and 11.7).

### COMBINATION OF STRESS FIELDS

Two or more stress fields of different origin may be superimposed to give a combined stress field. An example of such a combined field is shown in Figure 9.10. Stresses at any point may be combined by calculating each set of stresses in the form of stress components with reference to the same set of axes $x$, $y$ and $z$. The combined stress system is found by adding the components, e.g. $\sigma_x = \sigma_{x_1} + \sigma_{x_2}$, $\tau_{xy} = \tau_{x_1y_1} + \tau_{x_2y_2}$, etc. The new principal stresses may then be found by calculating positions for which $\tau = 0$. The method for calculating the principal stress axes given six stress components is given by Ramsay (1967, pp. 31–4).

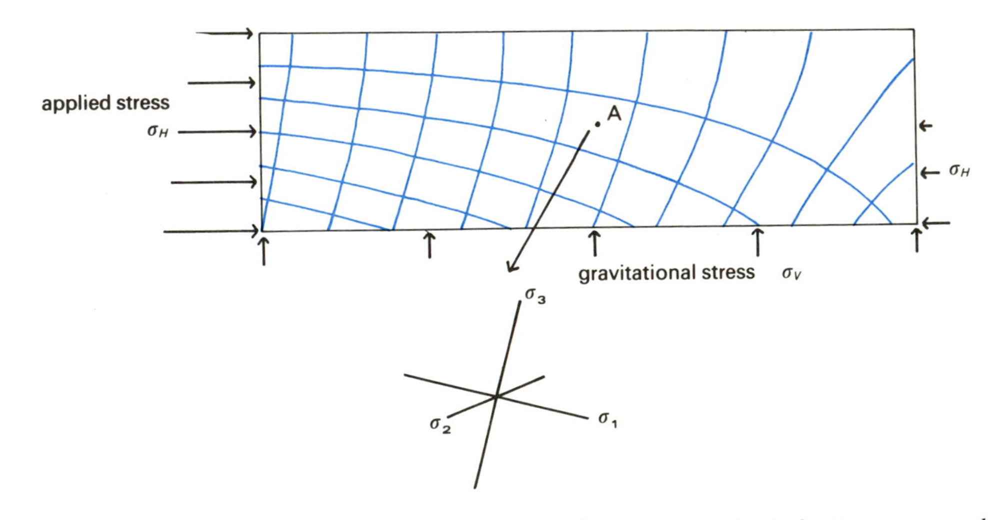


**Figure 5.11** Stress trajectories. The diagram shows theoretical stress trajectories (colour) in a rectangular block of crust subjected to a variable horizontal stress $\sigma_H$ applied to the sides of the block and a uniform vertical gravitational stress $\sigma_V$. The intermediate principal stress $\sigma_Z$ is perpendicular to the plane of the diagram. The stress axial cross at any point A can be found by interpolation. (After Hafner, W. (1951) *Bulletin of the Geological Society of America*, **62**, 373–98.

## FURTHER READING

Jaeger, I.C. and Cook, G.G.W. (1976) *Fundamentals of Rock Mechanics*, Chapman & Hall, New York. [Gives a comprehensive treatment of the physics of stress and the behaviour of materials.]

Means, W.D. (1976) *Stress and Strain*, Springer-Verlag, New York. [Gives a thorough account of stress as an aspect of continuum mechanics, but in a reasonably elementary way aimed specifically at the geologist. Easier to follow than Jaeger and Cook.]

Ramsay, J.G. (1967) *Folding and Fracturing of Rocks*, McGraw-Hill, New York.

# STRAIN 6

## 6.1 NATURE OF STRAIN

As explained in the previous chapter, **strain** is the geometrical expression of the amount of deformation caused by the action of a system of stresses on a body. We can thus define strain as the change in size and shape of a body resulting from the action of an applied stress field.

Strain is expressed as **dilation** (volume change) or **distortion** (shape change), or as a combination of these processes. In addition, it is often convenient to describe the distortion of a body in terms of a non-rotational shape change plus a rotational component (Figure 6.1).

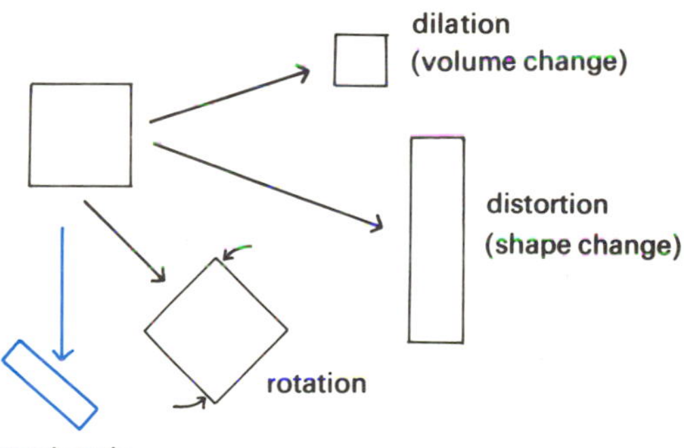


**Figure 6.1** The nature of strain: dilation, distortion and rotation.

### HOMOGENEOUS AND INHOMOGENEOUS STRAIN

If the amount of strain in all parts of a body is equal, the strain is said to be **homogeneous** (Figure 6.2A). The criteria for homogeneous strain are that straight lines remain straight and that parallel lines remain parallel. In the case of **inhomogeneous** (heterogeneous) strain, the strain in different parts of a body is unequal (Figure 6.2B). The criteria for inhomogeneous strain are thus that straight lines become curved and that parallel lines become non-parallel.

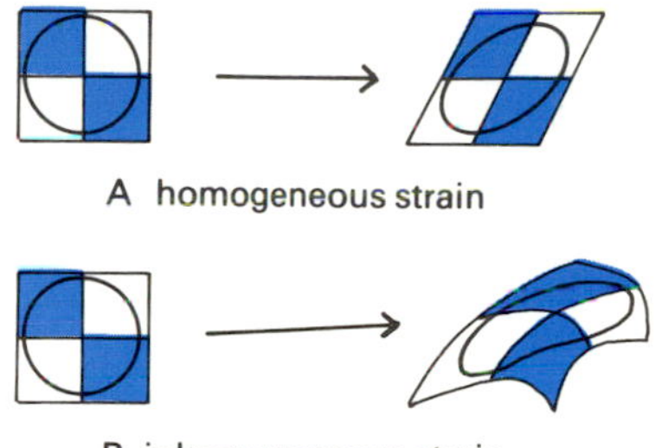


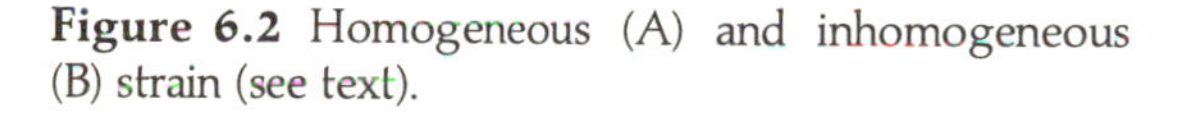

**Figure 6.2** Homogeneous (A) and inhomogeneous (B) strain (see text).

The difference between homogeneous and inhomogeneous strain can be illustrated simply by the folded layer of Figure 6.3. Taken as a whole, the fold exhibits inhomogeneous strain. However, the straight limbs of the fold taken separately exhibit homogeneous strain. This is an example of a very useful principle in strain analysis, which is that complex inhomogeneous strains are most conveniently studied by breaking them down into smaller homogeneous domains (cf. the study of fabric in section 4.4, and Figure 4.10).

## 6.2 MEASUREMENT OF STRAIN

Strain may be measured in two ways: either by a change in length of a line (linear strain, or **extension**) or by a change in the angle between two lines (angular strain, or **shear strain**) (Figure 6.4).

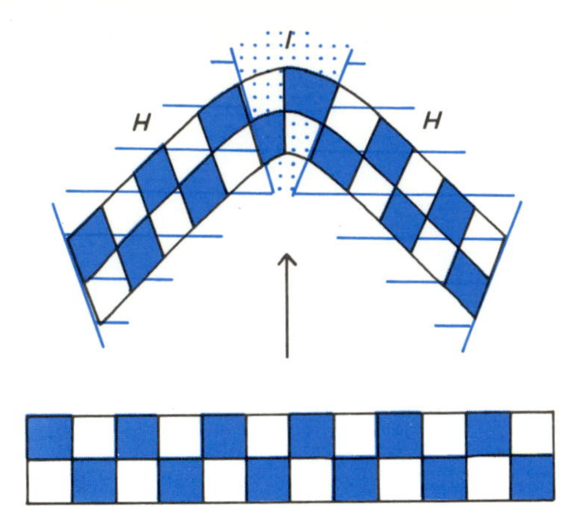


**Figure 6.3** Domains of homogeneous (H) and inhomogeneous (I) strain in a folded layer (see text).

Any strain geometry can be measured as a combination of these changes. They are defined as follows.

1. Extension

$$e = (l - l_0)l_0 \qquad (6.1)$$

where $l_0$ is the original length and $l$ the new length of a line. Note that a positive value of $e$ is an **elongation**, whereas a negative value of $e$ is a **shortening**.

Alternatively, the change in length of a line may be given by the **stretch**, which is defined as the ratio of the new length to the old length. Thus:

$$\theta = l/l_0 = (1 + e) \qquad (6.2)$$

For many purposes this measure of strain is more convenient. In the study of large crustal deformations, the term stretch is replaced by the **$\beta$-factor**.

2. Shear strain

$$\gamma\,(\text{gamma}) = \tan\psi\,(\text{psi}) \qquad (6.3)$$

where $\psi$ is the deflection of an originally right angle.

$e$, $\theta$ and $\gamma$ are all dimensionless quantities measuring the strain in a particular direction.

STRAIN IN TWO DIMENSIONS

Consider a circle of unit radius deformed into an ellipse with major axis $\theta_1$ and minor axis $\theta_2$ (Figure 6.4C). This ellipse is known as the **strain ellipse**. The point P $(x, y)$ on the unit circle is transferred to P′ $(x_1, y_1)$ on the ellipse. If $\alpha$ is the angle made by OP and the $x$-axis before deformation, and $\alpha'$ the

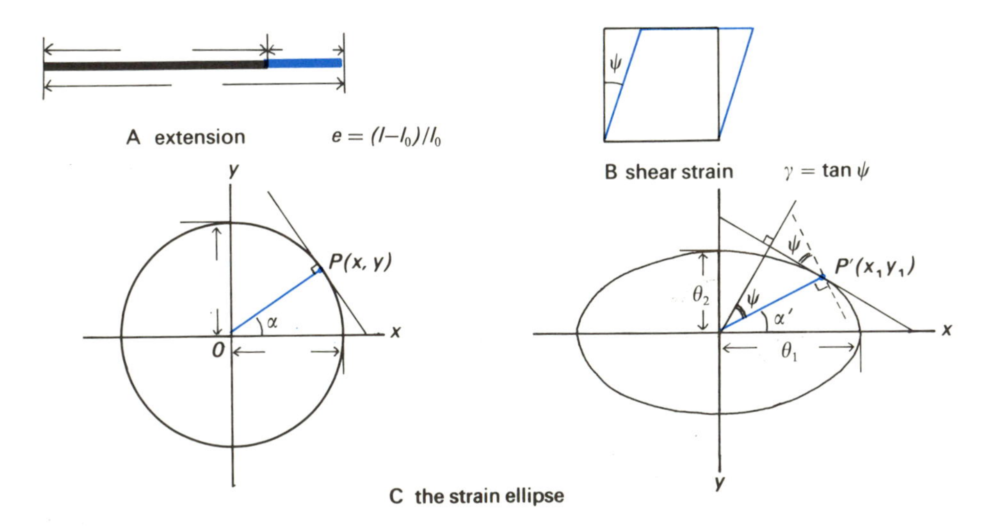


**Figure 6.4** Extension, shear strain and the strain ellipse. A. Extension $e = (l - l_0)/l_0$. B. Shear strain $\gamma = \tan\psi$. C. The strain ellipse (see text for explanation).

angle after deformation, then, since $x_1 = x\theta_1$ and $y_1 = y\theta_2$,

$$\tan\alpha' = \frac{y_1}{x_1} = \frac{y\theta_2}{x\theta_1} = \tan\alpha\,\frac{\theta_2}{\theta_1} \qquad (6.4)$$

and therefore,

$$\tan\alpha'/\tan\alpha = \theta_2/\theta_1 \qquad (6.5)$$

The length of the line OP is changed by an amount $\theta$, such that

$$\theta^2 = x_1^2 + y_1^2 = \theta_1^2\cos^2\alpha + \theta_2^2\sin^2\alpha \qquad (6.6)$$

### STRAIN IN THREE DIMENSIONS

The strain of a body can be measured in three dimensions with reference to three arbitrarily chosen coordinate axes *a*, *b* and *c* in the same way as for stress (see section 5.4). By taking an infinitesimal cube with sides parallel to *a*, *b* and *c*, we can describe the strain 'at a point' with reference to the change in shape of the cube, which becomes a parallelepiped (Figure 6.5). The infinitesimal strain can thus be measured by a set of extensions and deflections with reference to axes *a*, *b* and *c* necessary to transform the cube into the strain parallelepiped.

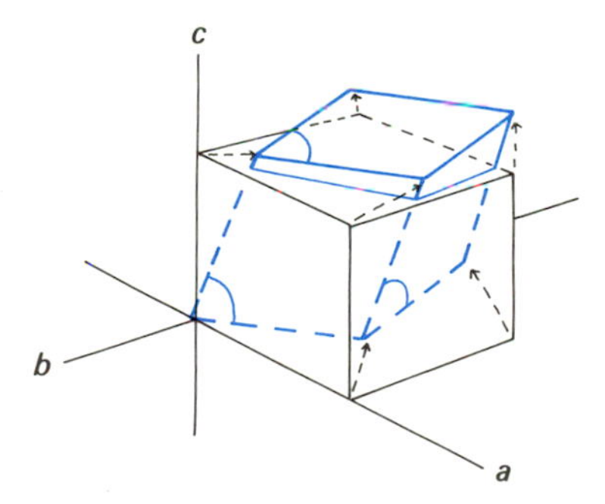


**Figure 6.5** Strain in three dimensions: deformation of a cube with sides parallel to orthogonal axes *a*, *b* and *c* (see text for explanation).

Alternatively, the strained body could be described by reference to a set of displacement vectors by which the eight corners of the cube were displaced to a set of new positions.

## 6.3 PRINCIPAL STRAIN AXES AND THE STRAIN ELLIPSOID

An alternative and more useful way to describe the strain is to select three mutually perpendicular axes *x*, *y* and *z* such that they are parallel respectively to the directions of greatest, intermediate and least elongation of the strained body. These axes *x*, *y* and *z* are known as the **principal strain axes**. They may be conveniently regarded as the axes of an ellipsoid, the **strain ellipsoid**, which is the shape taken up by a deformed sphere of unit radius (Figure 6.6). The maximum, intermediate and minimum axes, *X*, *Y* and *Z*, of this ellipsoid represent respectively the stretches $\theta_1$, $\theta_2$ and $\theta_3$ along *x*, *y* and *z*, and are known as the **principal strains**. To complete the description of the geometry of the strain, the orientations of *X*, *Y* and *Z* with respect to reference axes *a*, *b* and *c* have to be given in addition.

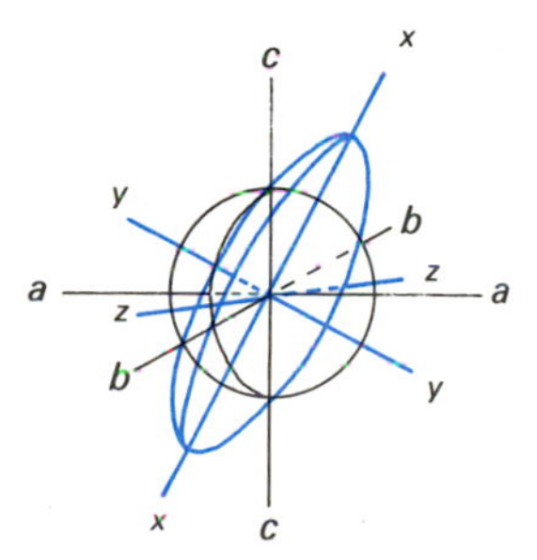


**Figure 6.6** The strain ellipsoid: principal strain axes *x*, *y* and *z* (see text for explanation).

## 6.4 PURE SHEAR AND SIMPLE SHEAR (DISTORTION AND ROTATION)

If the orientations of the principal strains *X*, *Y* and *Z* have not changed during the deformation, the strain is non-rotational, and is described as **coaxial**. Such a strain is generally known as **pure shear** (Figure 6.7A). Where a change in orientation has occurred, the strain is described as rotational, or non-coaxial, and this process is known as **simple shear** (Figure 6.7B). The difference is more easily portrayed in two dimensions.

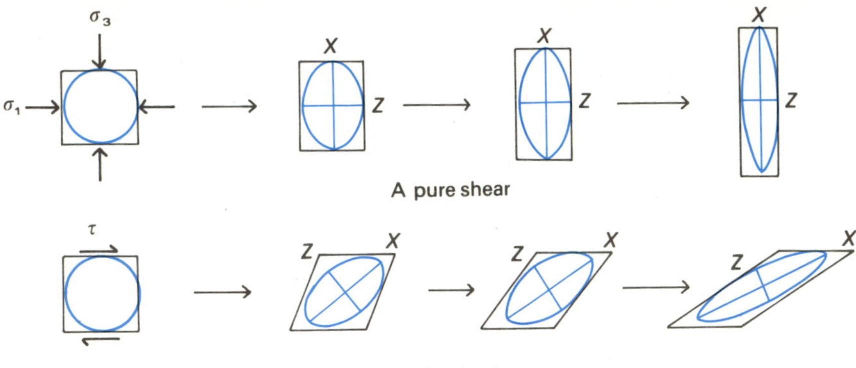


**Figure 6.7** Pure shear and simple shear. In the process of pure shear (A), which involves coaxial or non-rotational strain, the orientations of the principal strains $X$ and $Z$ do not change during progressive deformation. In simple shear (B), which involves rotational strain, principal strains $X$ and $Z$ rotate in a clockwise manner during progressive deformation.

A strain may thus be described in terms of a distortional component, which measures the ellipsoid shape, plus a rotational component, which measures the rotation of the principal strain axes from their original attitudes in the unstrained state.

## 6.5 SPECIAL TYPES OF HOMOGENEOUS STRAIN

It is convenient to recognize three special cases of homogeneous strain which can be distinguished by particular ratios of the principal strains $X$, $Y$ and $Z$. In the general case, the three are unequal and $X > Y > Z$. The special cases, shown in Figure 6.8, are as follows.

1. AXIALLY SYMMETRIC EXTENSION ($X > Y = Z$) (Figure 6.8A)

This is a constrictional type of strain which involves uniform extension in the $X$ direction and equal shortening in all directions at right angles to it. The deformed shape corresponds to a **prolate** type of strain ellipsoid, i.e. like a rugby ball or cigar.

2. AXIALLY SYMMETRIC SHORTENING ($X = Y > Z$) (Figure 6.8B)

This is a flattening type of strain which involves uniform shortening in the $Z$ direction and equal extension in all directions at right angles to it. The deformed shape corresponds to an **oblate** type of strain ellipsoid, i.e. like a pancake.

3. PLANE STRAIN ($X > Y = 1 > Z$) (Figure 6.8C)

This type of strain is distinguished by the intermediate principal strain axis remaining unchanged (i.e. $Y$ has unit length). $X$ is extended and $Z$ is shortened. Thus plane strain is a special type of triaxial ellipsoid.

## 6.6 VOLUME CHANGE DURING DEFORMATION

Changes in volume commonly accompany shape changes during deformation, and if these are not recognized they can cause misleading estimates of the principal strain ratios. The volume change, termed the **dilation**, Δ (delta), is given by:

$$\Delta = (V - V_0)/V_0 \qquad (6.7)$$

where $V$ and $V_0$ are the volumes in the deformed and undeformed states respectively.

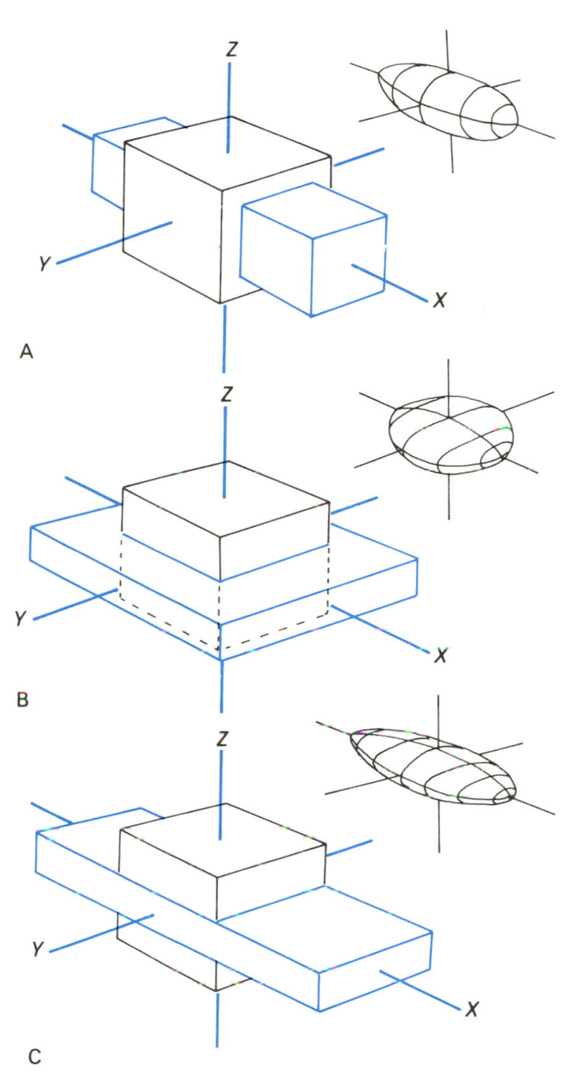


**Figure 6.8** Special types of homogeneous strain. A. Axially symmetric extension ($X > Y = Z$). This is a prolate uniaxial ellipsoid. B. Axially symmetric shortening ($X = Y > Z$). This is an oblate uniaxial ellipsoid. C. Plane strain ($X > Y = 1 > Z$). This is a triaxial ellipsoid, in which the intermediate axis is unchanged.

Since the volume of the strain ellipsoid derived from a unit sphere of volume $\frac{4}{3}\pi$ is $\frac{4}{3}(X \times Y \times Z)$, the dilation is given by:

$$\Delta = (X \times Y \times Z) - 1 \tag{6.8}$$

or as

$$1 + \Delta = \theta_x \times \theta_y \times \theta_z \tag{6.9}$$

where $\theta$ is the stretch.

## 6.7 GRAPHICAL REPRESENTATION OF HOMOGENEOUS STRAIN

A convenient way of expressing the various strain states is to use the Flinn diagram (Flinn, 1962). In this diagram (Figure 6.9A), the ratios of the principal strains are taken, such that

$$a = X/Y = \theta_x/\theta_y \tag{6.10}$$

and

$$b = Y/Z = \theta_y/\theta_z \tag{6.11}$$

and $a$ is plotted against $b$.

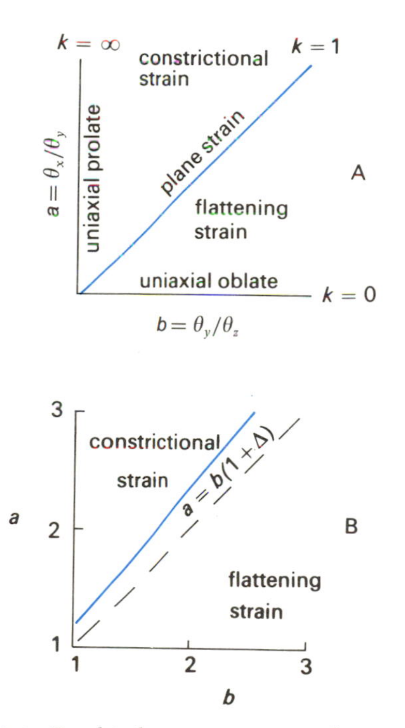


**Figure 6.9** Graphical representation of strain ellipsoids: the Flinn diagram. A. Different ellipsoids are described using the value $k = (a - 1)/(b - 1)$ (see text for explanation). B. If the volume is not constant, the line $a = b(1 + \Delta)$ divides the field of constrictional strain from the field of flattening strain. The diagram shows the effect of a 20% volume reduction. (A and B after Flinn, 1962, and Ramsay and Huber, 1983.)

The different shapes of ellipsoids are distinguished using the value $k$, such that

$$k = (a - 1)/(b - 1) \qquad (6.12)$$

The various strain states can be described as follows:

Axially symmetric extension: $k = \infty$
Constrictional strain (prolate ellipsoids): $1 < k < \infty$
Plane strain (at constant volume): $k = 1$
Flattening strain (oblate ellipsoids): $0 < k < 1$
Axially symmetric flattening: $k = 0$

In this way the shape of an ellipsoid can be described using only the value of parameter $k$, and constrictional and flattening strains can immediately be distinguished by whether $k$ is greater or less than 1.

Figure 6.9A is constructed assuming a constant volume, since the line $k = 1$ will pass through the origin only when the volume change $\Delta = 0$. When $\Delta \neq 0$, equation (6.12) reduces to

$$\begin{aligned} 1 + \Delta &= \theta_x \times \theta_z \\ &= a/b \end{aligned} \qquad (6.13)$$

since $\theta_y = 1$ for $k = 1$ ellipsoid.

Therefore

$$a = b(1 + \Delta) \qquad (6.14)$$

Thus for a volume change of $\Delta$, the line $a = b(1 + \Delta)$ represents plane strain and divides the constrictional from the flattening fields (Figure 6.9B).

## 6.8 PROGRESSIVE DEFORMATION AND FINITE STRAIN

The strained body at the time of measurement represents the total strain acquired by the body up to that time, and is produced by adding a series of strain increments as the body takes up a succession of different shapes and positions in response to the applied stress (Figure 6.10). This process, from initial to final position, is termed **progressive deformation**, and the final strain at the time of measurement is termed the **finite strain**. It is important to recognize that the nature of the finite strain is not necessarily a reliable guide to the various intermediate states of strain. Even quite simple cases of progressive strain display marked changes in strain pattern with time.

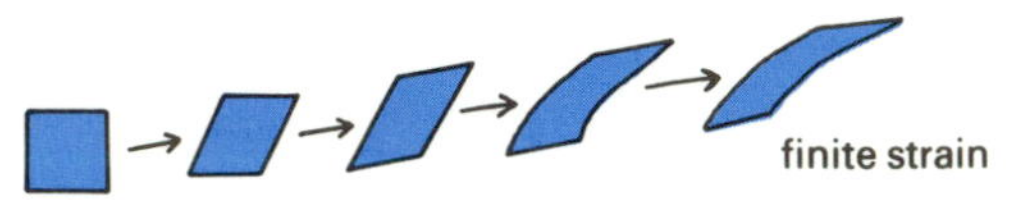


**Figure 6.10** Progressive deformation. The finite strain is achieved by adding successive strain increments to the initial unstrained shape.

At any given instant during progressive deformation, it is theoretically possible to distinguish the finite strain (total strain up to that time) from the **infinitesimal strain** at that point in time. In two dimensions, the finite strain ellipse (Figure 6.11A) can be divided into sectors of elongation and contraction separated by lines of no finite longitudinal strain (i.e. zero extension). Suitably oriented layers will exhibit boudinage in elongation sectors and folding in contraction sectors. The infinitesimal strain ellipse (Figure 6.11B) at that instant will show sectors whose lines are currently elongating or contracting.

By superimposing the two ellipses, we can distinguish four zones:

zone 1: continued elongation (boudins);
zone 2: contraction followed by elongation (unfolded or boudinaged folds);
zone 3: elongation followed by contraction (folded boudins);
zone 4: continued contraction (folds).

The distribution of these zones or sectors will depend on the strain history and, in particular, on whether the strain is coaxial (i.e. pure shear – Figure 6.11C) or rotational (i.e. simple shear – Figure 6.11D). Observations of the orientations of folded and boudinaged layers are therefore of great value in investigating progressive deformation.

### GROWTH FIBRE ANALYSIS

Another very useful method of investigating the strain history is to analyse the orientation of quartz

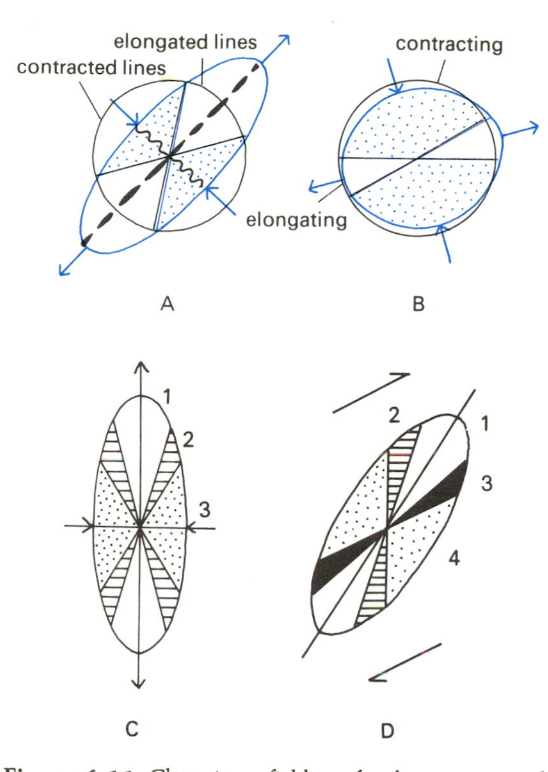


**Figure 6.11** Changing fields of elongation and contraction during progressive deformation (in two dimensions). A. The finite strain ellipse. Fields of elongated lines (boudinage) and contracted lines (folds) are separated by lines of zero extension which retain their original length (= radius of undeformed circle). B. The infinitesimal strain ellipse. Fields that at any instant during progressive deformation are contracting or elongating. C. Superimposition of ellipses A and B for pure shear will produce three zones: 1, continued elongation; 2, contraction followed by elongation; and 3, continued contraction. D. Superimposition of ellipses A and B for simple shear will produce an asymmetric arrangement of four zones: 1, continued elongation; 2, contraction followed by elongation; 3, elongation followed by contraction; and 4, continued contraction. (After Ramsay, 1967, figures 3.56 and 3.62.)

or calcite **growth fibres** commonly found in extensional cracks or veins. These growth fibres are crystals oriented parallel to the direction in which the walls of the crack have opened and thus mark the direction of extension in the rock (Figure 6.12). Changes in fibre orientation mark changes in the extension direction. Often quite complex changes in direction are recorded in this way. The process of periodic opening of cracks and filling with vein material is known as the **crack–seal mechanism** and is described in detail by Ramsay and Huber (1983).

## 6.9 RELATIONSHIP BETWEEN STRESS AND STRAIN

Since a strain results from the action of a stress, there must always be a direct geometrical relationship between the two. However, since the geometry of both stress and strain fields changes with

**Figure 6.12** Growth fibres in an extensional crack. The crystals are aligned in the extension direction. Quartz–calcite vein in siliceous limestone, Gemmipass, Switzerland. (From Ramsay and Huber, 1987, figure 13.7.

time, the relationship may not be a simple one. We can illustrate this principle with reference to the two cases of homogeneous strain shown in Figure 6.7 – pure shear and simple shear.

In the case of pure shear (Figure 6.7A), the orientation of the principal finite strain axes and the principal stress axes correspond, with $X \parallel \sigma_3$, $Y \parallel \sigma_2$ and $Z \parallel \sigma_1$; that is, the direction of greatest extension corresponds to the direction of minimum stress (which in many cases would be negative, or tensional) and the direction of least extension or greatest shortening corresponds to the direction of maximum stress.

However, in the simple shear case (Figure 6.7B), only the intermediate stress and strain axes correspond (since it is plane strain), and the X and Z axes rotate progressively clockwise with increasing strain away from their initial position; they will not in general correspond with the orientations of the $\sigma_1$ and $\sigma_3$ stress axes.

Since only the strain axes are observed in natural deformation, the position of the stress axes cannot immediately be deduced. It can be seen from Figure 6.7 that the magnitudes of the strain axes at stages 2, 3 and 4 of A are respectively identical with those of B although they have been produced differently. Thus the position of the stress axes can only be reconstructed by reference to the strain history of the deformed body.

Examples of how the stress field may be reconstructed from fault displacement data are discussed in section 9.2.

## FURTHER READING

Flinn, D. (1962) On folding during three-dimensional progressive deformation. *Quarterly Journal of the Geological Society of London*, **118**, 385–433.

Means, W.D. (1976) *Stress and Strain*, Springer-Verlag, New York.

Ramsay, J.G. (1967) *Folding and Fracturing of Rocks*, McGraw-Hill, New York.

Ramsay, J.G. and Huber, M.I. (1983) *The Techniques of Modern Structural Geology, Vol. 1: Strain Analysis*, Academic Press, New York.

Skjernąa, L. (1980) Rotation and deformation of randomly oriented planar and linear structures in progressive simple shear. *Journal of Structural Geology*, **2**, 101–9.

# STRESS AND STRAIN IN MATERIALS 7

The way in which individual minerals respond to stress varies widely, depending on the physical conditions under which the deformation takes place, and also on the compositional and mechanical properties of the material. Before considering the behaviour of rocks under stress, we shall discuss various 'ideal' types of response.

## 7.1 IDEAL ELASTIC AND VISCOUS STRAIN

### ELASTIC STRAIN

In ideal **elastic strain**, the removal of the deforming stress causes an immediate return of the body to its original unstrained shape. This type of strain is therefore temporary, and recoverable, and can be demonstrated by the compression and release of a spring. It corresponds to the type of strain associated, for example, with the propagation of seismic waves through the Earth, or with the passage of sound waves through any medium.

The behaviour of perfectly elastic bodies is governed by **Hooke's law**, which states that

$$e = \sigma/E \qquad (7.1)$$

where $e$ is the extensional strain, $\sigma$ the applied stress and $E$ a constant known as **Young's modulus** or the **elasticity** of the material. Thus stress/strain is a constant.

Also

$$e_V = P/K \qquad (7.2)$$

where $e_V$ is the dilational strain $= (V - V_0)/V_0$, where $V$ is the new volume, $V_0$ the original volume, $P$ the hydrostatic pressure, and $K$ a constant termed the **compressibility**. Thus for elastic shape and volume changes, the strain is directly proportional to the stress, i.e. stress and strain have a linear relationship (Figure 7.1A). Rocks exhibit perfectly elastic behaviour only under certain restricted conditions.

### VISCOUS STRAIN

In ideal **viscous strain**, there is no recovery after removal of the deforming stress, i.e. all the movement is permanent. Ideal viscous or 'Newtonian' behaviour is exhibited by the flow of fluids and is governed by the equation

$$\sigma = \eta\dot{e} \qquad (7.3)$$

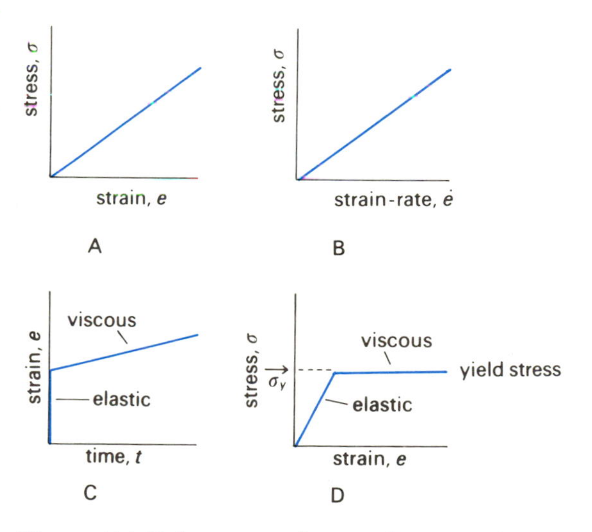


**Figure 7.1** Deformation of material: types of ideal strain. A. Ideal elastic strain: $\sigma \propto e$. B. Ideal viscous strain: $\sigma \propto \dot{e}$. C. Ideal elastoviscous strain at constant stress – an instantaneous elastic strain is followed by viscous strain for a stress of longer duration. D. Ideal plastic strain – elastic strain at low values of stress is replaced by viscous strain above the yield stress.

where $\eta$ (eta) is a constant termed the **viscosity** of the material and $\dot{e}$ is the strain rate (rate of change of shape with time). Thus in ideal viscous strain, the stress is linearly related to the strain rate (Figure 7.1B), so the higher the applied stress, the faster the material will deform, and the total strain is dependent on both the magnitude of the stress and the length of time for which it is applied. That is, stress = viscosity × strain rate.

For a constant stress, the strain will increase linearly with time $t$, since, integrating equation (7.3),

$$e = \sigma t/\eta \qquad (7.4)$$

For more realistic equations of flow, the 'effective viscosity' varies with both temperature and stress.

## 7.2 ELASTOVISCOUS, PLASTIC AND VISCOELASTIC BEHAVIOUR

Real rock materials combine the properties of ideal elastic and viscous bodies, and the strain in such materials may be regarded as having both elastic and viscous components. One approximation to the total strain may be obtained by adding equations (7.1) and (7.4) to give

$$e = \sigma/E \quad \text{(elastic component)}$$
$$+ \sigma t/\eta \quad \text{(viscous component)} \qquad (7.5)$$

at constant stress. However, this equation over-simplifies the relationship and several other factors have to be taken into account.

### ELASTOVISCOUS BEHAVIOUR

A material that basically obeys the viscous law (equation 7.3) but that behaves elastically for stresses of short duration is termed **elastoviscous** (Figure 7.1D). Pitch is a good example of such a material. When stressed, it will show elastic strain which is completely recoverable if the stress is rapidly removed. The material will flow, however, exhibiting perfectly viscous behaviour, for a stress held for any length of time.

### PLASTIC BEHAVIOUR

A **plastic** material is one that behaves elastically at low values of stress, but above a certain critical value of stress (the **yield stress** – see section 7.4) it behaves in a perfectly viscous manner (Figure 7.1D).

### VISCOELASTIC BEHAVIOUR

A material that, for a given stress, exhibits a basically elastic type of strain but which takes a certain time to reach its limiting value is said to show **viscoelastic** behaviour (Figure 7.2). Conversely, the removal of the stress does not cause an immediate return to the undeformed state, but there is a **delayed recovery** of the elastic strain. Delayed recovery is responsible for certain earthquake after-shocks which represent continued movements following the main release of strain represented by the earthquake (see section 9.3). Most rocks exhibit viscoelastic behaviour at low values of stress.

The behaviour of rock materials under stress cannot be described in terms of simple models of elastic, viscous or plastic behaviour but includes their characteristics in combination. Figure 7.2 shows diagrammatically how the strain–time diagram of a typical viscoelastic/plastic material (like most rocks) exhibits elements of all the above models. The term viscoelastic is also used in a rather looser sense to describe any combination of elastic and viscous behaviour. In practice, the relationship between stress and strain in rock deformation varies both with the magnitude of the stress and with the strain rate.

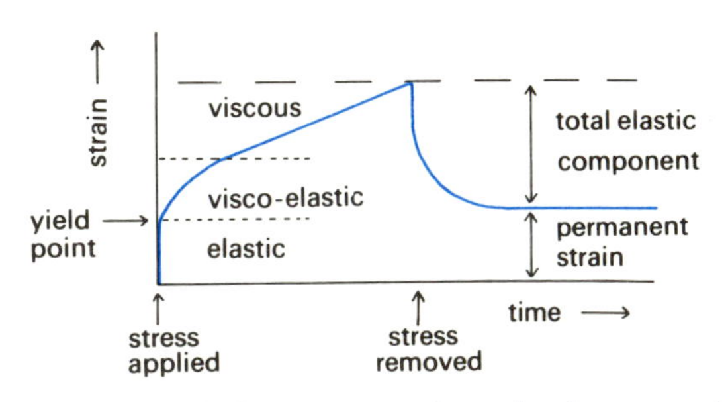


**Figure 7.2** Ideal strain–time relationship for a typical plastic material deformed above its yield point.

## 7.3 BRITTLE AND DUCTILE BEHAVIOUR

Where deformation leads to failure, the material loses cohesion by the development of a fracture or fractures, across which the continuity of the material is broken. This type of behaviour is called **brittle** behaviour and governs the development of faults and joints (see Chapter 9). **Ductile** behaviour, in contrast, produces permanent strain that exhibits smooth variations across the deformed sample or rock without any marked discontinuities. Most rock materials are capable of exhibiting either brittle or ductile behaviour, depending on such factors as the size of the differential stress ($\sigma_1 - \sigma_3$), the hydrostatic pressure, the temperature, the fluid pressure and the strain rate. In sections 7.4–7.8 we consider the effect of each of these factors on the strain, based on the results of laboratory experiments on rock materials.

## 7.4 THE EFFECTS OF VARIATION IN STRESS

The effect of the differential stress ($\sigma_1 - \sigma_3$) is considered separately from that of the hydrostatic part of the stress. Figure 7.3 summarizes diagrammatically the effects of increasing differential stress on the strain–time curve. For a low value of stress $\sigma_A$, the material may exhibit entirely elastic behaviour. For slightly larger values of stress $\sigma_B$ the strain may be partly viscoelastic, but there is a limiting value of strain which is basically elastic. Above a certain critical value of stress known as the 'yield stress' $\sigma_Y$ (see 'yield strength', below), the material exhibits essentially viscous behaviour for successively higher values of stress $\sigma_C$, $\sigma_D$ after an initial viscoelastic strain. Above a second critical value of stress known as the 'failure stress' $\sigma_R$ (see 'failure strength', below), the material exhibits accelerated viscous flow for higher values of stress $\sigma_E$, $\sigma_F$ leading to failure after the initial viscoelastic and viscous stages (cf. the creep curve, Figure 7.4).

We may distinguish three main fields on a typical strain–time diagram, corresponding to a progressive increase in stress – elastic, viscous and failure. In the case of a ductile substance, the viscous field is enlarged at the expense of the elastic and failure fields, and $\sigma_R \gg \sigma_Y$.

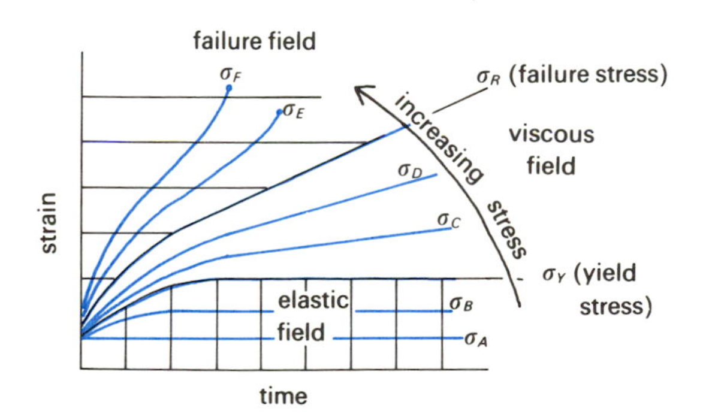


**Figure 7.3** Diagrammatic representation of the effect of increasing stress on a strain–time curve (see text for explanation).

### STRENGTH OF MATERIALS

The measured strength of a material is simply the value of the applied stress at which failure occurs. Many materials possess both a **yield strength**, defined as the limiting stress above which permanent deformation occurs, and a **failure strength**, or **ultimate strength**, above which failure occurs. The values of compressive and tensile strength are normally different (see section 9.1) – i.e. the yield strength and failure strength are usually higher under a compressive differential stress ($\sigma_1 - \sigma_3$ positive) than under a tensile differential stress ($\sigma_1 - \sigma_3$ negative).

Values of strength measured (or extrapolated) over long time periods are much smaller than those measured for short periods of time (the 'instantaneous' strength). The long-term strength of most rocks is only in the range 20–60% of their instantaneous strength (see effect of strain rate, section 7.9).

## 7.5 THE EFFECT OF HYDROSTATIC PRESSURE

In considering the effect of an applied stress field during deformation, the hydrostatic and deviatoric stress components must be considered separately (see section 5.6). Rocks at depth in the crust are

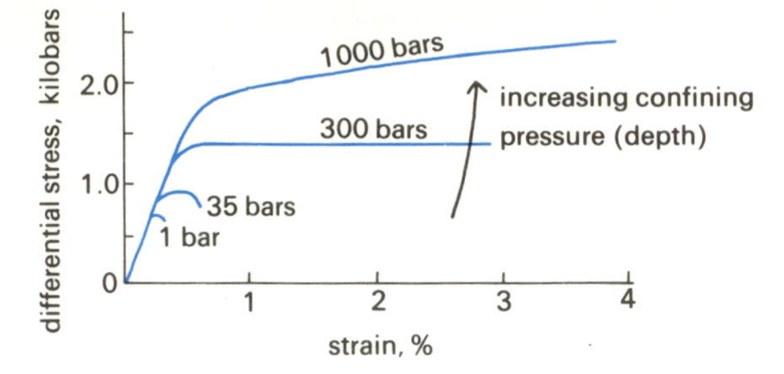


**Figure 7.4** Effect of increasing confining pressure on stress–strain curves for the experimental deformation of Wombeyan marble. Increase in confining pressure raises the yield stress or strength of the rock. (After Paterson, M.S. (1958) *Bulletin of the Geological Society of America*, **69**, 465–76.)

subjected to the gravitational load pressure of the overlying column of rock. This pressure can be assumed to be effectively hydrostatic, and is simply related to the thickness and mean density of the overlying material. The pressure at the base of a 35 km thick crust is about 10 kilobars, and realistic pressures for most naturally deformed crustal rocks range from several hundred bars upwards.

The hydrostatic pressure causes elastic volume changes that depend on the compressibility of the material (see equation (7.2)). The size of these volume changes is unimportant except at great depth. Primary seismic waves propagate by means of elastic volume changes in the material through which the waves are transmitted.

A more important aspect of hydrostatic pressure in the study of strain is its effect on the strength of materials. In laboratory experiments, the effect of hydrostatic pressure is investigated by varying the **confining pressure**, which is the radial stress applied to a deforming rock specimen subjected to a uniaxial compression or extension. With increasing confining pressure, both the yield stress $\sigma_Y$ and the failure stress $\sigma_R$ are raised, giving the material a higher effective strength. Figure 7.4 illustrates the effect of confining pressure on the stress–strain curves for the experimental deformation of marble. At low pressures the response is basically elastic and failure occurs at low values of $\sigma$. At 300 bars confining pressure, the yield stress is raised to around 1400 bars and is followed by a steady viscous increase in strain, indicating that the material has become ductile at high pressure.

## 7.6 THE EFFECT OF TEMPERATURE

With increasing temperature, the yield stress $\sigma_Y$ is lowered and the failure stress $\sigma_R$ is raised, which has the effect of enlarging the viscous field of deformation at the expense of the elastic and failure fields (cf. Figure 7.3). Consequently the material may be said to show an increase in ductility.

The effect of temperature on the stress–strain curves for the experimental deformation of marble is shown in Figure 7.5. The yield stress at 800 °C is less than 0.5 kilobar, about one-sixth of its value at room temperature. Moreover, the strain range of viscous behaviour is considerably increased.

These observations are consistent with our geological experience of metamorphic rocks deformed at elevated temperature and pressure. These exhibit much more ductile types of deformation than do the equivalent rocks at the surface.

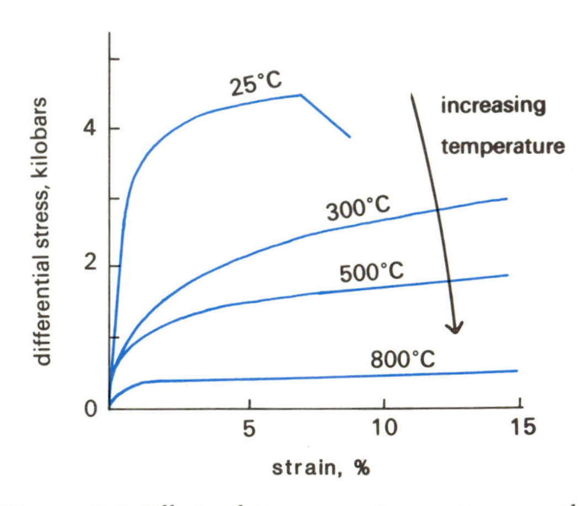


**Figure 7.5** Effect of increasing temperature on the stress–strain curves for the experimental deformation of Yule marble at 5 kilobars confining pressure. Increase in temperature decreases the yield stress or strength of the rock. (After Griggs, D.T., Turner, F.J. and Heard, H.C. (1960) in *Rock Deformation* (eds D.T. Griggs and J. Handin), *Memoirs of the Geological Society of America*, **79**.)

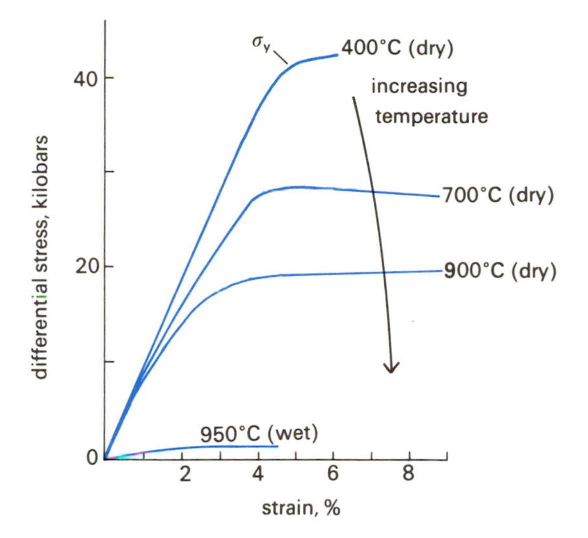


**Figure 7.6** Effect of pore-fluid pressure. Stress–strain curves for the deformation of wet and dry natural quartz crystals at 15 kilobars confining pressure, various temperatures and a strain rate of $0.8 \times 10^{-5}$/s. See text for explanation. (After Griggs, D.T. (1967) *Geophysical Journal of the Royal Astronomical Society*, **14**, 19–31.)

## 7.7 THE EFFECT OF PORE-FLUID PRESSURE

The presence of a fluid phase in rocks undergoing deformation is important in several ways. The most important effect is to facilitate deformation by reducing resistance to slip along planes of potential movement within a rock, from grain boundaries to major fault planes (see also section 9.1). The fluid pressure has the effect of reducing the shear stress required for slip, i.e. it reduces the shear strength of a rock. This is because the direct pressure between adjoining grains caused by the lithostatic or hydrostatic pressure is countered by the effect of the pore-fluid pressure. Fluid pressure may also promote mineralogical reactions, particularly at elevated temperatures, that reduce the mechanical strength of the rock.

The mechanical effect of the pore-fluid pressure is given by

$$P_e = P - P_f \quad (7.6)$$

where $P_e$ is the effective pressure on the solid material, $P$ the hydrostatic pressure, and $P_f$ the fluid pressure.

For saturated rocks, where the pore-fluid pressure may be very high compared with the hydrostatic pressure, the effect of the high hydrostatic pressure is cancelled out and the strength of a rock reduced to near-surface conditions. The effect of high fluid pressure on rock at elevated temperature is illustrated by the stress–strain curves for wet and dry quartz crystals (Figure 7.6). The yield stress at 950 °C in wet quartz is only about one-tenth of that required for the dry material at the same temperature. Clearly the ductility in this case is greatly increased, which explains why certain materials, normally strong even at high temperature, can flow under metamorphic conditions in the presence of aqueous fluids.

## 7.8 THE EFFECT OF TIME: STRAIN RATE

The relationship between stress and strain for real materials, which exhibit a combination of elastic, viscous and plastic properties, depends critically on the length of time for which the stress is applied. In laboratory experiments on rocks, with short durations of seconds or minutes, the behaviour of the material is effectively instantaneous in geological terms and differs significantly from that of the same material under stresses with more geologically realistic durations of months or years. The long-term strain behaviour of materials is termed **creep**. The important characteristic of creep behaviour is that viscous strain is produced over long periods of time under low stresses that would produce only elastic effects if applied over short periods. Another way of expressing this difference in behaviour is to state that the strain rate has a decisive influence on the type of strain behaviour: that is, materials deforming under low strain rates that are geologically realistic exhibit creep, whereas high strain rates (e.g. of experimental deformation) are associated with geologically instantaneous strain.

### THE TYPICAL CREEP CURVE

Most rocks when subjected to a constant low stress over a period of, say, months exhibit a strain curve that consists of three stages (Figure 7.7): an initial transient stage, termed **primary creep**, where the

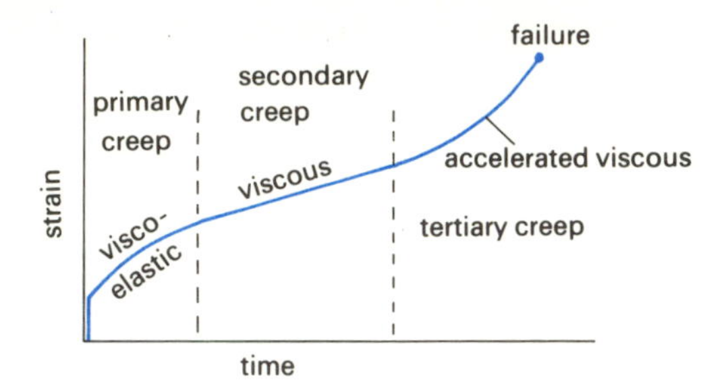


**Figure 7.7** Creep: strain–time diagram for long time periods.

material behaves viscoelastically, followed by a steady-state stage termed **secondary creep,** where the material exhibits essentially viscous flow, and a final stage of **tertiary creep**, where the material exhibits accelerated viscous strain leading ultimately to failure. Creep at constant stress may be represented by an equation of the form

$$e = A + B \log t + Ct \qquad (7.7)$$

where $A$, $B$ and $C$ are constants reflecting the physical properties of the material and the values of the stress and temperature (cf. equation (7.5)). Note the resemblance of the typical creep curve to the 'ideal' strain–time curve of Figure 7.2.

The importance of strain rate in rock deformation is exemplified by the fact that values of the yield and ultimate (dry) strengths of rocks are much higher if measured over short time periods than over geologically significant time periods. Deformation experiments lasting for days rather than seconds have been designed to investigate the effect of lower strain rates. A succession of stress–strain curves for Yule marble (Figure 7.8A) at strain rates ranging from $2 \times 10^{-3}\ s^{-1}$ (= 20% in 100 s) to $2 \times 10^{-7}\ s^{-1}$ (= 20% in 12 days) demonstrate a gradual reduction of strength and increase in ductility with decrease in strain rate.

A strain rate of $2 \times 10^{-7}\ s^{-1}$ is still much greater than geological strain rates, which are around $10^{-14}\ s^{-1}$, i.e. 10% in one million years. In order to extrapolate the experimental results to such a low value, a log stress/log strain rate plot has been used (Figure 7.8B). This plot indicates that at a temperature of 300 °C the differential stress required to maintain steady ductile flow at a strain rate of $10^{-14}\ s^{-1}$ is only about 400 bars (40 MPa), or 160 bars (16 MPa) at 400 °C.

## 7.9 SUMMARY: PHYSICAL CONTROLS ON STRAIN BEHAVIOUR

Brittle failure is typical of rocks at low confining

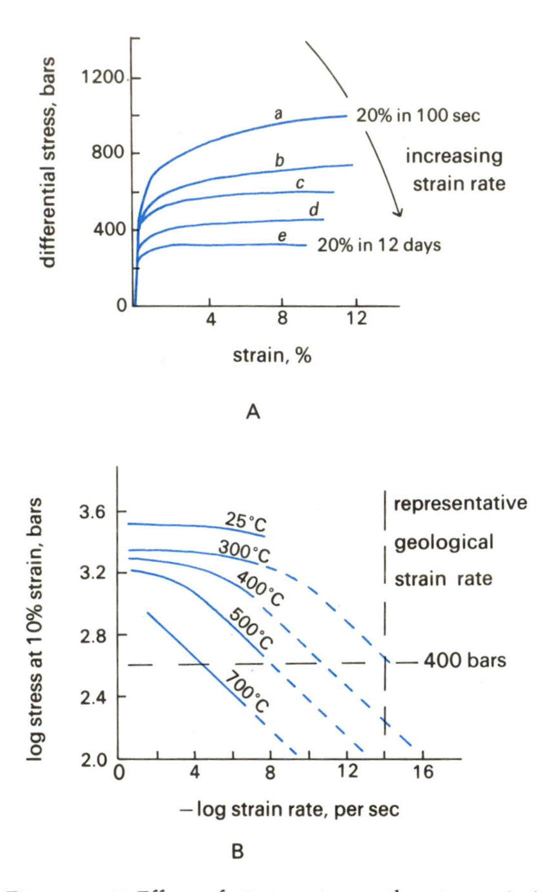


**Figure 7.8** Effect of strain rate on the stress–strain relationship. A. Stress–strain curves for Yule marble deformed by extension at 600 °C at strain rates ranging from $2 \times 10^{-3}\ s^{-1}$ to $2 \times 10^{-7}\ s^{-1}$ (a, $2 \times 10^{-3}$; b, $2 \times 10^{-4}$; c, $2 \times 10^{-5}$; d, $2 \times 10^{-6}$; e, $2 \times 10^{-7}\ s^{-1}$). B. Stress–strain curves for deformation of Yule marble at various temperatures, plotted as log stress at 10% strain ($\simeq$ yield stress) against log strain rate. The curves representing the experimental data are extrapolated (dashed lines) to show the effect of lower strain rates (see text for further explanation). (A and B after Heard, H.C. and Raleigh, C.B. (1972) *Bulletin of the Geological Society of America*, **83**, 935–56.)

pressure and low temperature appropriate to near-surface conditions. In the temperature–pressure range found throughout the greater part of the crust (e.g. hydrostatic pressures of 0.1–3 kilobars (10–300 MPa) and temperatures of 100–500 °C), however, most rocks show at least some ductile flow before failure. The yield strength, the critical value of stress difference required to initiate this ductile behaviour, is rather high in many rocks and would effectively inhibit yield under surface conditions. However, the yield strength is dramatically reduced by pore-fluid pressure (particularly at high temperatures) and, even more important, by decreasing the strain rate to geologically appropriate rates of about $10^{-14}\,s^{-1}$. Thus, given the physical conditions existing at some depth within the crust, and several million years under stress, most rocks will exhibit the kind of ductile behaviour familiar to all geologists who have studied folded rocks in metamorphic terrains. The same rocks under higher stresses and more rapid strain rates, however, will fracture and generate earthquakes.

## 7.10 MECHANISMS OF ROCK DEFORMATION

Since rocks consist of aggregates of individual crystal grains, normally of several different mineral species, the way that they deform depends partly on the properties of the individual crystals and partly on the texture of the rock as a whole. An igneous rock with an interlocking crystalline texture will clearly be stronger than a sandstone with a weak carbonate cement, and stronger in turn than a rock that is cut by pervasive planar fractures, regardless of the nature of the actual minerals. Much can be learned about the nature of rock deformation by studying the microscopic fabric of deformed rocks (see section 4.4) and it is possible in favourable cases to reconstruct in detail how the final strained shape of a deformed rock has been achieved by successive changes in crystal shapes and interrelationships.

The purely elastic properties of a rock are conferred on it by elastic distortions of the lattice of individual crystals. When the lattice is subjected to a differential stress, the atomic spacing is slightly changed by an amount that is proportional to the size of the stress and which also depends on the interatomic bonding force – a characteristic property of the crystal, and hence of a particular rock type. This mechanism is responsible for the primary elastic stage of the typical creep strain curve.

Permanent viscous strain is produced by various deformation mechanisms that operate on a microscopic scale. There are three main types of process: 'cataclasis', 'intracrystalline plasticity' and 'diffusive mass transfer'; these will now be discussed.

### CATACLASIS

**Cataclasis** is the process of fracture and sliding of rigid particles. It includes **grain boundary sliding**, which is one of the most common mechanisms of deformation, producing a parallel alignment of grain boundaries and rectangular grain shapes. Individual particles are undistorted. This process is therefore characterized by a shape fabric, but not by a crystal orientation fabric. It operates at low hydrostatic pressures and low temperatures, and requires a high differential stress.

### INTRACRYSTALLINE PLASTICITY

**Intracrystalline plasticity** comprises the two processes of **dislocation glide** and **dislocation creep**, both of which involve the movement of dislocations through the crystal lattice. These processes result in changes to the crystal shape, and give rise to a variety of characteristic crystal features, including undulose extinction, deformation bands, deformation twins, kink bands and deformation lamellae (see section 4.4). These features give rise to a preferred orientation of crystal domains, which, unlike grain boundary sliding, produce an oriented crystal fabric.

Dislocation glide becomes more difficult as deformation proceeds, since the dislocations intersect and become entangled. Increasing stress is therefore required for deformation to proceed at the same strain rate; this condition is known as **strain hardening**. Dislocation glide is replaced at higher temperatures by dislocation creep, where continued

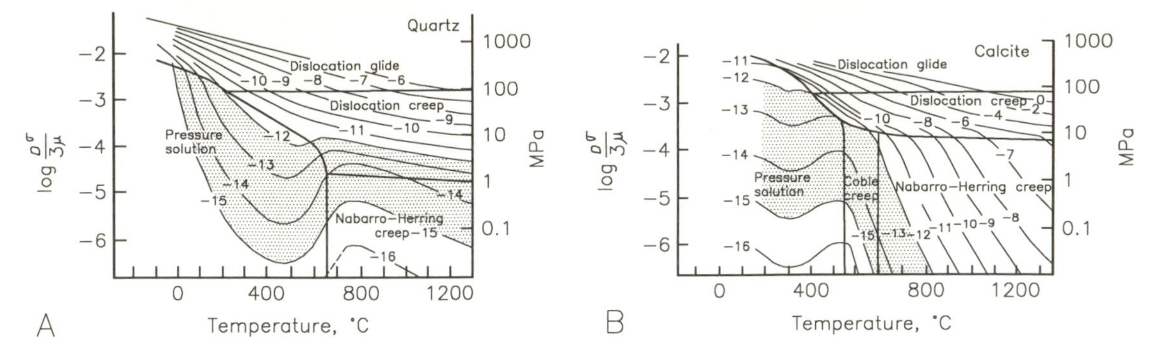


**Figure 7.9** The deformation map. Plots of normalized differential stress against temperature showing fields in which the different mechanisms dominate the strain rate; at field boundaries, the strain rates of the two adjacent mechanisms are approximately equal. Fine lines are contours of strain rate. Fields of geologically likely strain rates are shaded. A. Map for wet quartzite with 100 m grain size; stress scale is evaluated for shear modulus $\mu$, at 900 °C. B. Map for marble with 100 m grain size; stress scale is evaluated for shear modulus, $\mu$, at 500 °C. (After Twiss & Moores, 1992, figure 19.1.)

strain is aided by climbing of dislocations out of their original plane, and by the elimination of dislocations by recrystallization. These processes require intermediate to high stress levels, and operate at relatively high strain rates throughout a large temperature range.

DIFFUSIVE MASS TRANSFER

**Diffusive mass transfer** involves the movement of molecular-scale material, which may be in either the solid state or solution. In **solution creep** (often termed **pressure solution** or **crystal plasticity**), deformation takes place by the transfer of material in solution. This process is important in the formation of structures such as certain spaced cleavages and stylolites (see section 4.1). In **superplasticity**, the transfer of material is in the solid state, either via grain boundaries (**Coble creep**) or within the crystal lattice (**Nabarro–Herring creep**). All these processes take place at low stresses and low strain rates, and with increasing temperature, solution creep is replaced in turn by Coble creep and Nabarro–Herring creep. The above mechanisms may be related by means of a **deformation map** (Figure 7.9), which shows in simplified fashion how stress, strain rate, and temperature determine which process is likely to operate.

SUMMARY

Most rock deformation involves a combination of two or more of the above types of process. For example, diffusive mass transfer (superplasticity) accompanied by grain boundary sliding is thought to be responsible for the formation of very fine-grained mylonites (section 2.3). Since most rocks are polycrystalline aggregates, different crystal phases may exhibit different mechanisms; for instance, under greenschist facies conditions, feldspar may deform cataclastically whereas quartz may exhibit intracrystalline plasticity. The most important factors controlling the deformation mechanism are (1) the physical properties of the material, (2) the magnitude of the deviatoric and hydrostatic stress components, (3) the pore-fluid pressure, (4) the temperature, and (5) the grain size. All deformation processes may be enhanced by chemical reactions.

These processes are irreversible and under appropriate levels of stress and physical conditions give rise to uniform viscous flow – the secondary or steady-state stage of the creep strain curve of Figure 7.7. Accelerated viscous flow is usually caused by the spread of microfractures or slip surfaces through the rock in such a way that they link up to form continuous pervasive cracks, causing loss of cohesion and failure.

The relative importance of slip-type displacements

and recrystallization in promoting viscous flow is governed primarily by rock composition, temperature and the presence of pore-fluids. Low-temperature deformation involves mainly cataclasis and intracrystalline plasticity (or **coldworking**). Strain fabrics produced by both cataclasis and cold-working may be removed at elevated temperature by recrystallization. New unstrained and generally polygonal sub-grains develop from the original strained grains, which eventually become replaced. This process of strain recovery is known as **polygonization**, **hotworking** or **annealing**. The behaviour of crustal rocks below about 10–15 km (and also of upper mantle rocks) is largely governed by steady-state viscous flow at very low strain rates giving rise to a continuous and progressive cycle of distortion and annealing recrystallization. Under such conditions the strain rate is governed only by the effective viscosity (cf. equation (7.3)) of the material, and the size of the differential stress.

## FURTHER READING

Heard, H.C., Borg, I.Y., Carter, N.I. and Raleigh, C.B. (eds) (1972) *Flow and Fracture of Rocks*, American Geophysical Union, *Geophysical Monograph* 16.

Jaeger, J.C. and Cook, N.G.W. (1976) *Fundamentals of Rock Mechanics*, Chapman & Hall, New York.

Twiss, R.J. and Moores, E.M. (1992) *Structural Geology*, Freeman, New York. [See chapters 18 and 19 for a comprehensive treatment of the theoretical and experimental aspects of rock deformation.]

# DETERMINATION OF STRAIN IN ROCKS 8

The quantitative evaluation of the total or bulk finite strain in a given area resulting from a deformation is an important objective for the structural geologist. If we wish to evaluate the effects of orogenic compression, for example, we would ideally need to know the magnitudes and orientations of the three principal strain axes in all parts of the area. Once the strain distribution is known, in the form of a set of strain trajectories, or as a representative bulk strain ellipsoid, we can try to explain it in terms of stress or movement models.

There are three quite different approaches to the problem of quantifying strain. The first and most obvious is to measure individual strain ellipsoids using various kinds of strained objects, and sum the results over the area in question. The second approach is to estimate the total shortening or elongation by examining the geometry of folds or faults (this method is, however, difficult to apply in three dimensions). The third, and in some ways the simplest, approach is to assume that the strain on a large scale is essentially homogeneous statistically, and that the statistical arrangement of all planar and linear structural elements throughout the area reflects both the orientation and the size of the bulk finite principal strains. This approach has been particularly useful in dealing with highly deformed zones in certain Precambrian gneiss terrains.

## 8.1 FINDING THE PRINCIPAL STRAIN AXES

The problem of determining the strain in a deformed rock is made much easier if the principal strain axes can be found first. This is possible if the deformed rock possesses a new planar or linear fabric that reflects the finite strain geometry. Thus a planar flattening fabric (slaty cleavage, schistosity or gneissosity) will lie in the $XY$ plane of the strain ellipsoid (see sections 4.1 and 6.3), enabling the orientation of the $Z$ strain axis to be found immediately. In certain circumstances, the orientation of $X$ may also be found by observing the 'stretching' direction (i.e. an elongation lineation) in the plane of the foliation, and in this way all three strain axes may be uniquely determined. However, strongly linear fabrics ($X > Y \sim Z$) or strongly planar fabrics ($X \sim Y > Z$) will yield only one obvious principal strain axis.

Even if the orientation of all three principal strain axes can be measured, however, the complete determination of the strain requires the measurement of the extensions or stretches along each. This demands the use of some object or objects of known initial shape. Such objects are termed **strain markers**. A wide variety of strain markers has been used in strain studies. Much of the early work on strain determination was done on deformed fossils, but the simplest methods involve the measurement of initially spherical objects which assume an ellipsoidal form after deformation.

## 8.2 INITIALLY SPHERICAL OBJECTS AS STRAIN MARKERS

The strain ellipsoid assumed by objects which initially had a spherical shape may be measured directly if the objects can be seen in three dimensions. Such objects include the ooids in oolitic limestones, the spherulites and vesicles found in volcanic rocks, reduction spots in slates, recrystallization spots in hornfels and various kinds of concretions. Since such objects are rarely perfectly spherical, some allowance has to be made for the initial variation in shape. This can be easily done if the shape variation is random by taking the geometric mean of a number of observations, but if there is some non-random variation in initial shape, controlled by bedding for instance, a more sophisticated technique must be employed (see later).

Another problem concerns the degree of homogeneity of strain through the rock. Many strain markers are made of material different from their matrix, so the measured strain may not accurately reflect that of the whole rock. Thus hard siliceous or calcareous concretions would show much lower strains than a more ductile shaly matrix.

## DIRECT MEASUREMENT OF STRAIN AXES

Any three non-parallel sections through the rock are sufficient to determine the strain axes, but the calculations are easier if the three sections are mutually perpendicular, especially if they are parallel to the principal strain planes. Measurements can be made directly from thin sections or polished slabs, or on enlarged photographs if the objects are small. To obtain an accurate result, a number of measurements should be made. If these are plotted on a graph of long-axis length against short-axis length, the slope of the best-fit straight line through the origin gives the mean value of the strain ratio $Y'/X'$ (ratio of minimum to maximum stretch) in that plane (Figure 8.1). The visual method of plotting is preferable to simply calculating the arithmetic mean, since deviations from the straight line may be easily observed. Sometimes the larger particles give a different result from the smaller ones, perhaps because of differences in ductility.

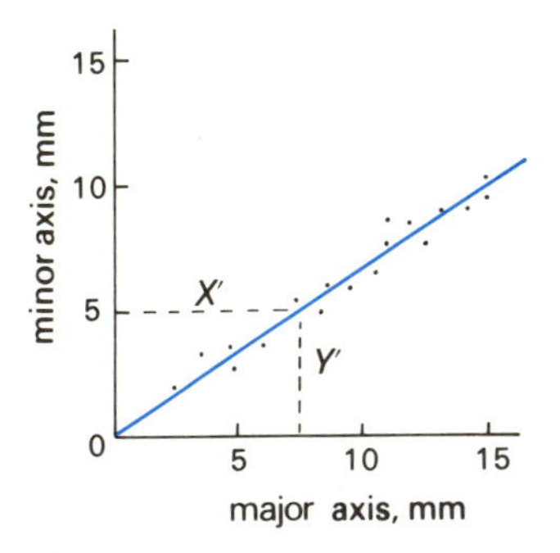


**Figure 8.1** Finding the principal strain ratio in two dimensions from deformed circular objects. The long axis and short axis of a number of ellipses (e.g. Figure 8.2A) are plotted. The slope of the coloured line gives the ratio $Y'/X'$. (After Ramsay and Huber, 1983, figure 5.10A.)

## CENTRE-TO-CENTRE METHOD

If there is a large difference in ductility between the measured objects and their matrix, the measured strain will obviously not apply to the whole rock. In the case, say, of quartz pebbles in a mudstone matrix, much higher strains would take place in the matrix than in the pebbles. To deal with this problem, the centre-to-centre method may be used. This method relies on the fact that the distances between the centres of randomly arranged spheres

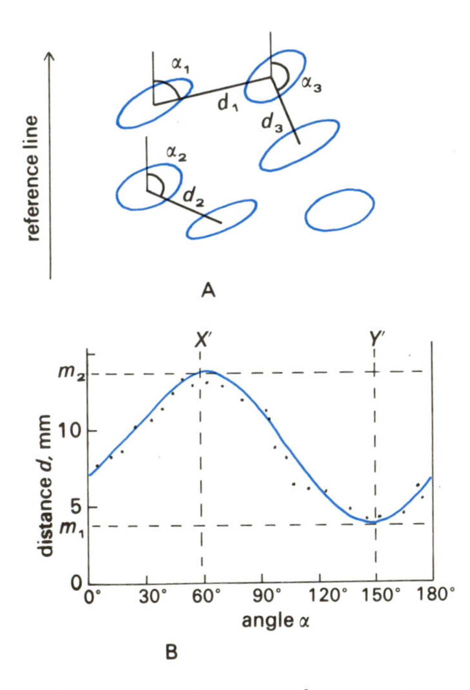


**Figure 8.2** Finding the principal strain ratio in two dimensions from the spacing of deformed objects. A. Distance $d$ between adjacent centres and angle $\alpha$ made with a reference line are measured for a number of adjacent centres. B. Plot of $d$ against $\alpha$ yields a minimum value of $d$ ($m_1$) and a maximum value ($m_2$) which represent $Y'$ and $X'$ respectively. The orientation of $X'$ and $Y'$ in relation to the reference line is given by $\alpha_X$ and $\alpha_Y$. (After Ramsay and Huber, 1983, figures 7.17 and 7.18.)

are systematically altered during strain in such a way that the changes in distance are related to direction. The ratio between the minimum and maximum mean distances is equal to the strain ratio $Y'/X'$.

To determine this ratio, a plot is made of the distances between adjacent centres against the orientation of the line between the centres (Figure 8.2). The plot gives a minimum and a maximum value ($m_1$ and $m_2$ respectively) for the distance, which is used to calculate the strain ratio, and two corresponding values for $\alpha$ ($\alpha_Y$ and $\alpha_X$) for the orientation of $Y'$ and $X'$.

The above method has the advantage that it can be applied to any rock with randomly distributed particles regardless of their respective ductilities or whether they form perfect ellipses. The main problem lies in determining accurately the positions of the centres of the particles.

## 8.3 DEFORMED CONGLOMERATES AS STRAIN MARKERS

Deformed conglomerates are much more common than other kinds of strain marker, and have therefore been frequently used by structural geologists for assessing strain in the field. Unfortunately deformed conglomerates suffer from the serious disadvantage that their initial shape is generally not spherical. Moreover, they may possess a primary sedimentary orientation which, when added to the deformation fabric, produces a compromise fabric that cannot be simply analysed. Added to these complications is a further factor – the effect of ductility, both between different types of pebble and between pebble and matrix. The centre-to-centre method described above would circumvent some of these problems, but centres of pebbles with a primary orientation fabric are unlikely to possess a random distribution – an essential precondition for this method. Much effort has been expended by structural geologists in devising satisfactory methods of analysing pebble strain. In general it is more useful to obtain a large number of approximate results from different localities and rock types than to concentrate effort on obtaining mathematically precise results from a single pebble bed that might be highly atypical.

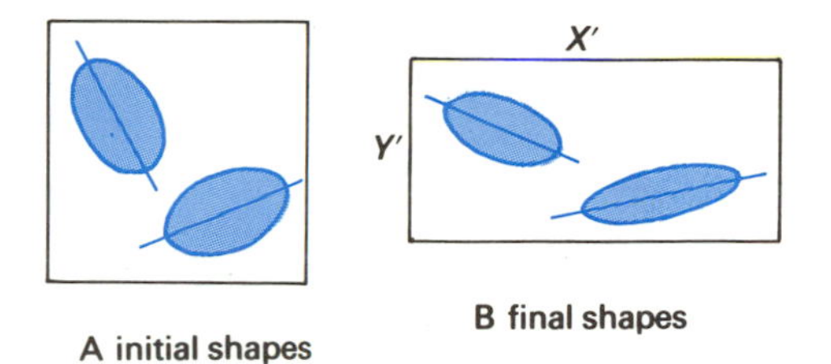


**Figure 8.3** Effect of an initial elliptical shape on the strain ratio and orientation. A. Initial shapes. B. Final shapes, after a homogeneous strain $X'$, $Y'$.

If we can assume that the pebbles originally possessed an ellipsoidal shape, the final shape as measured in two dimensions will be a compromise between two ellipses both in direction and in axial ratio (Figure 8.3). The fluctuation in observed axial directions and axial ratios will thus have a systematic relationship to the tectonic axes.

Figure 8.4 shows a graphical method of finding the tectonic strain ratio and orientation of the major axis of the strain ellipse by plotting $R$, the observed finite strain ratio $Y'/X'$, against $\alpha$, the angle that the long axis of the ellipse $X'$ makes with reference line $\alpha_0$. If enough measurements are made, the plot yields maximum and minimum values of $R$ (Figure 8.4A). The maximum value records the case where the initial elongation is parallel or near-parallel to

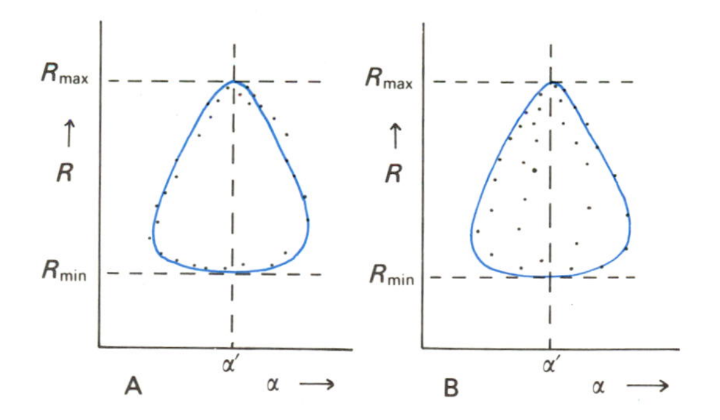


**Figure 8.4** Finding the principal strain ratio in two dimensions using pebbles with an initial elliptical shape (see text). A. Plot of $R = X'/Y'$ (observed ratio) against angle $\alpha$ with reference line for a constant initial shape ratio (A) and a variable initial shape ratio (B). The angle which $X'$ makes with the reference line is $\alpha'$. (After Ramsay, J.G. (1967), figures 5.27 and 5.28.)

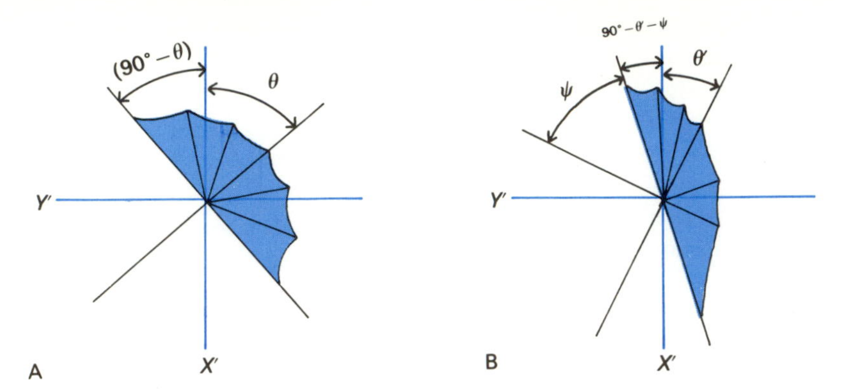


**Figure 8.5** Finding the principal strain ratio in two dimensions using the angular shear strain in a bilaterally symmetrical fossil. A. Before strain. B. After strain. The angular shear strain is $\psi$ and the strain ratio is given by $Y'/X' = \sqrt{(\tan\theta'/\tan(\theta'+\psi))}$ (see text for explanation.)

the tectonic strain elongation and the minimum value records the case where the initial elongation is perpendicular to the tectonic strain elongation. Therefore,

$$R^2_{max} = R_T R_0 \tag{8.1}$$

and

$$R^2_{min} = R_0/R_T \tag{8.2}$$

where $R_0$ is the original ratio $Y'_0/X'_0$ and $R_T$ the tectonic strain ratio $Y'_T/X'_T$; the tectonic strain ratio may be found by dividing (8.1) by (8.2), giving:

$$R_T = R_{max}/R_{min} \tag{8.3}$$

If the original ratio $R_0$ is variable but smaller than $R_T$, the relationship still holds but $R_0$ will represent the maximum value of the original ratio and the points on the graph will be distributed throughout the area bounded by the envelope shown in Figure 8.4B. If the orientation of the maximum strain axis $X'$ is not already known, it is given by the symmetry axis of the plot ($\alpha'$ in Figure 8.4A, B). This method is often referred to as the $R_f/\phi$ method, where $R_f$ is the finite strain and $\phi$ the angle with the reference line.

## 8.4 BILATERALLY SYMMETRICAL FOSSILS AS STRAIN MARKERS

Two-dimensional strain may also be determined by measuring the angular distortion of an original right angle in a bilaterally symmetrical fossil such as a brachiopod or a trilobite (Figure 8.5). The divergence from an original right angle will give the angular shear strain $\psi$.

To obtain the ratio of the principal strains in the plane of the fossils, only one measurement of $\psi$ is needed, provided that the orientations of the principal strain axes in the plane are already known. The axis of greatest extension $X'$ is commonly visible as a stretching lineation. If $\theta$ is the original angle between the axis of symmetry of the fossil and $X'$, then $90° - \theta$ is the complementary angle between the perpendicular to the symmetry axis and $X'$. Both these angles change as a result of the angular shear strain $\psi : \theta$ to $\theta'$ and $(90° - \theta)$ to $(90° - \theta' - \psi)$. From equation (6.4),

$$\frac{Y'}{X'} = \frac{\tan\theta'}{\tan\theta} \quad \text{and} \quad \frac{Y'}{X'} = \frac{\tan(90° - \theta' - \psi)}{\tan(90° - \theta)} \tag{8.4a}$$

where $X'$ and $Y'$ are the principal strains in the plane concerned. Thus, multiplying,

$$\left(\frac{Y'}{X'}\right)^2 = \frac{\tan\theta' \tan(90° - \theta' - \psi)}{\tan\theta \tan(90° - \theta)}$$

$$= \tan\theta' \tan(90° - \theta' - \psi)$$

$$(\text{since } \tan\theta \tan(90° - \theta) = 1)$$

$$= \frac{\tan\theta'}{\tan(\theta' + \psi)} \tag{8.4b}$$

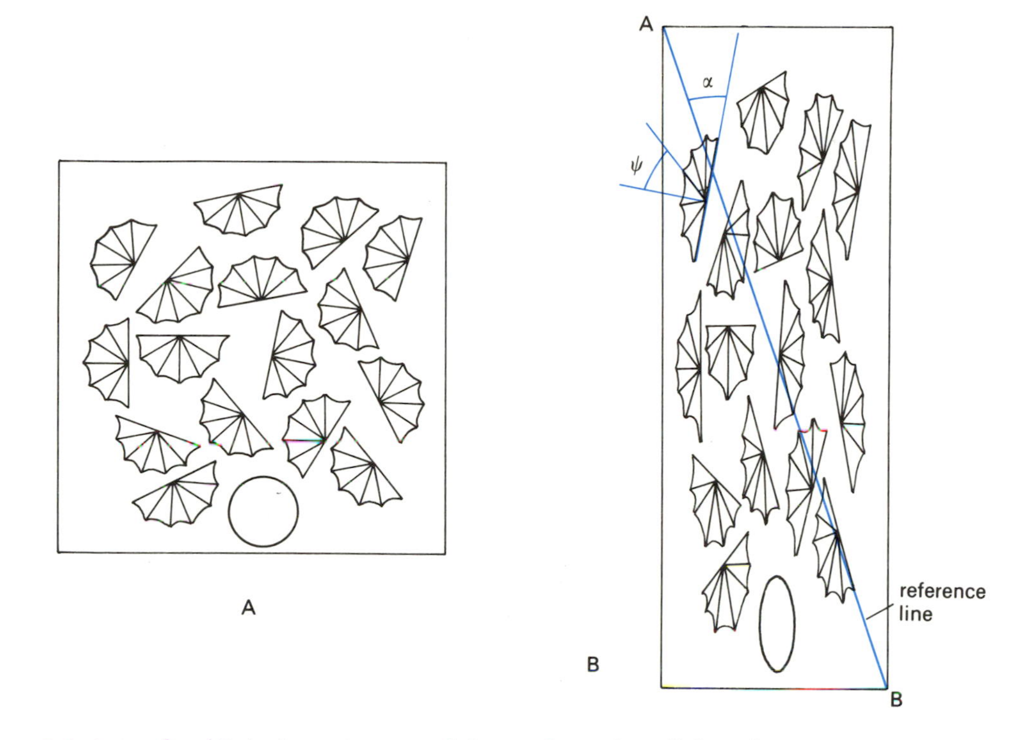


**Figure 8.6** A, initial and B, final, orientations and shapes of a number of bilaterally symmetrical fossils with varying orientation after a homogeneous strain of $Y'/X'$. (After Ramsay, J.G. (1967), figures 5.60 and 5.62.)

The above method is of course dependent on knowing the direction of $X'$ and $Y'$ in the plane of the fossil, and these may not be immediately obvious from the stretching lineation. However where there are a number of fossils in the same plane showing variation in angular shear strain (Figure 8.6), it is possible to measure both the strain ratio and the orientation of the strain axes by finding the maximum angular shear strain. It is not necessary that the fossils all be of the same species.

The angular shear strain $\psi$ (measured as in Figure 8.5) may be plotted against the angle $\alpha$ between a standard identifiable line in the fossil (e.g. a hinge line) and an arbitrary reference line on the plane (Figure 8.6B). If the section shows sufficient variation in orientation, the resulting graph will take the form of a curve cutting the $\alpha$ axis at two points (i.e. where $\psi = 0$), which give the directions of the maximum and minimum strain axes. The strain ratio is found by estimating the maximum angular shear strain from the curve and using the formula:

$$Y'/X' = \tan\theta'/\tan\theta \quad \text{(see equation 6.4)} \quad (8.5)$$

Since the maximum shear strain occurs when $\theta = 45^\circ$ and $\tan\theta = 1$,

$$Y'/X' = \tan\theta'_{max} \quad (8.6)$$

The directions of $X'$ and $Y'$ in the plane of the fossils may also be found by observing shortened or elongated forms of deformed fossils that have suffered no angular shear strain (Figure 8.7). The axes of bilateral symmetry in these cases will be parallel to the maximum and minimum strain axes in the plane and a simple comparison between them will yield the strain ratio directly. Thus, assuming that the length/breadth ratio of both samples was originally identical, the strain ratio can be found simply (see Figure 8.7).

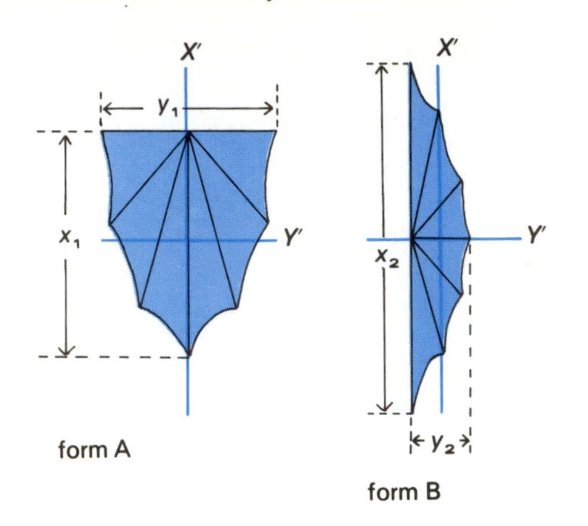


**Figure 8.7** Finding the principal strain ratio in two dimensions using short and long fossil shapes showing no angular shear strain. If the lengths of the short and long forms are respectively $x_1$ and $x_2$ and the breadths $y_1$ and $y_2$, then the strain ratio $Y'/X' = \sqrt{(y_1y_2/x_1x_2)}$.

## 8.5 STRAIN DETERMINATION IN THREE DIMENSIONS

The methods outlined in the preceding sections are designed to discover the strain ratio $Y'/X'$ in a single plane. If this plane possesses two of the principal strain axes, e.g. $X$ and $Y$, it is a comparatively simple matter to obtain a second strain ratio in a plane normal to the first, containing say $Y$ and $Z$. The two ratios thus obtained, $Y:X$ and $Z:Y$, give a complete description of the three-dimensional strain. This is usually given in the form $X:Y:Z = 5:2:1$, for example, i.e. taking $Z = 1$ (Figure 8.8).

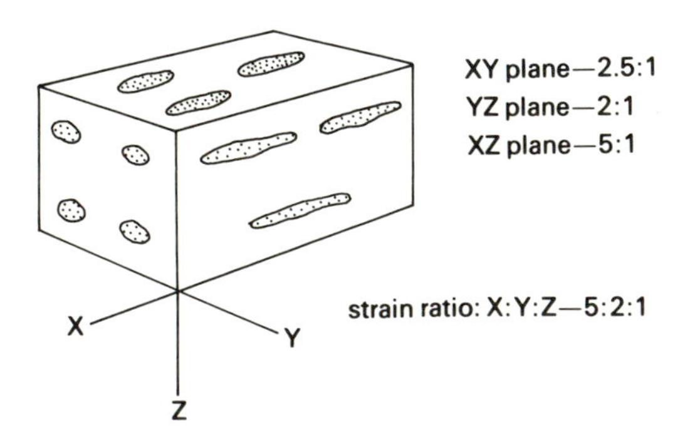


**Figure 8.8** Principal strain ratios in three dimensions. If the strain ratios are known for any two of the three principal strain planes (say $XY$ and $YZ$), the strain ratios can be calculated in three dimensions.

Where the positions of the principal strain axes $X$, $Y$ and $Z$ cannot be directly observed from the rock fabric, the strain may be determined using any three mutually perpendicular sections on which the strain ratio $Y'/X'$ is measured. The method is described by Ramsay (1967, pp. 142–7).

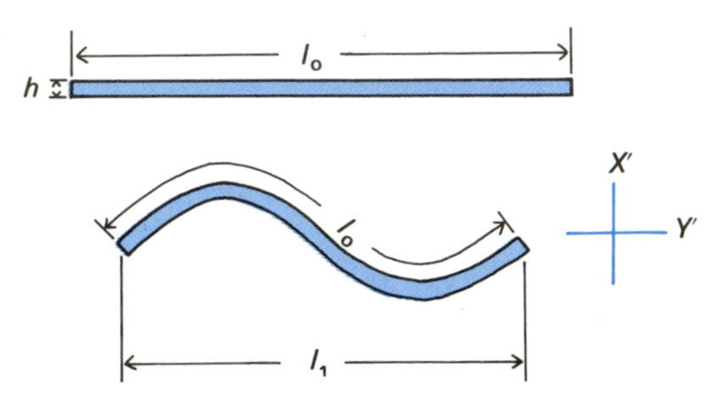


**Figure 8.9** Determination of shortening from a folded layer. The strain ratio $Y'/X' = l_1^2/l_0^2$, assuming the area of the layer $hl_0$ is constant.

## 8.6 USE OF FOLD SETS IN STRAIN DETERMINATION

It is possible to measure the strain represented by a set of folds by comparing the final and the (presumed) initial shapes of a layered sequence. Fold geometry is often complex and depends partly on the mechanism of formation. The relationship between fold geometry and strain is discussed in more detail in Chapter 10; however, for our present purpose, we can assume that a simplified analysis can be made, in the plane perpendicular to the fold axis, by comparing the initial and final lengths of a folded layer (Figure 8.9). If $l_0$ is the original length of a folded layer of thickness $h$ and $l_1$ is the new length, the stretch measured perpendicular to the axial plane is

$$Y' = l_1/l_0 \tag{8.7}$$

The stretch parallel to the axial plane is found by assuming that the area $hl_0$ is constant and dividing by the new length, thus:

$$X' = \frac{hl_0}{l_1} \bigg/ h = \frac{l_0}{l_1} \tag{8.8}$$

The strain ratio is thus

$$Y'/X' = l_1^2/l_0^2 \tag{8.9}$$

This method takes no account of possible layer-parallel shortening, but this effect can be minimized by comparing several layers of different thickness and physical properties and taking the maximum value as an estimate of the total strain in that plane. A more serious restriction is that the method can only measure two-dimensional strain and takes no account of strain parallel to the fold axes. On a large scale, however, estimates of strain based on the geometry of major folds are likely to be more accurate than observations based on strain markers from specific lithologies that may be biased in favour of ductile lithologies and more highly strained zones.

## 8.7 TWO-DIMENSIONAL STRAIN FROM BALANCED SECTIONS

Section balancing has been developed as a method of unravelling complex thrust and extensional fault zones by restoring them to their original lengths, in order to measure the fault displacement and to reconstruct the sequence of movements responsible for their often complex geometry (see section 2.6).

Two-dimensional strain can be estimated using the same method as that described in the previous section, except that, instead of considering a single layer, we are concerned with the change in dimensions of a planar section consisting of a number of layers (Figure 8.10). The area of the section is assumed to be constant, and the length $l_1$ and thickness $t_1$ of a deformed section are compared with the original 'stratigraphic' thickness $t_0$ of the same sequence in an undeformed section. The original length of the deformed section $l_0$ is then found by

$$l_0 = l_1 t_1 / t_0 \tag{8.10a}$$

and the strain by

$$Y'/X' = l_1^2/l_0^2 = t_0^2/t_1^2 \tag{8.10b}$$

The section must be taken perpendicular to the 'orogenic strike', (i.e. the main fold trend). A three-dimensional study of fault displacements can be made if sufficient data are available and, except in the simplest types of fault belts, strain in the third dimension should be considered.

## 8.8 BULK HOMOGENEOUS STRAIN

A quite different approach to the problem of strain measurement is to ignore the detail of local strain patterns, which are typically highly variable in any case, and treat the strain as statistically homogeneous over a given area. All the measured structural elements (foliations, fold axial planes, fold axes, etc.) in the area are plotted on a stereogram and compared with the ideal distributions that would be obtained with various types of strain ellipsoid. In this treatment we regard the bulk strain pattern as a large-scale strain ellipsoid, wherein planar and linear

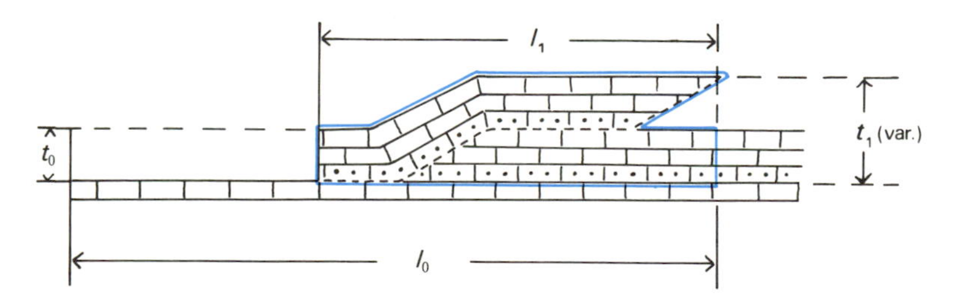


**Figure 8.10** Two-dimensional strain from balanced thrust sections. The section balances if all the restored layers are of equal length (see section 2.6). The area of the section is assumed to be constant, and the length $l_1$ and thickness $t_1$ of a deformed section are compared with the original 'stratigraphic' thickness $t_0$ of the same sequence in an undeformed section. The original length of the deformed section, $l_0$, is then found by measuring the restored length of the section (see text).

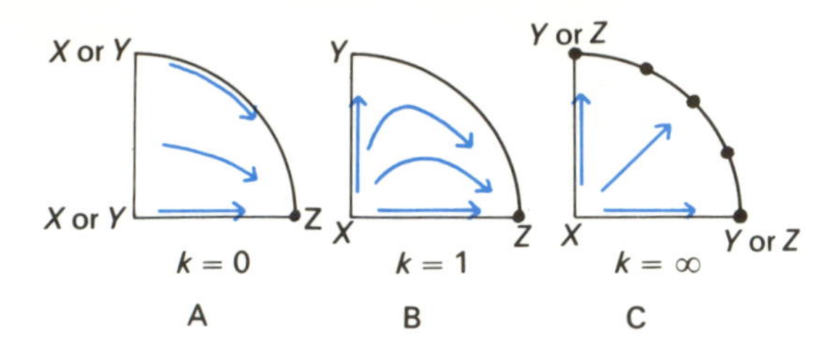


**Figure 8.11** Stereographic plot (one quadrant) showing movement paths taken by poles to planes undergoing passive rotations during progressive homogeneous deformation. The dots show the positions of poles at infinite strain. A, $k=0$, oblate strain; B, $k=1$, plane strain; C, $k=\infty$, prolate strain (see text for explanation). (After Flinn, D. (1962) *Quarterly Journal of the Geological Society of London*, **118**, 385–433).

elements rotate passively in response to progressive strain towards the theoretical positions in the strain ellipsoid that would be taken up at infinite strain.

In this model, the planar structural elements are regarded as passive rather than active; that is, the strain pattern is caused not by the active rotations of planar surfaces around fold axes, for example, but by the passive rotations of a collection of randomly-oriented planar markers situated within a ductile matrix.

The bulk homogeneous strain model is most appropriate in studying areas of rather high strain where there was a considerable variation in attitude of pre-strain surfaces. These conditions apply particularly within major shear zones in certain Precambrian gneiss complexes which have been studied using this approach. Figures 8.11 and 8.12 show how the model may be applied.

## STRAIN FIELDS AND MOVEMENT PATHS

If we refer back to section 6.7, the various strain states were expressed in the Flinn diagram in terms of the value $k$. Figure 8.11 shows stereogram quadrants representing the three special cases: oblate uniaxial strain ($k = 0$), plane strain at constant volume ($k = 1$) and prolate uniaxial strain ($k = \infty$). The arrows show the movement paths taken by the poles to the planes, which rotate passively under progressive strain. For oblate uniaxial strain, where $X = Y > Z$, the planes will rotate towards the $XY$ plane (the plane of flattening) and the poles will rotate towards $Z$; for plane strain, where $X > Y > Z$, the planes will again rotate towards the $XY$ plane but will first move towards $XZ$ if their poles lie in or near $XY$ and the poles will again rotate towards $Z$; and for prolate uniaxial (constrictional) strain, where $X > Y = Z$, the poles to the planes will migrate towards the circumference of the stereogram and the planes will intersect along the $X$-axis. As a consequence of the behaviour of planes under constrictional strains, fold axes will migrate into parallelism with $X$ (since the fold axes represent the intersections of the planes of the fold surfaces). Directions of extension indicated by boudinage structures (section 4.3) may also be plotted and will similarly migrate towards 'ideal' positions parallel to $X$, or into the $XY$ plane where $X = Y$.

## MEASUREMENT OF HOMOGENEOUS STRAIN

In order to make use of these properties in providing a quantitative estimate of the bulk strain, it is necessary to measure the extent to which a strain ellipsoid departs from sphericity. The amount of deformation can be defined by $r$, where

$$r = a + b - 1 \qquad (8.11)$$

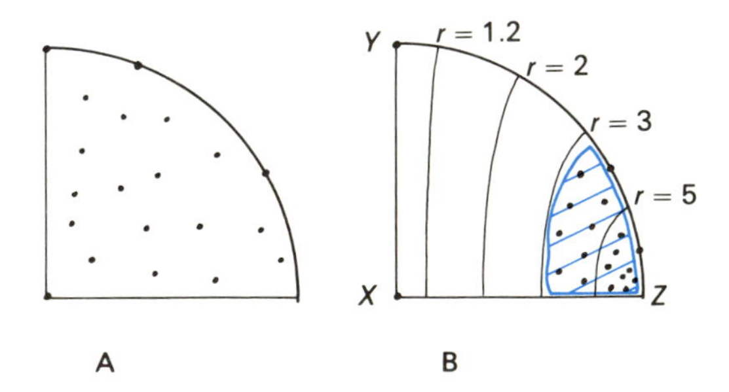


**Figure 8.12** Equal-area plot (one quadrant) showing an initially random distribution of poles to planes (A), and (B) the distribution of the same poles at a strain level of $r = 3$ (where $r = a + b - 1$). Limits of distribution for various other $r$ values are shown. (Modified from Watterson, J. (1968) *Meddelelser om Grønland*, **175**, figure 41.)

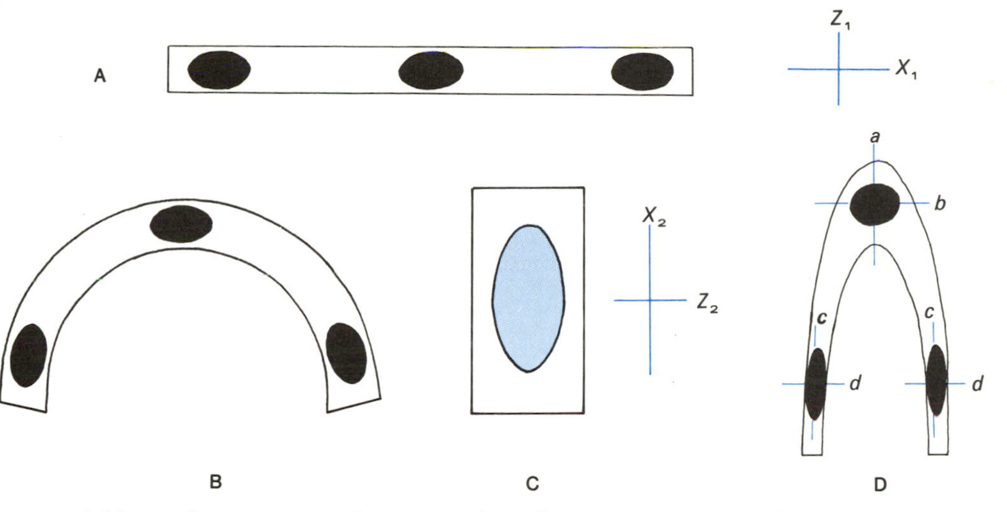


**Figure 8.13** Addition of strains in two dimensions where the strain axes are parallel. A shows the initial strain $X_1$, $Z_1$ in a layer before folding. During folding (B) a second homogeneous strain $X_2$, $Z_2$ is added (C). The final strain (D) varies – at the hinge $a = Z_1X_2$ and $b = X_1Z_2$, whereas on the limbs $c = X_1X_2$ and $d = Z_1Z_2$ (see text)

given that $a = X/Y = \theta_x/\theta_y$ and $b = Y/Z = \theta_y/\theta_z$ (see equations (6.10) and (6.11)).

Figure 8.12 shows the distribution of poles to planes in a stereogram at a strain level of $r = 3$ for a particular type of ellipsoid ($k = 1$). The stereogram is also contoured to show the limits of distribution of poles for different values of $r$.

Stereograms can be constructed for several different $k$-values. Each will have a different set of $r$ contours, depending on the ratios of $X/Y$ and $Y/Z$. The correct $k$-value could thus be found by comparison with these ideal distributions. Given a sufficient number of measurements, and provided that our initial assumptions of random orientation are justified, it is possible to specify completely the shape of the strain ellipsoid and consequently to quantify the amount of strain.

## 8.9 SUPERIMPOSITION OF STRAINS

As for stresses, it is possible to add or subtract strains, and two strain ellipsoids of whatever orientation may be added to produce a 'compromise' ellipsoid. The principle may be illustrated by considering the effect of adding stretches $\theta_1$, $\theta_2$ and $\theta_3$ of a strain ellipsoid B along the axes $x$, $y$ and $z$ of another strain ellipsoid, A, with principal strains $X$, $Y$ and $Z$. The new stretches along $x$, $y$ and $z$ are obtained by multiplying the respective stretches, giving $X_1\theta_1$, $Y_2\theta_2$ and $Z_3\theta_3$. To determine the geometry of the new finite strain ellipsoid, the strains should be described in tensor notation as described by Means (1976).

Figure 8.13 illustrates the principle of superimposition of strain in two dimensions in the special case where the axes of the two strain ellipsoids correspond. The first strain is represented by three ellipses (with axes $X_1$, $Z_1$) in the plane of the layer (Figure 8.13A). The layer is isoclinally folded (Figure 8.13B) and a second homogeneous flattening strain imposed, represented by an ellipse with axes $X_2$, $Z_2$ (Figure 8.13C), giving the superimposed strain pattern of Figure 8.13D. The original strain ellipses are now oriented in such a way that $X_1 \parallel Z_2$ in the fold hinge, whereas $X_1 \parallel X_2$ on the fold limb. The final strain in the hinge is therefore $a = Z_1X_2$, $b = X_1Z_2$ and on the limbs it is $c = X_1X_2$, $d = Z_1Z_2$, giving final strain ratios in the hinge of $X_1Z_2/Z_1X_2$ and on the limbs of $Z_1Z_2/X_1X_2$.

## FURTHER READING

Hossack, J. (1979) The use of balanced cross-sections in the calculation of orogenic contraction: a review. *Journal of the Geological Society of London*, **136**, 705–11.

Means, W.D. (1976) *Stress and Strain*, Springer-Verlag, New York.

Ramsay, J.G. (1967) *Folding and Fracturing of Rocks*, McGraw-Hill, New York.

Ramsay, J.G. and Huber, M.I. (1983) *The Techniques of Modern Structural Geology, Vol. 1: Strain Analysis*, Academic Press, New York. [This work contains a comprehensive and rigorous treatment, with examples, of all the main methods of strain analysis. It is clear and easy to read, and is highly recommended to students who wish to pursue this topic.]

The morphology of faults has been described in Chapter 2. In this chapter we shall discuss faulting as a process or mechanism, and how faults may be related to stress.

## 9.1 STRESS CONDITIONS FOR BRITTLE FAILURE

When a material fractures under conditions of brittle deformation (see section 7.3), it is said to exhibit **brittle failure**. The stress conditions at the point of failure are known as the **stress criteria of brittle strength**. These criteria include both the shear stress and the hydrostatic pressure, and vary with rock composition, temperature, etc.

When rocks fail under compression in experimental conditions, it is found that, in general, two sets of planar shear fractures are formed which intersect in a line parallel to the intermediate principal stress axis $\sigma_2$ (Figure 9.1A). Moreover, the acute angle between the shear fractures is bisected by the maximum principal stress axis $\sigma_1$.

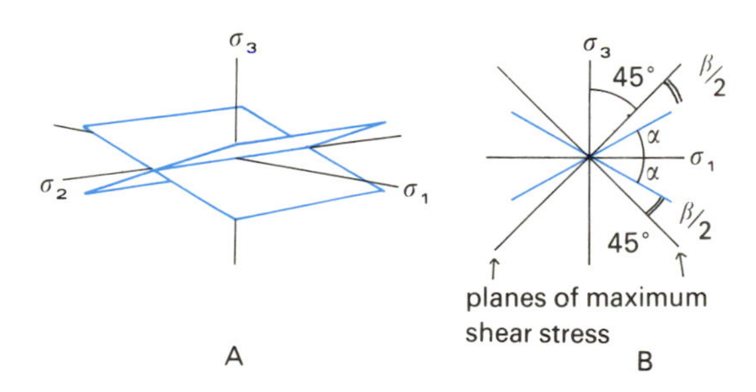


**Figure 9.1** Relationship between shear fractures and principal stress axes. A. Shear fractures ideally intersect in $\sigma_2$ and make an acute angle with $\sigma_1$. B. Plane perpendicular to $\sigma_2$. Shear fractures make an angle $\alpha$ with $\sigma_1$, and $\beta/2$ with the planes of maximum shear stress. Thus $2\alpha + \beta = 90°$.

The actual fracture planes do not correspond to the planes of maximum shear stress, which make angles of 45° with $\sigma_1$ (see section 5.5). If the size of the acute angle between the fracture planes is $2\alpha$, then the difference between this angle and the 'ideal' angle made by the planes of maximum shear stress is $\beta = 90° - 2\alpha$ (Figure 9.1B). The angle $\beta$ is sometimes referred to as the **angle of internal friction** of the material and it is different for different stress states.

### THE MOHR STRESS DIAGRAM

This diagram (Figure 9.2A) is a convenient way of portraying the relationship in two dimensions between shear stress, hydrostatic pressure and the angle of failure at the point where failure occurs. Each state of stress is represented by a circle with centre $(\sigma_1 + \sigma_3)/2$ (= the mean stress or hydrostatic component) and radius $(\sigma_1 - \sigma_3)/2$ (= the stress difference) that intersects the $\sigma$ axis in two points, $\sigma_1$ and $\sigma_3$. It is assumed for convenience that $\sigma_1 > \sigma_2 = \sigma_3$. Let the values of stresses $\sigma_1$ and $\sigma_3$ at failure be represented by the circle shown in Figure 9.2A and the angle between the shear fractures be $2\alpha$, then the stress conditions at failure are represented by the point X. The shear stress at failure, $\tau_R$, is given by

$$\tau_R = \tfrac{1}{2}(\sigma_1 - \sigma_3)\cos\beta \qquad (9.1)$$

from equation (5.4), since $\beta = (90° - 2\alpha)$. $\beta$ is the angle that the tangent to the circle at X makes with the horizontal. The hydrostatic pressure at failure is given by

$$P = \tfrac{1}{2}(\sigma_1 + \sigma_3) \qquad (9.2)$$

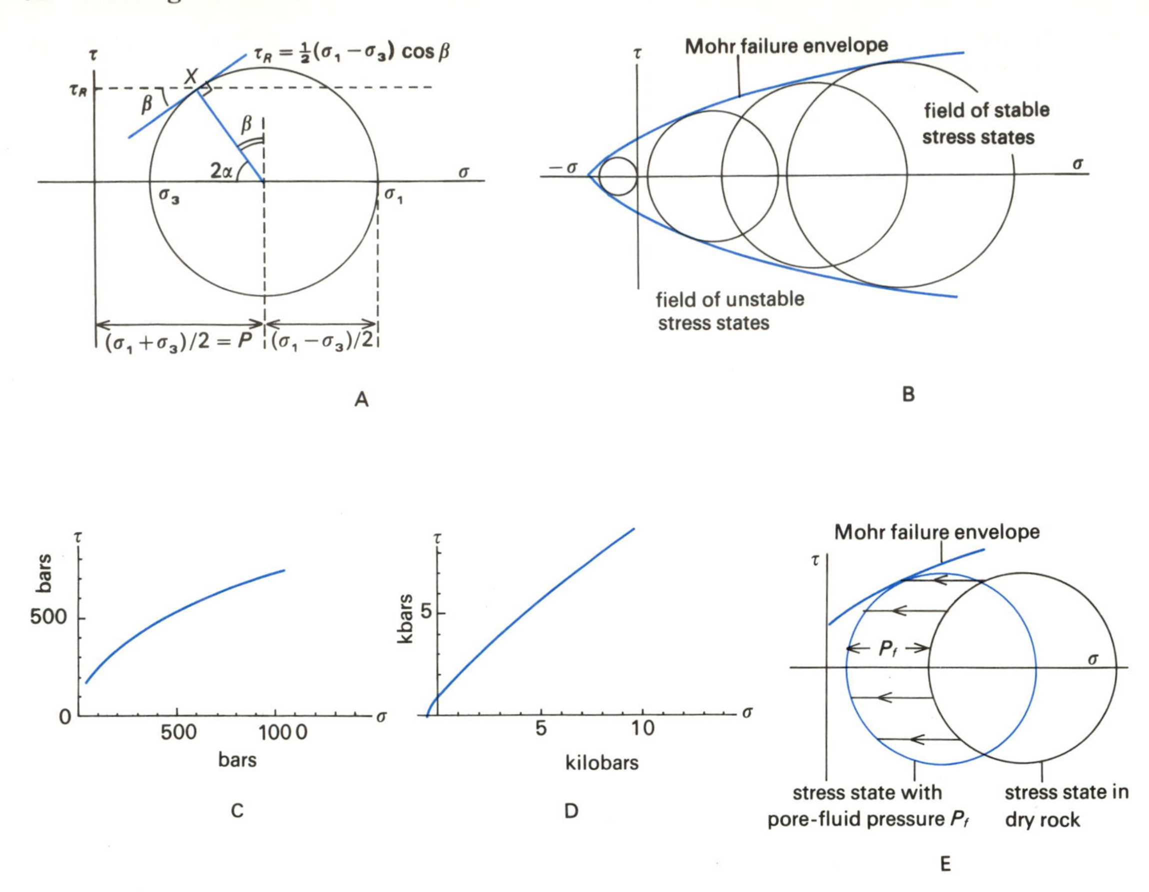


**Figure 9.2** The Mohr stress diagram: failure criteria in two dimensions. A. Stress conditions at failure for a shear fracture making an angle $\alpha$ with $\sigma_1$ (see text for explanation). B. The Mohr failure envelope joining points of failure for different stress states separates the field of stable stress states from the failure field. Note that the shape of the Mohr curve implies an increase in the values of the shear stress $\tau$ and fracture angle $\alpha$ with an increase in the mean (hydrostatic) stress $(\sigma_1 + \sigma_3)/2$. C, D. Mohr envelopes derived from the experimental deformation of Wombeyan marble (C) and Frederick diabase (dolerite) (D). (After Paterson, M.G. (1958) *Bulletin of the Geological Society of America*, **69**, 465–76 (C), and Brace, W.F. (1964) in *State of Stress in the Earth's Crust* (ed. W.R. Judd), Elsevier, Amsterdam, figure 22 (D).) E. Effect of pore-fluid pressure. The black Mohr circle represents a stress state where the shear stress is too low for failure to occur. The effect of pore-fluid pressure $P_f$ is to reduce the normal stress to the value represented by the coloured circle, which intersects the failure envelope, indicating that failure would occur.

and the normal stress across the fracture plane is given by

$$\sigma = \tfrac{1}{2}(\sigma_1 + \sigma_3) - \tfrac{1}{2}(\sigma_1 - \sigma_3)\sin\beta \quad (9.3)$$

(cf. equation (5.3)).

The effect of varying the stress conditions at failure is shown in Figure 9.2B, where a number of circles are drawn to show various values of $\sigma_1$ and $\sigma_3$ at which failure occurred for a particular rock specimen. The curve joining the points of failure for the different stress states is called the **Mohr failure envelope**, and divides the field of stable stress states within the envelope from the failure field outside. This curve illustrates the general principle that the value of the shear stress $\tau$ required to produce failure increases as the hydrostatic pressure $(\sigma_1 + \sigma_3)/2$ increases, i.e. as the size of the Mohr circle increases.

The diagram also shows the effect of negative compressive stress, or tensile stress, represented by that part of the Mohr envelope to the left of the axis. It is clear from the shape of the envelope that the value of shear stress required to produce failure under tension is much smaller than that required under compression, which is in agreement with the observation that rocks are much stronger under compression. Moreover, although most rocks have a finite tensile strength, their compressive strength is effectively infinite if the shear stress is below the required shear strength. It may also be observed from the shape of the Mohr envelope in the tensional field that tensile shear fractures make smaller angles with $\sigma_1$ ($\beta$ is large), and that where the shear stress is zero, the value of $\sigma$ corresponds to the tensile strength of the material.

The rock composition has a marked effect on the general shape of the failure envelope. Figures 9.2C and D show two failure envelopes derived respectively from the experimental deformation of marble and diabase (dolerite). Note that the shear strength of the diabase increases more rapidly with increase in compressive stress than is the case for the marble.

## FAILURE CRITERIA

The shear stress $\tau$ acting along the fracture plane to promote failure is opposed by the compressive stress $\sigma$ acting across the fracture plane which tends to close the crack and prevent failure (Figure 5.3). The simplest relationship between shear stress $\tau$ and normal stress $\sigma$ at failure is given by

$$\tau = c + \mu\sigma \qquad (9.4)$$

where $c$ and $\mu$ are constants. This relationship is called the **Coulomb failure criterion** and gives a linear Mohr envelope with a slope of $\mu$. However, very few materials behave in this way. A more realistic interpretation is given by the **Griffith failure criterion**, which is based on the suggestion that failure results from the propagation and linking of minute defects ('Griffith cracks') in the material (cf. section 4.4). Stress concentration occurs around the ends of the cracks, which spread spontaneously above a certain critical stress. The Griffith crack theory leads to the relationship

$$\tau^2 = |4\sigma_t(\sigma_t + \sigma)| \qquad (9.5)$$

where $\sigma_t$ is the tensile strength (a negative value) of the material. This curve gives a parabolic Mohr envelope, which means that $\beta$ is large at low hydrostatic pressures and small at high hydrostatic pressures, rather than constant as in the Coulomb criterion, and corresponds quite closely to many experimentally derived failure curves (cf. Figure 9.2C, D).

These criteria are based on the two-dimensional Mohr diagram assuming that $\sigma_2 = \sigma_3$. It is likely that where $\sigma_1 > \sigma_2 > \sigma_3$, the value of $\sigma_2$ will have some effect on the criterion and therefore equation (9.5) can be modified to give a three-dimensional failure criterion of the form

$$\tau_{oct}^2 = |8\sigma_t P| \qquad (9.6)$$

where $P$ is the hydrostatic pressure (stress), equal to $\frac{1}{3}(\sigma_1 + \sigma_2 + \sigma_3)$, and $\tau_{oct}$ is the 'octahedral' or three-dimensional shear stress, such that:

$$\tau_{oct}^2 = \tfrac{1}{9}[(\sigma_1 - \sigma_2)^2 + (\sigma_2 - \sigma_3)^2 + (\sigma_3 - \sigma_1)^2] \qquad (9.7)$$

This is referred to as the **Griffith–Murrell failure criterion**.

The Griffith–Murrell ('open crack') failure criterion is a close approximation to the behaviour of rocks failing under low hydrostatic pressures, but under high pressures where cracks would be closed by higher normal stresses, the Coulomb criterion is thought to be more accurate. For a fuller treatment of this subject, the reader is referred to Jaeger and Cook (1976).

## EFFECT OF PORE-FLUID PRESSURE

As we saw in section 7.7, the effect of pore-fluid pressure is to reduce the effective hydrostatic pressure to a value $P_e = P - P_f$, where $P_f$ is the pore-fluid pressure. For saturated rocks, where $P_f$ approaches $P$, the value of the shear stress at failure is greatly reduced. The effect on failure can be

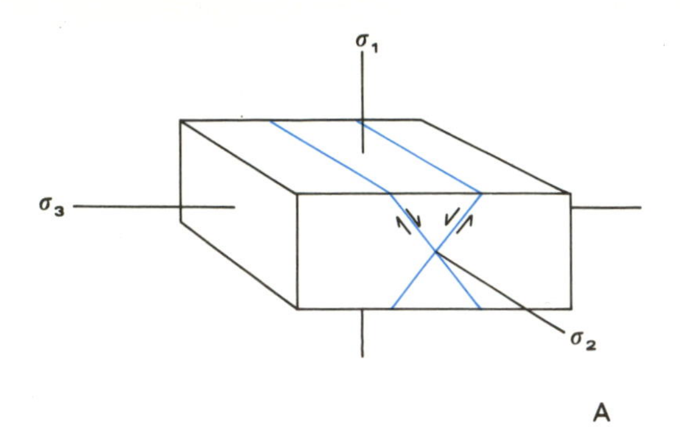


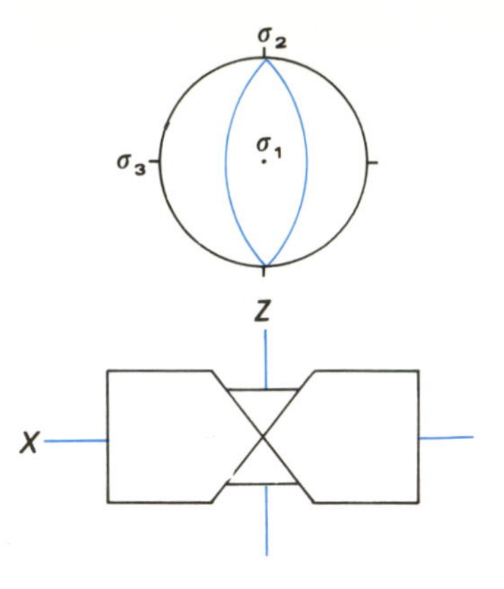


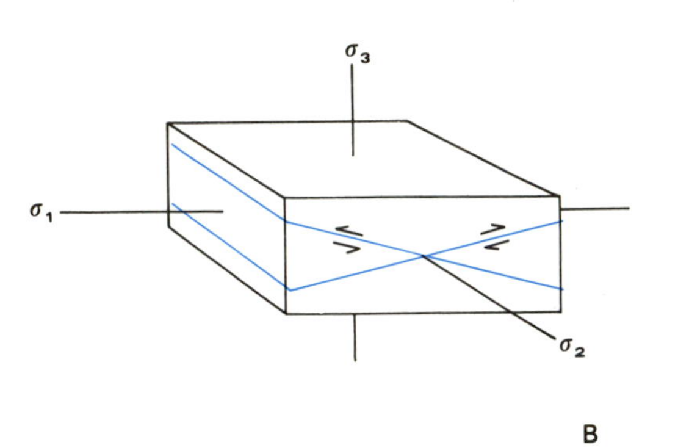


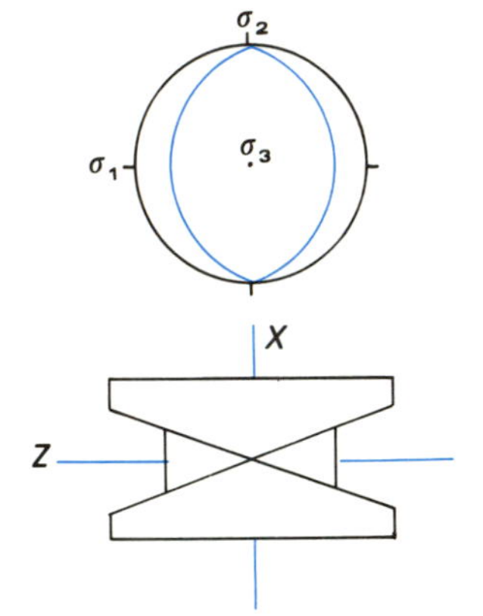


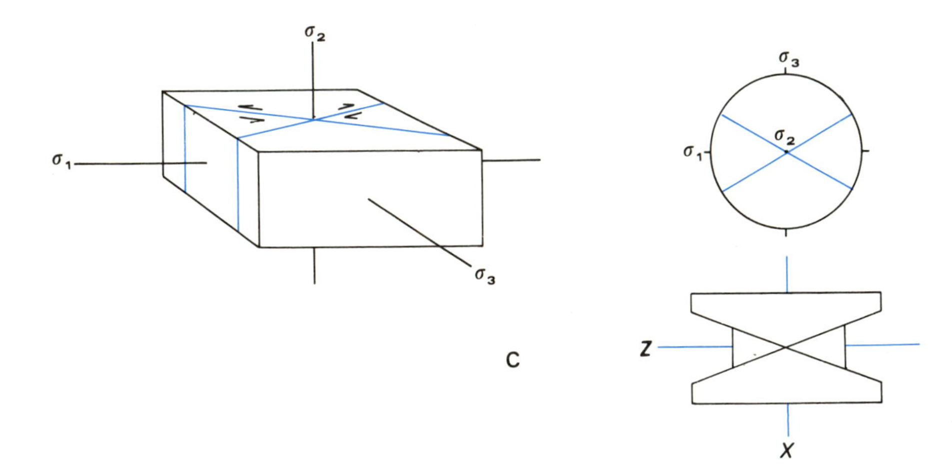


**Figure 9.3** Fault orientation in relation to principal stress and strain axes. A. Normal fault sets. B. Thrust fault sets. C. Strike-slip fault sets. (See text for further explanation.) Stereograms, plan view. Strain diagrams, $\perp Y \parallel \sigma_2$.

shown by a simple change in the Coulomb criterion (equation (9.4)) to

$$\tau = c + \mu(\sigma - P_f) \qquad (9.8)$$

which gives a reduction of $\mu P_f$ in the value of the shear stress required to produce failure. The effect can be demonstrated on the Mohr diagram as a leftwards shift of the Mohr circle by an amount $P_f$ (Figure 9.2E). If $P_f$ is large enough, the circle will intersect the Mohr envelope and failure will occur, although the size of the shear stress may be much too low to produce failure in dry rock.

High pore-fluid pressures are particularly important in the movement of thrust sheets. To move large thrust sheets in dry conditions requires impossibly large forces to overcome the friction along the base of the sheet. If the thrust plane is lubricated by water, however, the required shear stress is reduced to a reasonable level. If we consider the problem of a thrust sheet sliding under gravity, the critical inclination of the plane of sliding should be about 30° in dry rocks, whereas much smaller inclinations are commonplace in nature. W.W. Rubey and M.K. Hubbert have studied the effect of pore-fluid pressure in reducing the critical slope necessary for gravity sliding. They have shown that, for example, the critical inclination may be reduced to only 5° if the pore-fluid pressure is 85% of the hydrostatic pressure. Pore-fluid pressures as high as this, and even higher, have been measured in deep boreholes.

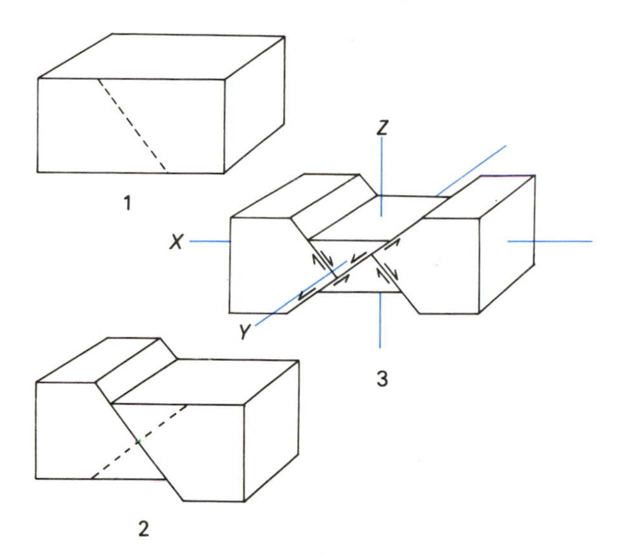


**Figure 9.4** Shortening and extension of a block by complementary shear displacements on normal faults. This diagram explains the relationship between fault orientations and strain axes as shown in Figure 9.3.

## 9.2 FAULT ORIENTATION IN RELATION TO STRESS AND STRAIN AXES

We have seen that both theory and experimental results predict a simple relationship between the orientation of the principal stress axes and the shear fracture planes where $\sigma_1$ bisects the acute angle between two sets of shear fractures (see Figure 9.1). This relationship can be used to investigate natural fault systems and in favourable cases to derive from them the orientation of the causal stress field. Since the theory applies to the initiation of fractures in completely homogeneous material, we would not expect very close agreement between theoretical and natural relationships, but nevertheless the theoretical model offers a very useful basis for understanding natural fault systems.

Since we can assume that shear stresses along the surface of the Earth are zero, it follows that one of the principal stress axes will be approximately vertical and the other two approximately horizontal. This leads to a simple threefold classification of fault sets based on the three possible orientations of the stress axes (Figure 9.3).

### NORMAL FAULT SETS (Figure 9.3A)

Here $\sigma_1$ is vertical and corresponds to gravitational load. The two sets of normal faults intersect parallel to $\sigma_2$ and dip more steeply than 45°. Actual values of $\alpha$ (the angle between the fault plane and $\sigma_1$) are in the range 25°–30°. The sense of simple shear movement on these faults implies a contraction of the block parallel to $\sigma_1$ and extension parallel to $\sigma_3$ (Figure 9.4). Thus the $X$ strain axis is parallel to $\sigma_3$, the $Z$-axis is parallel to $\sigma_1$ and $Y$ is unchanged (plane strain).

THRUST FAULT SETS (Figure 9.3B)

Here $\sigma_1$ is horizontal and $\sigma_3$ is vertical. Consequently two sets of thrust faults intersect along the horizontal $\sigma_2$ axis and dip less steeply than 45°. Actual values of $\alpha$ for thrusts are in the range 20°–25°. The sense of shear on the thrusts implies horizontal contraction parallel to $\sigma_1$ and vertical extension. It should be noted that the symmetrical arrangement of thrusts as illustrated in Figure 9.3B is rare in nature; thrust sets are typically strongly asymmetric, with one direction dominant. Since the least stress must be vertical, thrusts will form more easily at relatively high levels in the crust, where the lithostatic pressure is low.

STRIKE-SLIP FAULT SETS (Figure 9.3C)

Here $\sigma_1$ is again horizontal but, as $\sigma_3$ is also horizontal, two sets of strike-slip faults intersect along a vertical $\sigma_2$, each dipping vertically. Actual values of $\alpha$ for strike-slip faults are around 30°. The sense of shear on these faults implies horizontal contraction parallel to $\sigma_1$ and horizontal extension parallel to $\sigma_3$. As in thrust fault sets, one sense of movement is usually dominant.

FAULTING IN A NON-HOMOGENEOUS BODY

In a natural situation, where a stress field is applied to a rock body containing different rock types and planar discontinuities of different orientations, these simple rules break down. The effect of a plane of weakness (a previous fault, for example) oriented at the 'wrong' angle to $\sigma_1$ may be to cause faulting preferentially on that plane rather than to initiate a new fracture plane at the 'correct' angle. Both thrust and extensional fault systems (see sections 2.6 and 2.7) show examples of varying fault inclination (e.g. from thrust to reverse fault) controlled by movement rather than by initial stress state.

Where conjugate fault sets are found, the orientation of the principal stress axes may be determined by assuming that they intersect along $\sigma_2$ and that $\sigma_1$ bisects the acute angle, even if there is a certain amount of variation in the orientation of the faults. Where, however, there is no obvious conjugate set, and there is a range of orientations, any individual fault may make an angle with all three principal stress axes. In this case the direction of displacement on the fault plane depends on both the orientations and the relative sizes of the three principal stresses and cannot be simply interpreted (see section 5.5 and Figure 5.9).

Where the sense of displacement is known on a number of planes of various orientations, it is often possible to find the approximate positions of the principal stress axes. This is most easily done on a stereogram by plotting the positions of lines across which a change in sense of displacement occurs. These lines will contain the principal stress axes (Figure 9.5).

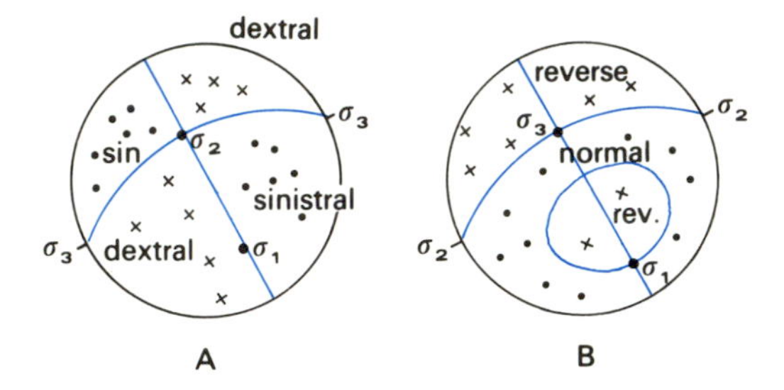


**Figure 9.5** Finding approximate orientations of the principal stress axes from the sense of displacement on variably oriented faults. The principal stress planes shown on the stereogram divide faults with sinistral displacement from faults with dextral displacement (A) and faults with reverse displacement from faults with normal displacement (B). (After Davidson, L. and Park, R.G. (1978) *Quarterly Journal of the Geological Society of London*, **135**, 282–9, figure 6.)

STRESS TRAJECTORIES AND FAULT ORIENTATION

The simple relationship between the orientation of the stress axes and the horizontal, as shown in Figure 9.3, may hold true generally near the surface, but at depth a variety of factors give rise to variations in orientation of the stress axes which are reflected in turn in variations in the orientation of faults. In favourable circumstances it may be possible to use such a variation to draw stress trajectories (section 5.7) showing the variation in

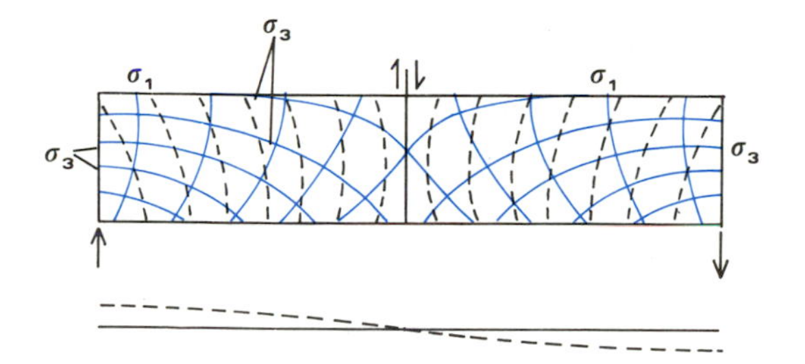


**Figure 9.6** Stress trajectories (coloured lines) and normal fault orientations (dashed lines) resulting from an elastic upwarp in two dimensions. Theoretical model. (After Sanford A.R. (1959) *Bulletin of the Geological Society of America,* **76**, 19.)

stress orientation which can then be related to theoretical models. Figure 9.6 shows an example of a set of theoretical stress trajectories obtained for an elastic upwarp, together with the set of curved normal faults that would be predicted from such a stress distribution.

A variation in the angle at which a fault is initiated may be predicted, even under a uniform stress orientation, due to the effect of the downward increase in hydrostatic pressure. This changes the fault angle $\alpha$ as shown by the change in the slope of the Mohr failure envelope (Figure 9.2B), particularly where the horizontal stress changes from tensional to compressional with depth (Figure 9.7). Note that this effect need not result in curved (listric) faults. A fault initiated at depth at an angle of 30° from the vertical may propagate to the surface without significant change in inclination. It is thought that most large faults probably initiate at depths of 10–15 km, in the strongest part of the crust.

### DISPLACEMENT VERSUS STRESS IN FAULT ORIENTATION

Once a fault or set of faults has been initiated, the subsequent development of the fault or faults is governed by the displacement geometry of the deforming system, rather than by the initiating stress field.

Thus the evolution of thrust fault systems, for example, is controlled by the movement of large thrust sheets, which is in turn governed by the presence of easy slip horizons and by other local geometrical considerations. Major strike-slip fault systems are often linked kinematically to plate-boundary movement vectors. In both thrust and extensional fault systems, the original fault planes become deformed by subsequent movements and their present orientation no longer provides a reliable guide to the initial stress state.

## 9.3 FAULTING AND EARTHQUAKES

### FAULT INITIATION AND PROPAGATION

When a brittle rock is compressed, certain strain effects take place before fracturing. These changes are important in earthquake prediction. The initial strain is elastic, but when the shear stress reaches a value of about half the shear strength, the rock begins to show some permanent strain due to the opening and propagation of small cracks in the zone of greatest strain within which the fault movement eventually occurs. The intensity of this microfracturing increases as the shear strength is approached. The opening of the cracks causes an

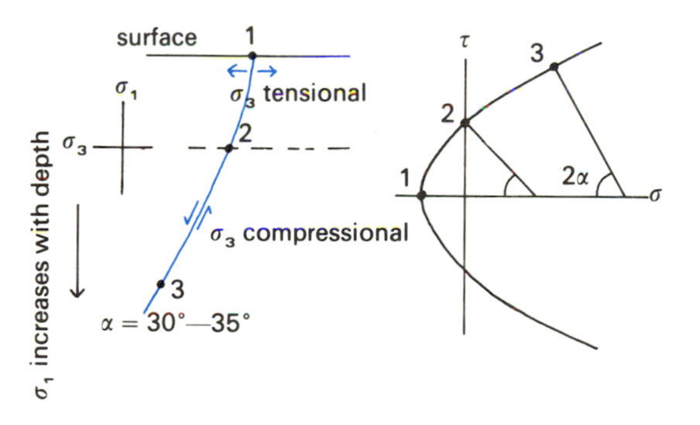


**Figure 9.7** Change of fault inclination with depth of initiation due to increase in hydrostatic pressure, as predicted by the Mohr diagram (cf. Figure 9.2). The hade of the fault ($\alpha$) changes from 0° at point 1 (pure extension with no shear stress) to around 30° at point 3 within the compressional field. Note that faults initiated at a particular angle at depth may, in practice, propagate to the surface maintaining the same inclination, i.e. this diagram does not necessarily predict listric faults (see text for further explanation).

increase in volume or dilation in the rock which is associated with an increase in the fluid content, as ground water migrates into the cracks. The rise in pore-fluid pressure has a significant effect in weakening the rock, as noted earlier (section 9.1).

The rate of propagation of the microfractures is the critical factor in determining the strain rate, and hence the time taken for fracturing (and the resulting earthquake) to occur. The influence of pore-fluid pressure on the strain rate led to experiments at the Rangely Oil Field in Western Colorado, where the level of earthquake activity was artificially increased by pumping water into a fault zone. It has been proposed that large and potentially destructive earthquakes could be prevented by generating smaller ones by this method, and thus releasing the stress more safely.

## RECOGNITION OF IMPENDING EARTHQUAKES

There are several features associated with the above changes that can be used to predict earthquakes. The dilation of the rock is accompanied by a decrease in P-wave earthquake velocities and also by an uplift of the ground around the area of the fault. Another change that has been noticed is an increase in the amount of radon (an inert gas) in the atmosphere, presumably because of its release during the microfracturing process. The increase in pore-fluid pressure causes an increase in electrical resistivity, which can be easily measured. All these changes are reversed in the period of rapid deformation leading up to the earthquake. Lastly, all large earthquakes appear to be preceded by a number of small shocks that increase in frequency immediately before the main earthquake.

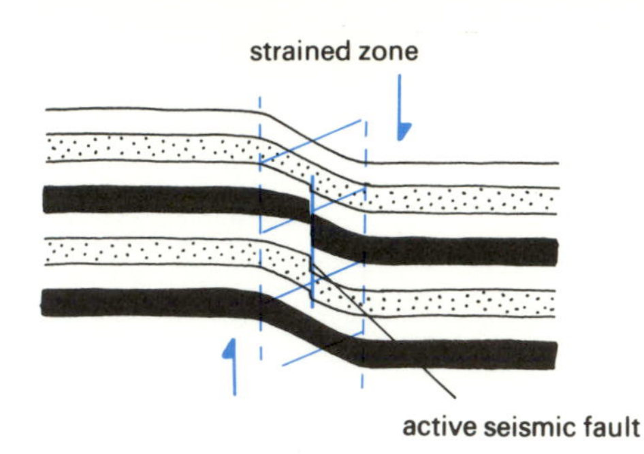


**Figure 9.8** Lateral replacement of seismic slip by aseismic strain (see text for explanation).

## FAULT DISPLACEMENT

When a fault plane has become established, further strain effects are partly in the form of very fast movements (**slip**) along the fault plane, and partly of slow movements comparable to the pre-fracturing strains.

The former cause earthquakes (i.e. they are **seismic**), with displacement rates of the order of metres per second, whereas the latter are not associated with earthquakes (i.e. they are **aseismic**) and take place at velocities of the order of centimetres per year – similar to the movements involved in folding, etc. Only a limited section of a large fault will take part in a particular episode of seismic slip, and the slip displacement dies out gradually at both ends into regions of aseismic movement (Figure 9.8).

Any given section of a long-lived fault plane will exhibit short periods of fast seismic slip separated by long periods when that part of the fracture is inactive – i.e. the fault 'sticks'. This behaviour is known as **stick–slip** and is typical of the region

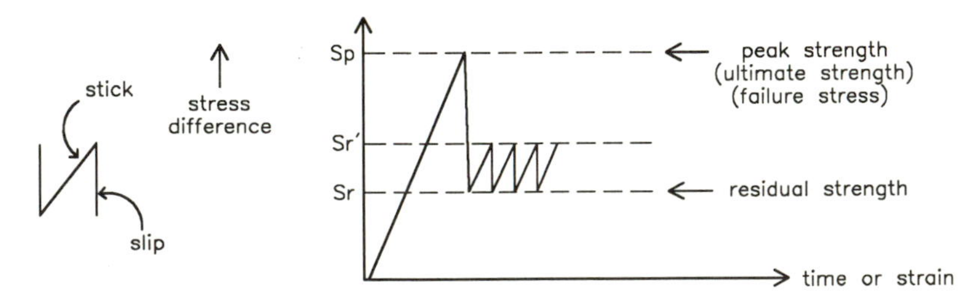


**Figure 9.9** Stress–strain (or stress–time) graph illustrating stick–slip behaviour on an active fault.

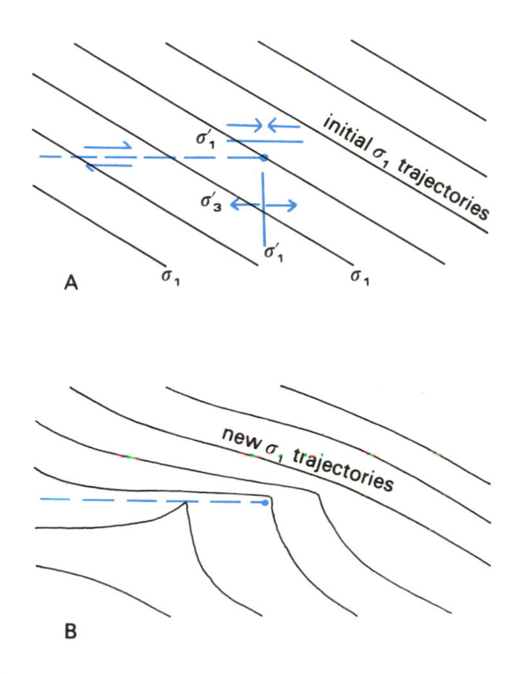


**Figure 9.10** Stress trajectories around the end of an active fault. A. The initial set of $\sigma_1$ trajectories together with the new stresses $\sigma_{1'}$ and $\sigma_{3'}$ arising from the additional compressive and extensional effects arising from the fault movement (see also Figure 9.11B). B. The addition of the two sets of stresses produces a new set of stress trajectories, i.e. a combined stress field. (After Chinnery, M.A. (1966) *Canadian Journal of Earth Sciences*, **3**, 163–74.)

between 5 km and 10 km in depth, where the majority of fault-generated earthquake foci occur.

The stress–strain graph for fault movement shows a characteristic sawtooth pattern (Figure 9.9). After reaching a peak value of stress $\sigma_p$, which corresponds to the failure strength of the rock, the stress drops instantaneously to a minimum value $\sigma_r$, which is the stress required to overcome the sliding friction on the fault surface. The stress then increases again to a value $\sigma_{r'}$, which is the stress required to overcome the static friction on the fault surface. Successive oscillations then take place involving the alternation of essentially elastic strain (stick) and very rapid sliding (slip). The difference between $\sigma_r$ and $\sigma_{r'}$ depends on the roughness of the fault surface, and on the extent of welding by vein material since the last slip episode. The value of the stress drop is typically 1–10 MPa. It is this periodic build-up and release of stress that causes the pattern of repeated earthquakes on existing large faults.

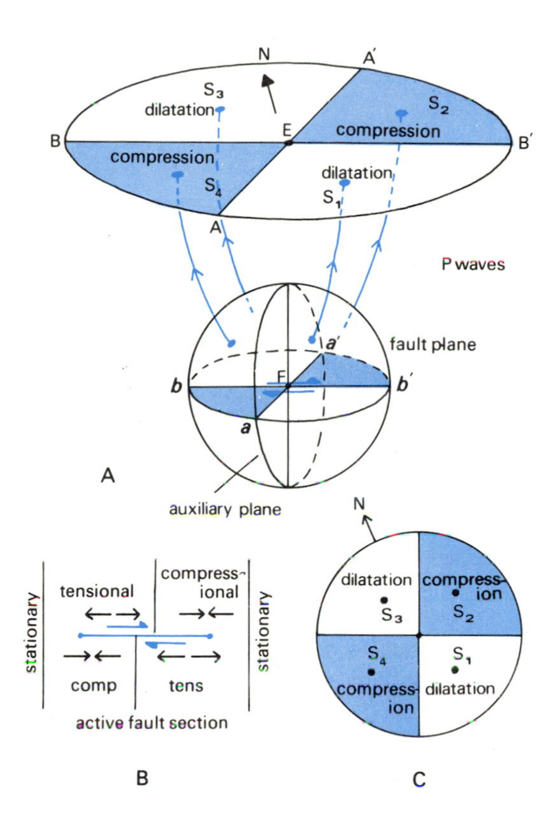


**Figure 9.11** Focal mechanisms of an earthquake: first-motion study. A. A small sphere has been drawn around the focus of an earthquake, F, resulting from movement along an active segment of fault *bb′*. The dextral fault movement results in compression and dilation in opposite quadrants (see also B). The P waves originating in the compressive quadrants will show compressive first motions on arrival at the surface at recording stations $S_2$ and $S_4$, and those originating in the dilational quadrants will show dilational first motion on arrival at $S_1$ and $S_3$. Given enough stations, the orientation of the planes dividing the quadrants can be determined (C). For simplicity, in this example the fault movement is horizontal and the fault plane vertical, but oblique fault movements can be reconstructed using the same method. The effect of any oblique fault displacement can be shown by tilting the sphere to the required orientation. (After Bolt, 1978, p. 67.)

Above 5 km, stable sliding takes place because the compressive stress across the fault plane is low. Below 10 km, because of the increase in confining pressure, there is a transition to a more ductile form of deformation (see section 10.6).

## SECONDARY STRESS-FIELDS

The process of faulting, by locally releasing stress in the strained zone, and by the lateral movement of blocks of rock along the fault, causes a modification of the stress field around the active region which may in turn influence further fault movements. Secondary stress fields are particularly important around the end of a line of active slip. Figure 9.10 shows an example of how a new set of stress trajectories may be derived by superimposing new compressional and tensional stresses $\sigma_{1'}$, $\sigma_{3'}$ parallel to the fault on an oblique set of pre-faulting stress axes $\sigma_1$. Complicated systems of branching or splay faults at the end of a major fault (Figure 2.9) may be explained in this way.

## EARTHQUAKE FAULT-PLANE SOLUTIONS

The orientation of fault planes and the displacement direction along them may be determined under favourable circumstances by a seismological method called **focal-plane** or **fault-plane solution**. This method is particularly useful in determining the origin of earthquakes relating to concealed faults, especially in the oceans, and has proved to be very important in the development of plate tectonic theory by enabling the relative motions of lithospheric plates to be determined.

The method is illustrated in Figure 9.11. If an earthquake originates by a shear displacement along a section of fault plane, the plane perpendicular to the displacement vector of the fault and midway along the displaced sector (the **auxiliary plane**) will divide regions of compression from regions of tension. These regions will be in opposite quadrants since the movement is in opposite directions on each side of the fault (Figure 9.11A, B). The pattern of compression and dilation is preserved in the seismic waves that are radiated from the earthquake source and the phase of the initial seismic wave received at the recording station reflects its origin. Thus if there are a sufficient number of seismograph stations in different directions from the earthquake, the orientation of the planes dividing the compressional and tensional quadrants can be determined (Figure 9.10C). One of these planes is the fault plane and the other is the auxiliary plane. It is not possible from the first-motion study alone to determine which of these planes is the fault, but if one is more likely, given our knowledge of the local geology, then the direction and sense of movement can be obtained as shown, the displacement vector being perpendicular to the line of intersection of the planes.

## FURTHER READING

Bolt, B.A. (1978) *Earthquakes: a Primer*, Freeman, San Francisco.

Jaeger, J.C. and Cook, N.G.W. (1976) *Fundamentals of Rock Mechanics*, Chapman & Hall, New York.

Price, N.J. and Cosgrove, J.W. (1990) *Analysis of Geological Structures*, chapter 5, Cambridge University Press, Cambridge.

# STRAIN IN FOLDS AND SHEAR ZONES 10

## 10.1 FOLDING MECHANISMS AND FOLD GEOMETRY

Several different mechanisms of fold formation have been discussed in section 3.8. The application of the concept of strain to the analysis of folds enables us to investigate the fold mechanism in greater depth. Figure 10.1 shows examples of five different mechanisms, which can be distinguished by the strain distribution in their respective fold geometries.

### BUCKLING

In a fold produced by buckling of a single layer under lateral compression, the layer maintains its thickness throughout so that a parallel or concentric fold (see section 3.4) is produced. The strain within the layer is dictated by extension around the outer arc and compression in the inner arc, separated by a neutral surface of no strain near the centre of the layer (Figure 10.1A). The geometry of natural buckle folds is typically much more complex, however, and is discussed in more detail in section 10.2.

### FLEXURAL SLIP

This process involves a shear displacement or slip between successive layers deformed by buckling (Figure 10.1B). This type of folding characterizes the deformation of relatively strong layers separated by planes or thin zones of weakness. In ideal flexural slip, the limbs would be unstrained and the strain would be concentrated at the hinge.

### FLEXURAL SHEAR

In this process, a fold produced by buckling is modified by simple shear acting parallel to the limbs of the fold, producing a strain distribution in which the long axes of the strain ellipses are divergent from the centre or core of the fold (Figure 10.1C).

### OBLIQUE SHEAR OR FLOW

If planes of simple shear are oblique or transverse to a layer and the amount and direction of shear displacement varies along the length of the layer, a fold will be formed by passive rotation of the layer (Figure 10.1D). This process has been termed 'heterogeneous simple shear' and is important in shear zones (see section 10.6). The strain distribution is similar to that of flexural shear. This mechanism produces an ideal similar fold (see section 3.4) and can be illustrated using a card-deck by drawing parallel lines on the edges of the deck and displacing the deck to make a fold shape (Figure 10.1E). The mechanism may also operate to modify the shape of an existing fold.

### KINKING

This process forms folds of the kink band or chevron type which typically have straight limbs and sharp hinges (section 3.4). The geometry is controlled by the rotation of sets of layers which remain planar between the kink planes, whereas rapid changes of orientation take place along the kink planes (Figure 10.1F). The limbs of the fold deform by flexural slip, and the process depends on the flow of highly ductile material separating the stronger active layers. Kinking ideally produces folds of overall similar profile, although individual layers exhibit different geometries (e.g. compare the white and black layers in the kink fold of Figure 10.16B).

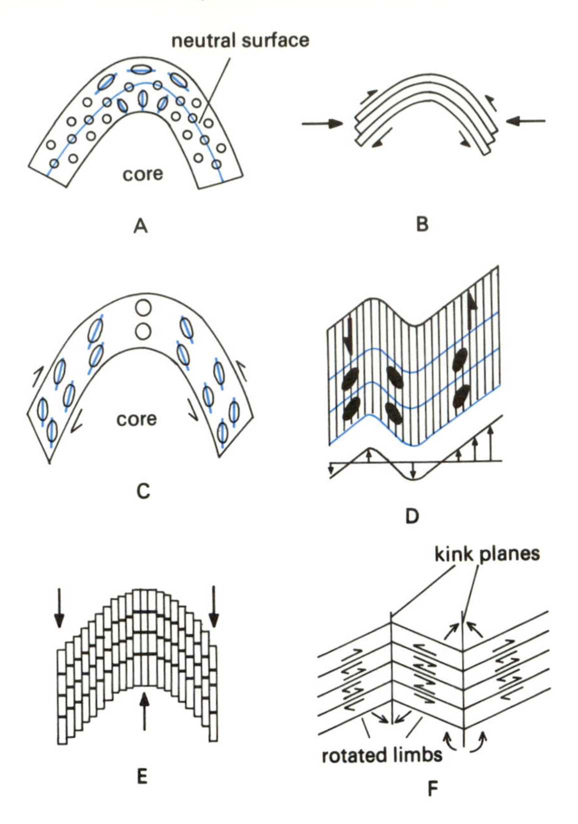


**Figure 10.1** Fold mechanisms. A. Buckling showing the strain distribution within the folded layer. The neutral surface of no strain separates the extensional strain at the outer arc of the hinge from the compressional strain at the inner arc. B. Flexural slip – successive layers are displaced upwards towards the antiform crest with respect to the layer below. Individual layers are relatively unstrained. C. Flexural shear – the limbs of the buckle fold are modified by oppositely directed simple shear acting parallel to the layers. The hinge area is unstrained. D. Oblique shear – the fold is the result of changes in the amount or direction of simple shear displacement. E. Card-deck model of an oblique-shear fold. F. Kinking – the fold is produced by the rotation of a set of layers on either side of a kink plane (axial surface). The layers deform partly by flexural slip (see text).

DIFFERENCES IN GEOMETRY

Each of the ideal mechanisms described above produces a characteristically different fold geometry. These differences lead to a few simple geometrical tests which can be applied to natural folds in order to find out how closely they match one of the ideal types.

1. Ideal buckling forms parallel folds; there is plane strain with the $Y$ strain axis parallel to the fold axis and a combination of extensional and shortening strains in the hinge area. An initially straight lineation lying in the plane of the layer becomes curved during folding and the angle made with the fold axis remains unchanged only on the neutral surface (Figure 10.2A). On surfaces above and below the neutral surface, the lineation distribution changes, depending on the amount of strain.

2. Ideal flexural shear also produces parallel folds and plane strain, with $Y$ parallel to the fold axis. The strain distribution defines a simple divergent fan of the $XY$ planes. Since there is no distortion within the folded surface, the angle made by a lineation within that surface with the fold axis is constant throughout the fold (Figure

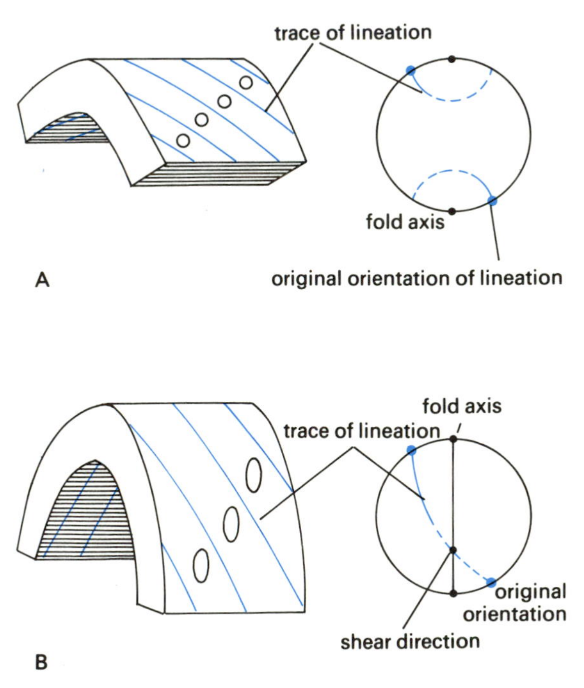


**Figure 10.2** Simple models of lineation reorientation during folding. A. Buckle folding of an initially straight lineation where there is no strain in the plane of the layer (e.g. at the neutral surface, or in flexural shear folding). B. Folding of an initially straight lineation during oblique shear folding.

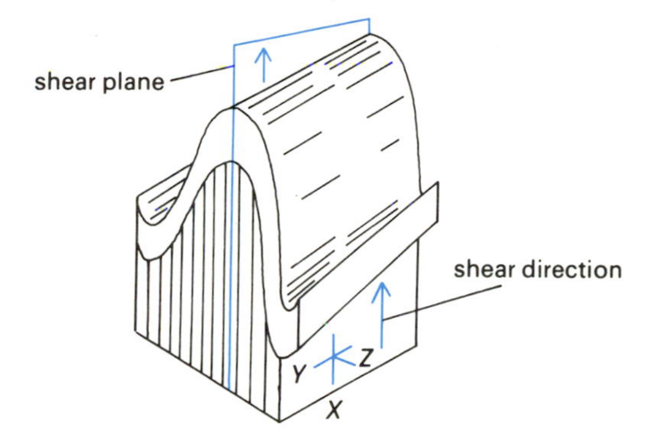


**Figure 10.3** Oblique shear fold showing shear plane and shear direction. Note that the fold axis is oblique to $Y$.

10.2A) and the trace makes a small circle on a stereogram.

3. Ideal oblique shear produces similar folds in which the thickness of the folded layer measured parallel to the shear plane is constant, but the orthogonal thickness varies systematically, thinning on the limbs. Again the process produces plane strain, but the shear direction need not be perpendicular to the fold axis. $Y$ is perpendicular to the shear direction and may or may not be parallel to the fold axis (Figure 10.3). An initially straight lineation is distorted in a systematic way such that it is rotated towards the shear direction. After folding, it therefore lies in a plane defined by the original orientation and the shear direction, i.e. it forms a great circle distribution on the stereogram (Figure 10.2B).

4. Ideal chevron folds (see section 3.4) have straight limbs and sharp angular hinges with fold angles of 60°. The competent layers maintain their thickness, whereas the intervening incompetent material exhibits extreme thickness variation and large strains. This leads to folds with alternating class 1C and class 3 geometry (Figure 3.12), which enables the folds to maintain an overall similar form.

## MODIFICATIONS DUE TO HOMOGENEOUS STRAIN

The geometry of a fold is considerably modified if a homogeneous strain is imposed on a layer before, during or after the folding. Figure 10.4A shows the strain distribution produced by layer shortening prior to folding, and Figure 10.4B the strain distribution due to the superimposition of a homogeneous

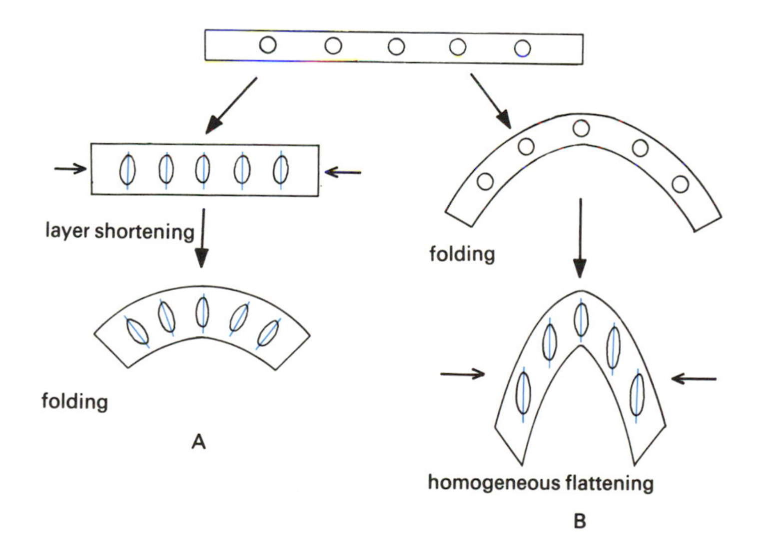


**Figure 10.4** Strain pattern produced by (A) folding a layer shortened before folding, and (B) superimposing a homogeneous strain on a pre-existing fold.

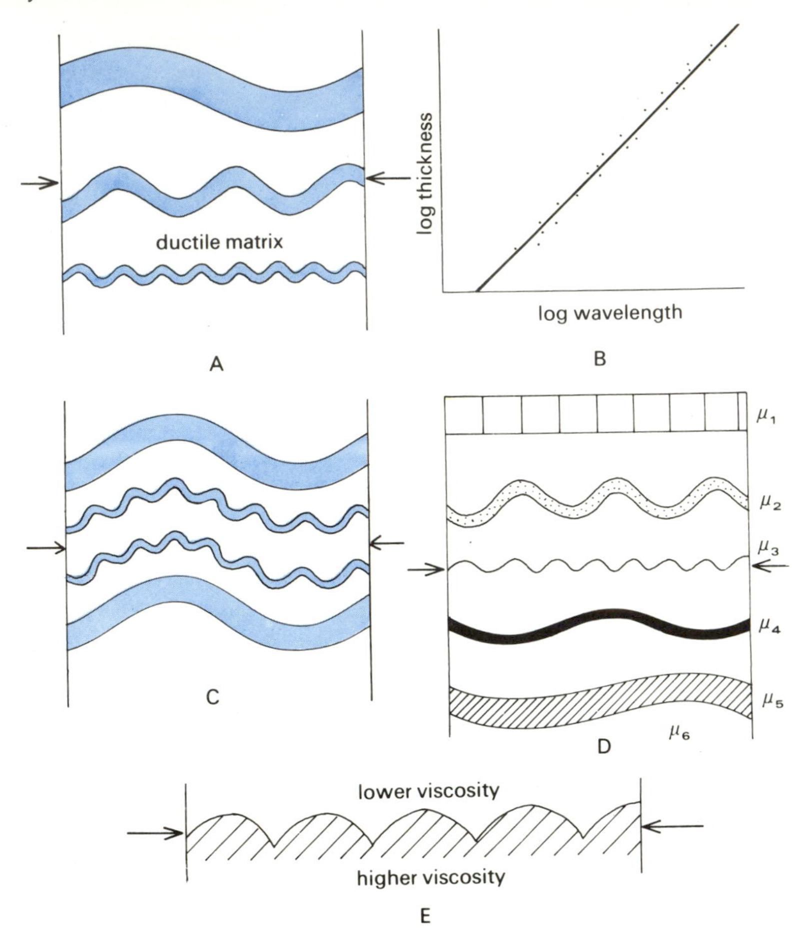


**Figure 10.5** Influence of layer thickness and viscosity contrast on the wavelength of buckle folds. A, B. There is a linear relationship between log layer thickness and log buckle wavelength for widely separated layers of constant viscosity in a ductile matrix of much lower viscosity. C. Buckle folds of different wavelength may be superimposed if the layers are close enough to interfere. D. Buckle folding produced by a number of layers of different viscosity $\mu_1$–$\mu_5$ and different thickness in a ductile matrix of much lower viscosity $\mu_6$. A, C and D are examples of disharmonic folding (see section 3.6). (After Ramberg, H. (1964) *Tectonophysics*, **1**, 307–41.) E. Buckle folding of an interface between two thick layers of contrasting viscosity. The cusps point towards the material of higher viscosity (cf. mullion structure, see section 4.2 and Figure 4.8B).

flattening strain on a fold. These strains have to be added to the strains produced by the folding process to give the total finite strain pattern.

## 10.2 CHARACTERISTICS OF BUCKLE FOLDS

The term **buckling** has traditionally been applied (see Ramsay 1967) to the process that produces (at least initially) a fold of approximately sinusoidal shape, or with relatively smooth changes of curvature, by compression acting parallel to the length of a layer. Folds formed in this way have a class 1 shape (Figure 3.12), often closely approximating class 1B (parallel folds). Both experimental and theoretical studies have indicated that buckle folds will only form in relatively strong layers in a more ductile matrix and that there is a limiting value of the viscosity ratio between the layer and the matrix below which buckling cannot be initiated. Many structural geologists now apply the term 'buckle folds' to any set of folds generated by active compression of a layer or multilayer, including folds of kink or chevron type.

## CONTROLS ON FOLD WAVELENGTH

Many natural sets of buckle folds exhibit a dominant wavelength that seems to be characteristic of a particular layer and is different from that of other layers in the same rock. Two key factors appear to control the wavelength: the layer thickness and the viscosity ratio between the layer and matrix. The relationship between layer thickness and wavelength can be easily demonstrated by studying natural folds of layers of the same composition and varying thickness in a ductile matrix of constant composition (Figure 10.5A, B). The layers have to be sufficiently far apart for their waveforms not to interfere. Where they are close together, we find that a larger wavelength may be superimposed on a smaller wavelength as shown in Figure 10.5C. Where buckling has affected layers of different viscosity as well, a more complicated arrangement results (Figure 10.5D), which cannot be simply interpreted. Buckling of layers of different thickness or properties is one of the commonest causes of disharmonic folding (Figure 10.6; see also section 3.6 and Figure 3.15).

**Figure 10.6** Disharmonic folding, shown by different fold wavelengths in layers of different thickness, in finely laminated sandstone with interlayered mudstone. Note interference between closely spaced layers with the same wavelength. New Harbour beds, Anglesey.

## BUCKLING OF AN INTERFACE

Buckle folding may also affect the planar interface between materials of contrasting viscosity. When this occurs, the folds have a characteristic form – those closing in one direction (e.g. the antiforms) have a broad rounded shape and those closing in the opposite direction (e.g. the synforms) have a narrow or cuspate shape (Figure 10.5E). The cusps always point towards the material with the higher viscosity. This is a common cause of mullion structure (see section 4.2 and Figure 4.8B).

## LAYER PARALLEL SHORTENING IN BUCKLE FOLDS

Comparing the lengths of folded layers of different wavelengths gives us a method of estimating the amount of shortening that has taken place in the layers prior to buckling. Figure 10.7A shows an example of three layers where the present lengths (after unfolding) differ significantly. Layer 1, because of its greater thickness, has not folded. The minimum amount of layer parallel shortening that has taken place is given by the difference in the length of layers 1 and 3. Thus by comparing the layers with the greatest and least unfolded lengths in a set of folds, an estimate of the layer parallel shortening may be made. This is very useful in interpreting the strain distribution within the folded layers (Figure 10.4A). Layer 2 will have experienced homogeneous strain before folding, whereas layer 3 may be almost unstrained before folding. Layer 1 might appear undeformed if the state of strain within the layer is ignored. It has been suggested that the amount of layer shortening is related to the viscosity contrast, in that the proportion of layer shortening to fold shortening

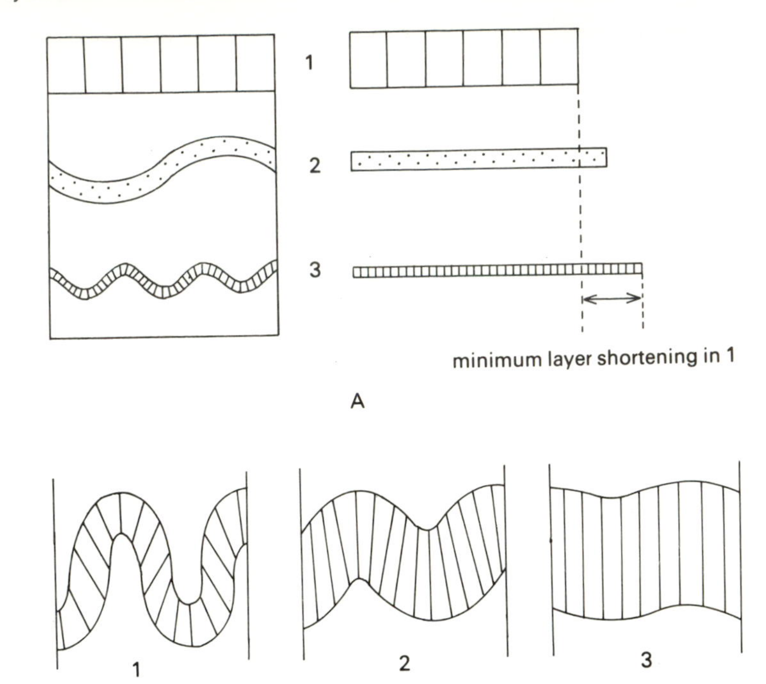


**Figure 10.7** Layer-parallel shortening in folds of different wavelength. A. Unfolding of layers 1, 2 and 3 shows three different apparent original lengths due to different amounts of layer-parallel shortening before folding. The minimum shortening in layer 1 is shown (i.e. assuming that there has been no layer-parallel shortening in layer 3). B. Diagrams 1–3 show the effects of different viscosity contrast on the extent of layer-parallel shortening relative to fold shortening for 63.2% total shortening. Viscosity ratios: (1) 42.1, (2) 17.5, (3) 5.2. (Experimental results from Dieterich, J.H. (1970) *Canadian Journal of Earth Sciences*, **7**, 467–76.)

increases with decrease in the viscosity ratio (Figure 10.7B). With a low viscosity ratio, of course, no folding will occur and the layer will deform solely by layer shortening.

Because of the effect of shortening on layer thickness, the relationship between dominant wavelength and layer thickness described earlier requires modification. As deformation proceeds, there will be a tendency for folds with larger thickness/wavelength ratios to be favoured.

## SHORTENING STRAIN IN PARALLEL FOLDS

The strain produced by parallel folding can be simply calculated by comparing the unfolded and folded lengths of a folded layer (section 8.6). The maximum shortening is achieved where the fold is concentric (Figure 10.8), in which case, given a layer thickness $t$, the initial length of the fold is $\pi t$

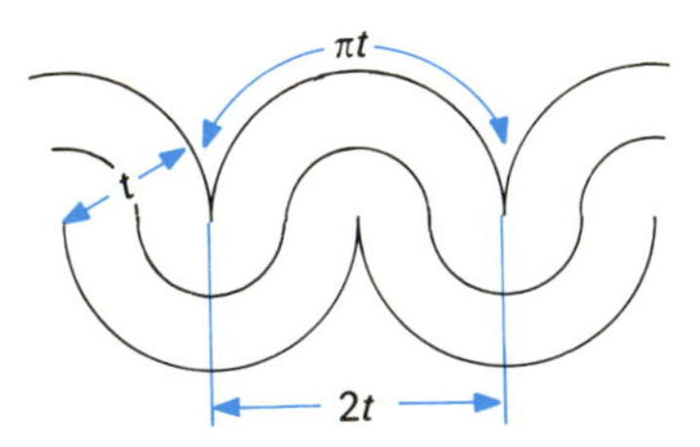


**Figure 10.8** Maximum shortening in a concentric fold. The shortening is given by $e = (\pi t - 2t)/2t = (\pi - 2)/2$, where $\pi t$ is the original length and $2t$ the new length of a folded layer. (After Ramsay, 1967, figure 7.48.)

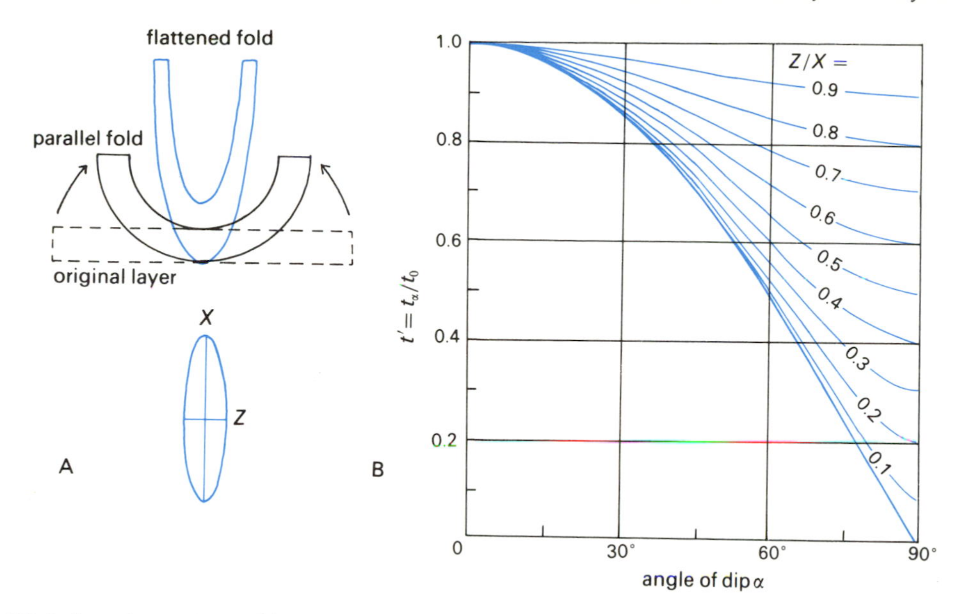


**Figure 10.9** Superimposition of homogeneous flattening on a parallel fold. A. Homogeneous strain $X$, $Z$ is superimposed on the parallel fold to produce a flattened parallel fold. B. Curves of $t'$ against $\alpha$ for the determination of the amount of flattening $Z'/X'$ in buckle folds. (After Ramsay, 1967, figure 7.79.)

and the new length $2t$. Thus the shortening $e$ is given by:

$$e = (\pi t - 2t)/2t$$
$$= (\pi - 2)/2 = 0.36(36\%) \qquad (10.1)$$

Further shortening strain can only take place by modifying the fold shape. This modification takes the form of a flattening of the fold perpendicular to the axial surface and an equivalent extension parallel to the axial surface, and produces a fold with class 1C geometry known as a 'flattened parallel fold'. The way in which a flattening strain is imposed on a buckle fold may be quite complex in detail, but in many cases the shape of the folded layer approximates closely to the result obtained if a homogeneous strain were simply added to the ideal parallel fold (Figure 10.9A). The addition of a homogeneous flattening strain with a strain ratio $Z/X$ can be demonstrated on the $t'/\alpha$ plot as a family of curves of $t'/\alpha$ for various values of $Z/X$ (Figure 10.9B), i.e. various degrees of flattening. The curves gradually approach the class 2 similar fold curve as the strain increases. When the limbs of an isoclinal fold become parallel and $\alpha = 90°$,

$$t' = Z/X \qquad (10.2)$$

That is, the strain ratio is given by the ratio between the thickness on the limbs and the thickness at the hinge. For a rapid estimation of the amount of flattening in a non-isoclinal fold, it is sufficient to measure $t/t_0$ for two or three values of $\alpha$ and plot them on the graph of Figure 10.9B.

## STRAIN WITHIN AND OUTSIDE THE FOLDED LAYER

The strain distribution within a buckled layer depends on the actual mechanism involved (Figure 10.1). Buckling that produces a neutral surface (Figure 10.1A) shows a type of strain distribution termed **tangential longitudinal strain**, where the strain axes are parallel to the layer boundaries. For certain types of fold profile there is a concentration of strain in the hinge zones, but the limbs may be relatively unstrained. Where there has been some flexural shear (Figure 10.1C) the strain ellipses (and any associated planar fabric) have a divergent fan arrangement. Many natural sets of folds involve alternating layers of contrasting viscosity, such as sandstone bands in shale (Figure 10.10). Where these deform by compression parallel to the layering, the more competent sandstone layers buckle to give, at least initially, a convergent fan arrangement of strain axes, owing to the predominance of tangential strain. The more ductile shale, however, deforms rather differently. Where a narrow band of shale is constrained by the movement of adjacent competent layers, it will deform by flexural shear, which will produce a divergent fan of strain axes. However, away from the influence of active buckling layers, it will tend to deform more by homogeneous flattening, producing parallel strain axes.

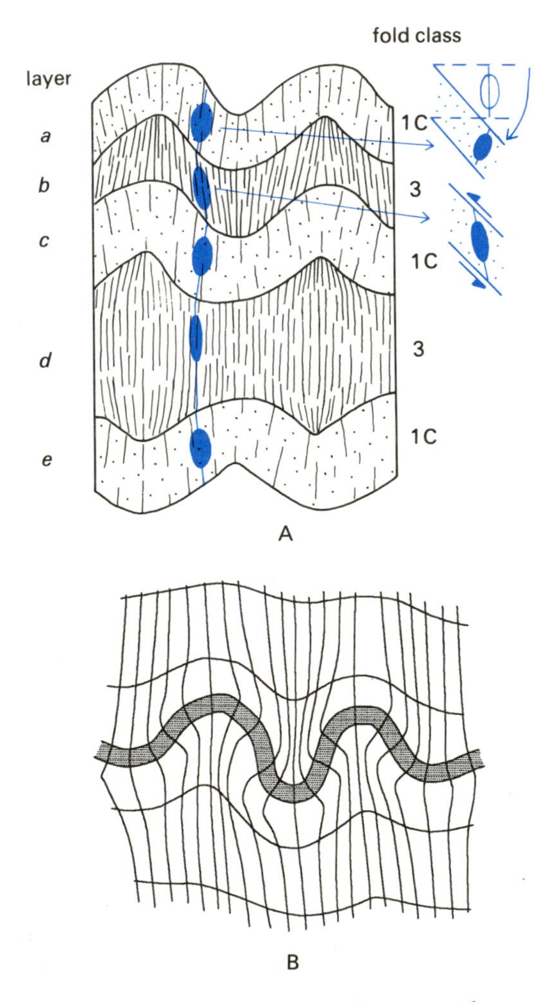


**Figure 10.10** A. Cleavage refraction. Alternating convergent and divergent slaty cleavage fans produced by the folding of alternate competent and incompetent layers (see text). B. Local distortions of cleavage pattern produced by 'contact strain' (see text).

These differences in strain pattern are very easy to detect where the rock contains a penetrative planar fabric, e.g. a slaty cleavage, that is parallel to the $XY$ plane. The layers *a–e* in Figure 10.10A illustrate the variation in attitude of a slaty cleavage under these conditions. Such alternation of convergent and divergent cleavage fans is an example of **cleavage refraction**, which is a change in the attitude of the cleavage on passing from one bed into another of different composition. Within a competent layer, the cleavage planes may be curved. This is usually associated with a change in physical properties and is particularly common in graded beds, where the degree of convergence decreases upwards with decrease in grain size.

The strain pattern near the interface between two layers of contrasting behaviour may be quite complex. A 'contact strain' effect may be produced by the displacements and rotations of an individual competent layer, which control the geometry of the adjoining less competent layers. This effect may be clearly seen by local distortions of the cleavage pattern (Figure 10.10B).

The pattern of alternating convergent and divergent cleavage fans is duplicated by the isogon geometry. Layers *a*, *c* and *e* of Figure 10.10A exhibit class 1C geometry, and layer *b* and the outer parts of layer *d*, class 3 geometry.

It is interesting to note that alternations of this kind may result in an overall class 2 geometry – that is, the dip isogons drawn from the top to the base of a set of layers (e.g. from the top of layer *a* to the base of layer *c*) would be approximately parallel, enabling the folding to continue indefinitely downwards.

In this way the individual buckled layers, with their rather complex strain geometry, when viewed on a larger scale, may become part of a much more homogeneous looking system where the total shortening strains measured at various levels

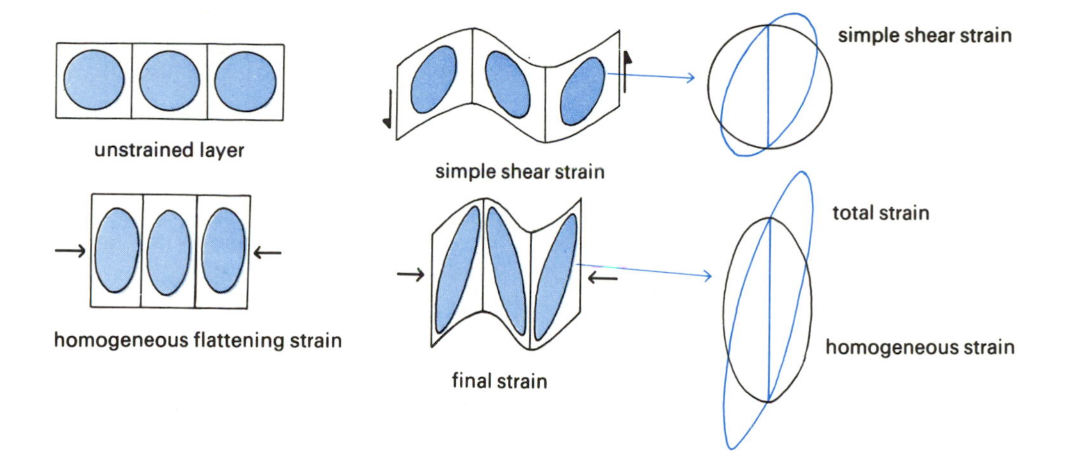


**Figure 10.11** Superimposition of variable simple shear strain and homogeneous flattening strain in a similar fold.

through the folded rock are approximately the same.

## 10.3 OBLIQUE SHEAR OR FLOW FOLDING

Folds may be formed by the passive rotation of a layer under simple shear directed obliquely to the layer. This process is often called **flow folding**, because of the analogy with the folding of marker layers in a flowing liquid. If the shear direction is constant through the fold, an ideal similar shape (class 2) is produced. In order to form a fold, the amount of shear displacement must vary along the layer, otherwise the layer would remain straight. Moreover, the layers must behave passively for ideal class 2 geometry; if there is any component of buckling (i.e. if the layers bend under lateral stress), the geometry will correspond to class 1C, closely approaching class 2 perhaps, but never reaching it.

Folds of similar type only form in rocks that are in a very ductile state during deformation, typically under medium- to high-grade metamorphism, although rocks of very low viscosity (such as salt and gypsum) behave in this way even at low temperature. Another essential condition is that the viscosity contrast between adjacent layers should be low – below the limit for the initiation of buckling.

### ROLE OF HOMOGENEOUS STRAIN

The imposition of a uniform homogeneous flattening strain on similar folds does not change the similar geometry, provided that one of the principal strain planes of the homogeneous deformation is parallel to the shear plane. The respective contributions of the two types of strain are illustrated in Figure 10.11. The simple shear strain only affects the fold limbs and produces a divergent fan of $XZ$ planes. The addition of a suitably oriented homogeneous component increases the strain on the limbs, decreases the divergence, and produces a strain in the hinge zone equivalent to the homogeneous strain ellipse. Homogeneous flattening of non-similar folds will tend to make them more like the ideal similar model and at high strains such folds may be indistinguishable from true similar folds.

### ORIGIN OF SIMILAR FOLDS

Many folds possessing similar geometry are probably the result of extreme flattening of buckle folds. Where similar folds are formed by variable oblique shear, we have to consider a mechanism for this variation. One likely cause of variation in simple shear is the influence of a fold shape generated by a buckling or bending mechanism. If we imagine a thick ductile layer bounded by two stronger, more competent layers, the buckling of the competent layers will transmit their shape through the ductile layer as a variable simple shear. Although the strain in the ductile layer near its margins may be influenced by contact strain induced by the buckling mechanism, a more uniform pattern would be expected in the central part of the layer.

An overall similar fold shape can be produced in a folded multilayer exhibiting alternating class 1C and class 3 shapes in such a way as to retain the same amount of shortening throughout the multilayer. This process is probably quite important in generating similar folds. Ideal chevron folds have this geometry (Figure 10.16).

## SUPERIMPOSITION OF OBLIQUE SHEAR (FLOW) FOLDS

It is geometrically much easier to deform an already folded surface by oblique shear than by buckling, since the shape of the passive layers is irrelevant and exercises no control over the second fold shape. The principle of superimposition of flow folds can be demonstrated by drawing a first fold shape on a card deck and then deforming the card deck by varying the amount of simple shear (Figure 10.12).

Complex interference patterns may be generated by superimposing a set of flow folds on a previously folded surface. The geometrical pattern depends on the spatial relationships between the

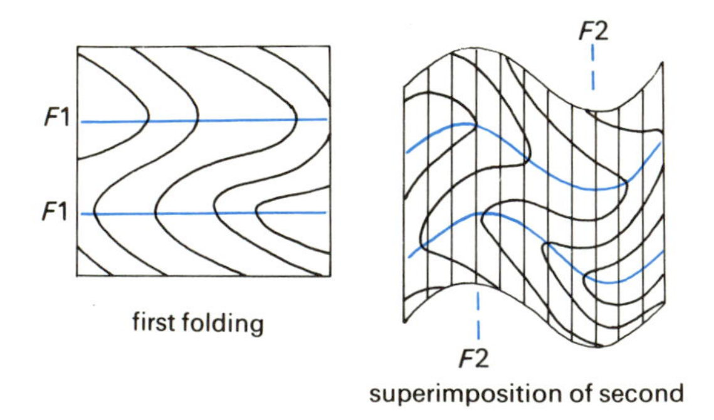


**Figure 10.12** Superimposition of oblique shear folds (flow folds) on a previously folded surface demonstrated by a card deck model.

two fold sets; in particular between the attitudes of the two fold axes and axial planes, and on the shear or flow direction of the second folds. Figure 3.18B (section 3.7) shows three common types of interference pattern produced on a flat surface by superimposition of fold sets – dome and basin, crescent and mushroom, and double zigzag. These three types can be regarded as 'end-members' in a continuous series of interference shapes. We may add a fourth type which produces no interference because the axial planes and axes of the two fold sets are parallel.

We can demonstrate how interference structures are formed by examining three cases (Figure 10.13). In each of these cases the second folds are upright with a vertical flow or shear direction. In the first case (Figure 10.13A) both the axial planes and axes of the F2 folds are near perpendicular to those of F1, and the flow direction of F2 lies within or near the axial plane of F1, which is also upright. This type of relationship produces a dome and basin pattern.

In the second case (Figure 10.13B) the F1 folds are recumbent and the axial planes and axes of F2 again make a large angle with those of F1. However, the flow direction of F2 also makes a large angle with the F1 axial planes. This relationship produces a crescent-shaped interference pattern.

In the third case (Figure 10.13C) the F1 axial planes are gently inclined and thus again make a large angle with the F2 axial planes. However, the fold axes of both sets of folds are subparallel. The F2 flow direction is oblique to the F1 axial planes. This relationship produces a double zigzag pattern.

As an example of how a mapped interference structure may be interpreted, let us return to the Loch Monar structure illustrated in Figure 3.18C. A three-dimensional model of this structure (Figure 10.14) shows that the plunge of the second folds is controlled by the attitude of the first fold limbs as well as by the F2 axial planes. Thus the F2 folds plunge vertically on the southern limb of the F1 synform, subhorizontally across the hinge zone, and have a moderate plunge to the south west on the northern limb. If we imagine the F2 folds to be unfolded, the shape of the F1 fold is revealed as an east–west trending synform with a moderately

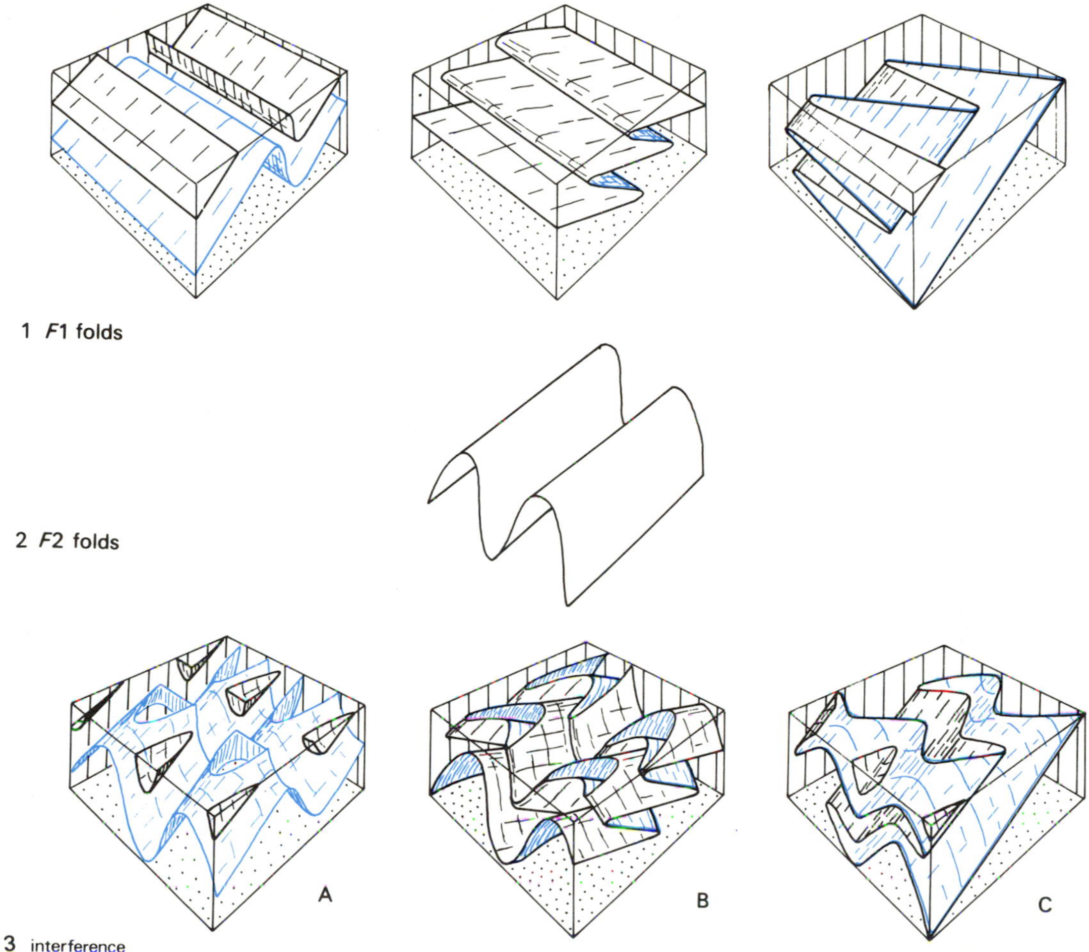


**Figure 10.13** Interference patterns generated by the superimposition of upright flow folds on previous folds of varying attitude (see text for details). (After Ramsay, 1967, figures 10.3, 10.8 and 10.15.)

south dipping northern limb, a vertical southern limb and a low plunge to the west.

FORMATION OF SHEATH FOLDS

At high levels of strain, very complex fold shapes may be produced by variable simple shear or flow superimposed on ordinary buckle folds. During progressive simple shear, the fold axes may behave passively and rotate towards the shear direction until at high strain ($\gamma \sim 15$) they will become subparallel to the shear direction (Figure 10.15). In this way it is possible to produce folds with strongly curved axes that ultimately develop very elongate dome and basin shapes. Such folds, termed **sheath folds**, are characteristic of strongly deformed ductile shear zones (see section 10.6).

## 10.4 KINKING AND FORMATION OF CHEVRON FOLDS

The characteristic features of kink bands and chevron folds (see section 3.4) are straight limbs and sharp angular hinges. Kink bands (Figure

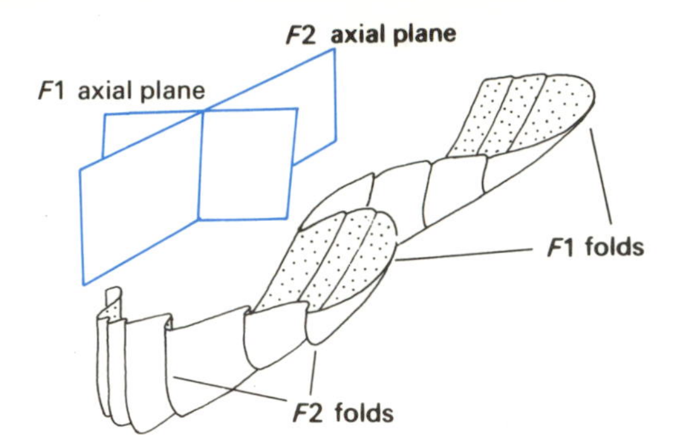


**Figure 10.14** Geometry of an interference structure in the Moine complex at Loch Monar, northwestern Scotland. (See text for details.) (After Ramsay, 1967, figure 10.25.)

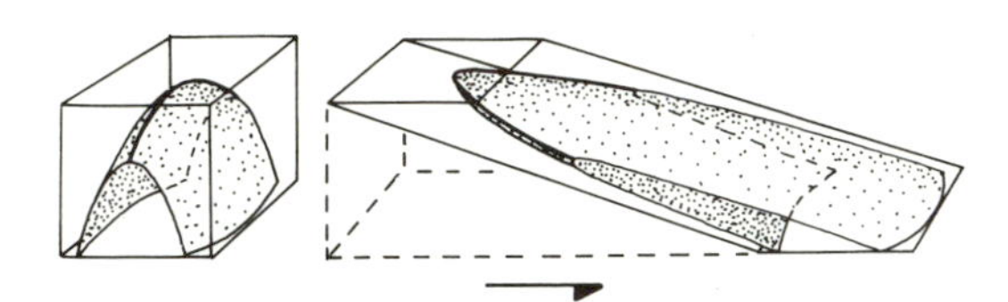

**Figure 10.15** Development of a sheath fold by the progressive simple shear of an initial non-cylindrical fold. (After Ramsay, J.G., 1980, figure 16.)

10.16A) are asymmetric structures where the deformation is essentially confined to the material within the kink band, whereas chevron folds are more symmetrical and continuous (Figure 10.16B, C). The formation of both kink bands and chevron folds appears to involve a combination of flexural slip between the competent layers, which maintain their thickness, and ductile flow in the intervening incompetent material. This leads to folds with alternating class 1B and class 3 geometry, which enables the folds to maintain an overall similar form.

The basic geometry of a kink band, as shown in Figure 10.16A, can be explained in terms of a rotation of the short limb of the asymmetrical fold couple so that the kink band behaves as a zone of simple shear. The rotation can take place with very little internal strain in the layers, provided that inter-layer slip can take place as shown. The amount of rotation of the short limb is expressed by the angle $\alpha$. The kink plane makes an angle $\beta_1$ with the short limb and $\beta_2$ with the long limb. If $\beta_1 > \beta_2$ and the layer thickness remains constant, dilation must take place between the layers within the kink band. In nature this results in extensional fissures which are often filled with quartz or calcite. Usually, however, the kink layers rotate until $\beta_1 = \beta_2$ when no fissure space exists and further movement would result in a contraction across the layers.

Kink bands usually occur in conjugate pairs or sets, but to achieve maximum shortening by kinking, the whole length of the layers must be involved in the folding. Experiments on the compression of phyllite have shown that, under continued compression, individual kinks grow by lateral expansion until neighbouring oppositely directed pairs merge completely to form chevron folds. This process is illustrated in Figure 10.16D. The development of chevron folds from kink bands will only take place when $\beta_1 = \beta_2$, i.e. when the axial surface of the kink band (or kink plane) bisects the fold angle, and if $\alpha = \beta_1 = \beta_2 = 60°$. 'Ideal' chevron folds which are produced from kink bands with this geometry will thus have a fold angle of 60° (Figure 10.16C).

Not all chevron folds may have formed in this way, however. Many natural fold sets show continuous variation between curved buckle folds and folds with chevron type geometry (Figure 10.17) and it is possible to straighten the limbs and increase the angularity of the hinges in buckle folds by progressive flattening (Figure 10.4B). It is therefore possible that many, if not all, natural chevron folds may form in this way rather than by the amalgamation of kink bands.

## SHORTENING AND INTERNAL STRAIN IN KINK FOLDING

Ideally the kinking process involves no internal strain in the layers on the limbs of the kinks and the strain is concentrated in the zone where the layer is bent around the hinge (Figure 10.18). The proportion of strained to unstrained layer is clearly greater when the layers are thicker in relation to the length of the limbs (compare Figures 10.18A

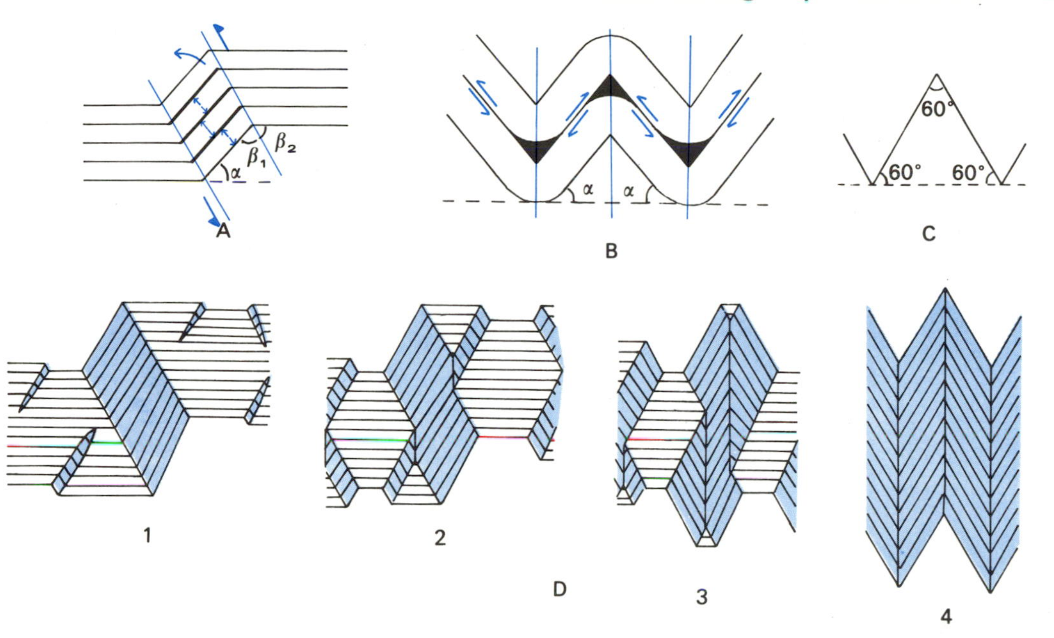


**Figure 10.16** Geometry of kink bands and chevron folds. A. Kink band – the inclination of the short (rotated) limb is $\alpha$ and the kink plane makes an angle $\beta_1$ with the short limb and $\beta_2$ with the long limb. If $\beta_1 > \beta_2$, dilation occurs between the layers of the short limb. B. Chevron fold – this is a symmetrical structure where the limb inclinations $\alpha$ are equal. If the layers maintain a constant thickness, gaps occur at the hinges. C. An ideal chevron fold possesses limb inclinations of 60° and therefore also an inter-limb angle of 60°. D. Progressive growth of conjugate kink bands ultimately forming symmetric chevron folds. (1) 12.5%, (2) 25%, (3) 40%, (4) 50% shortening. (From experiments on deformation of phyllite by M.S. Paterson and L.E. Weiss. A, B and D after Ramsay, 1967, figures 7.111, 7.120 and 7.124.)

and B). Thus the ratio between layer thickness $t$ and limb length $l$ is critical in determining the proportion of strained to unstrained material in the folds, e.g. the ratio of internal strain to total shortening is much greater in Figure 10.18A than in 10.18B.

The amount of shortening that can be achieved by chevron folding depends on the angle $\alpha$, but cannot be increased beyond a limiting value because of the 'locking' effect brought about by the progressive increase in internal strain. Actual 'locking' values of $\alpha$ vary with the $t/l$ ratio and the nature of the material. The relationship between the $t/l$ ratio, amount of shortening and $\alpha$ are shown in Figure 10.18C. This diagram shows that a fold with a low $t/l$ ratio can accommodate a given strain with a smaller limb dip (or fold angle) than one with a high $t/l$ ratio. We should expect therefore that chevron folds would form preferentially in thin-layered rocks. We find that in practice this $t/l$ ratio is generally less than 0.1 and is associated with shortening strains of 50–65%. Thus chevron folding is a more efficient method of shortening layers than buckling, since ideal buckles involve a maximum shortening of only 36% (Figure 10.8).

Further shortening, as in buckle folds, will be accommodated by a modification of the chevron geometry by flattening of the limbs and extension of the hinges and can be approximated by the addition of a homogeneous strain (cf. Figure 10.9).

## 10.5 CONDITIONS CONTROLLING THE FOLD MECHANISM

There is no satisfactory theory at present that enables us to predict which fold mechanism will affect a given layered rock material under compression.

**Figure 10.17** Chevron folds in Devonian phyllites, Boscastle, Cornwall. Note that ideal chevron folds should have straight limbs, whereas these are curved in places (see text).

However, there are certain generalizations which we may use as a guide. First, it is clear that a low-viscosity contrast between adjacent layers inhibits both buckling and kinking but favours flow. Second, a high-viscosity contrast between adjacent layers is necessary for both buckling and kinking, but buckling of an individual layer or set of layers will be inhibited if the nearest competent layers are too close. Thus we would expect kinking to take place in a relatively thick sequence of thin closely spaced competent layers and buckling where the competent layers are separated by thicker layers of more ductile material. Flexural slip and flexural shear are mechanisms that may operate together with either buckling or kinking.

Passive bending is the mechanism attributed to 'accommodation' folds associated with thrust and extensional fault systems (Figures 2.11 and 2.13). Such folds are produced essentially by the action of gravity as a layer accommodates to a change in shape caused by the sliding of a sheet over a surface of changing inclination. The layering will then typically also be subject to compression or extension. The bending process clearly requires some additional mechanism or mechanisms to enable the layering to change shape. These may include flexural slip, flexural shear or oblique shear; and if the layers are compressed, buckling or kinking may also be involved.

Finally, we must remember that the scale of observation is very important in deciding both geometry and mechanism. Thus a layer which might be regarded as a buckle fold or a flow fold on a large scale might well involve elements of different mechanisms on a smaller scale.

## 10.6 SHEAR ZONES

Brittle faulting is generally confined to the uppermost 10–15 km of the crust. Below this, owing to the change in physical properties of the rocks brought about by the increase in temperature and confining pressure, brittle behaviour gradually gives way to ductile flow. If we examine the structure of high-grade metamorphic terrains, representing uplifted deep crustal material, we find numerous examples of shear zones, which represent areas of ductile displacement analogous to faults but without discrete fracture planes. The existence of such structures at depth has led to the realization that

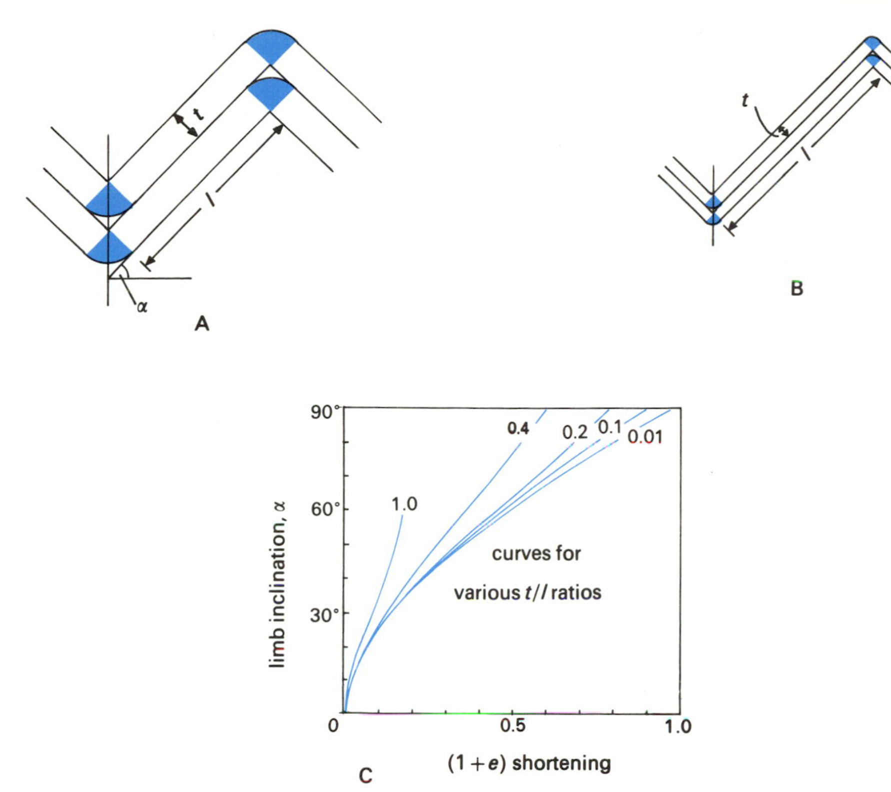


**Figure 10.18** Internal strain in kinked layers of thickness $t$ as a function of limb length $l$ and limb inclination $\alpha$. A. A high $t/l$ ratio produces a large strained zone. B. A low $t/l$ ratio produces a proportionally much smaller strained zone. C. Plot of shortening $(1 + e)$ against limb inclination for various values of $t/l$ (see text). (After Ramsay, 1967, figure 7.112.)

steep major faults or fault zones at the surface must be replaced by ductile shear zones at deeper levels in such a way that the total displacement of the blocks on either side is maintained (Figure 10.19). Major strike-slip fault zones, for example, particularly those that constitute plate boundaries, are presumed to continue at depth as strike-slip shear zones (see section 3.9). Studies of deeply eroded Precambrian shield regions demonstrate the importance of major strike-slip shear zones at deep crustal levels.

GEOMETRY OF SHEAR ZONES

Ideally, shear zones are contained between two parallel planar margins and are produced by simple shear stresses acting parallel to these margins (Figure 10.20). In practice, the margin may be very poorly defined, since the strain near the margin is usually low (Figure 10.22). An ideal shear zone represents plane strain and there is no displacement in the plane at right angles to the shear direction. The displacement plane therefore contains the $X$ and $Z$ principal strain axes and the $Y$ strain axis remains unchanged. The shear plane contains the shear direction, and is parallel to the shear zone margins and perpendicular to the displacement plane.

SHEAR ZONE DISPLACEMENT

There is a simple geometrical relationship between displacement, width and shear strain in ideal shear zones, and this is illustrated in Figure 10.20B. If we take a section parallel to the displacement plane,

$$d = x \tan \psi = x\gamma \qquad (10.3)$$

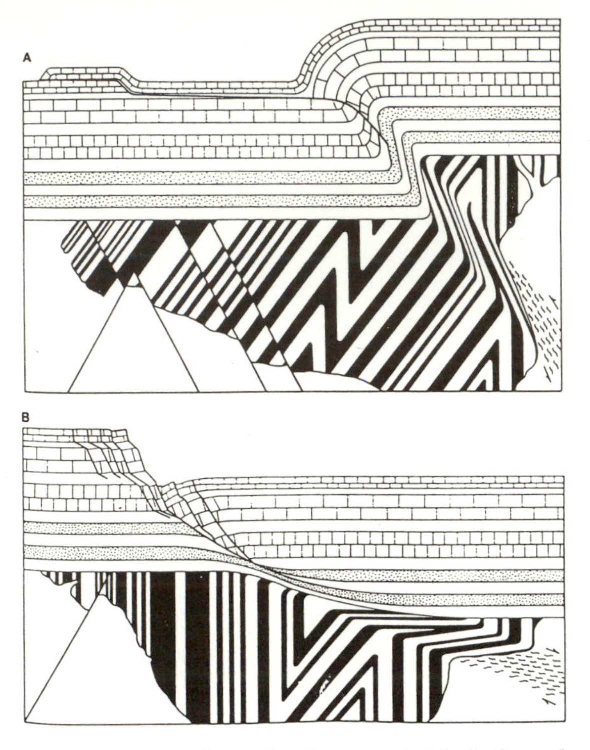


**Figure 10.19** Relationship between brittle faults and ductile shear zones: A, in compression, B, in extension. (From Ramsay, 1987, figure 26.30, with permission.)

where $d$ is the displacement, $x$ the width, $\psi$ the deflection of a right angle and $\gamma$ the shear strain (equation (6.3)). In the general case of any line making an angle with the shear plane, if the new angle is $\alpha'$,

$$\gamma = \tan(90° - \alpha') - \tan(90° - \alpha) = \cot\alpha' - \cot\alpha \qquad (10.4)$$

STRAIN WITHIN SHEAR ZONES

The mean shear strain within a shear zone can be simply calculated from equation (10.3), and equals the displacement $d$ divided by the width $x$. This strain can be represented by a strain ellipse which makes an angle $\theta'$ with the shear plane (Figure 10.21A) such that

$$\tan 2\theta' = 2/\gamma \qquad (10.5)$$

(See Ramsay, 1980, for a detailed analysis of shear zone geometry.)

Normally the shear strain varies continuously across a shear zone, commencing with low values near the margins and reaching a maximum in the centre (Figure 10.21B). The variation in the angle $\theta'$ and the strain ratio $X/Z$ with $\gamma$ may be conveniently expressed graphically (Figure 10.21C). Where the displacement cannot be directly measured for a shear zone with varying shear strain, it can be calculated using the relationship

$$d = \int_0^{x_1} \gamma \, dx \qquad (10.6)$$

by plotting $\gamma$ against $x$ and measuring the area under the curve (Figure 10.21D).

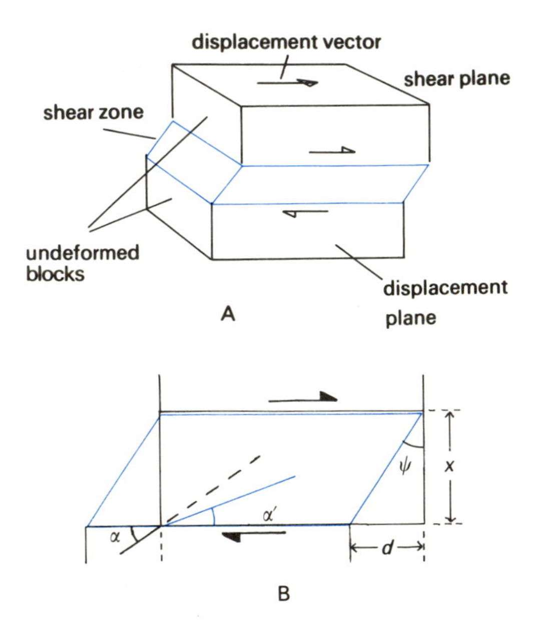


**Figure 10.20** Main elements of ideal shear zone geometry. A. The shear zone is a zone of deformation between two undeformed blocks that have been displaced relative to each other. Note the shear direction (displacement vector) lying within the shear plane, which is parallel to the margins of the shear zone. B. Profile view in the displacement plane. The displacement $d$ is given by $\tan\psi = x\gamma$, where $x$ is the width, $\psi$ the angular shear strain and $\gamma$ the shear strain. If a line makes an initial angle $\alpha$ with the shear plane and an angle $\alpha'$ after shearing, $\gamma = \cot\alpha - \cot\alpha'$. (After Ramsay, 1980.)

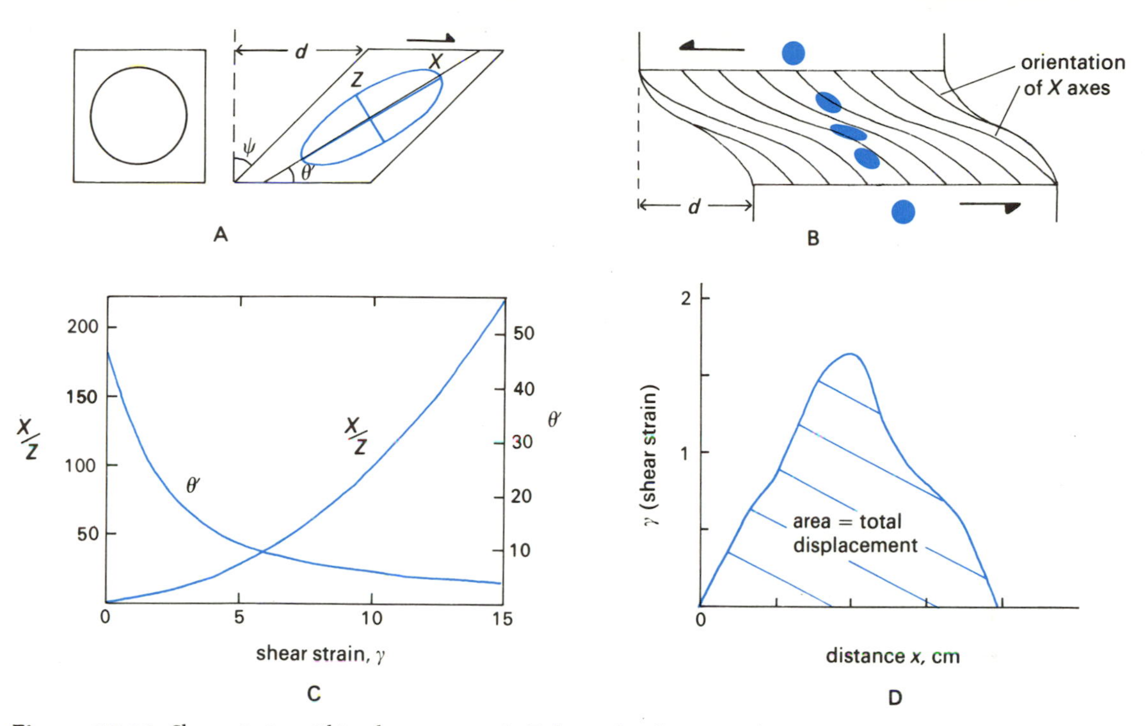


**Figure 10.21** Shear strain within shear zones. A. Relationship between shear strain and strain axes (see text). B. Continuous variation in shear strain across a shear zone shown by a change in the orientation of the $X$ strain axis and in the strain ratio $X/Z$. The angle $\theta'$ made by $X$ with the shear direction is 45° at the margin of the shear zone and decreases to a minimum in the centre. C. Plot showing the variation of the angle $\theta'$ and the strain ratio $X/Z$ with increase in the shear strain, $\gamma$. D. Plot of $\gamma$ against $x$ (distance across the shear zone) for the shear zone of B. The shear displacement $d$ is equal to the area under the curve. (A–D after Ramsay, 1980.)

## FABRICS WITHIN SHEAR ZONES

The variation of strain within shear zones is best studied by examining the fabrics developed in previously undeformed rocks, and many excellent examples may be found in deep-seated, coarse-grained, igneous rocks. Figure 10.22A shows a small shear zone in a gabbroic rock, where the strain is expressed by the deformation of originally equidimensional felsic and mafic grain aggregates into elliptical shapes, which become more elongate towards the centre of the shear zone. The long axes of these grain aggregates define a shape fabric or foliation that commences at the margins of the shear zone, making an angle of 45° with the shear direction. Towards the centre of the shear zone this fabric becomes more intense, and the angle that it makes with the shear direction decreases to a minimum. Clearly this fabric offers two methods of measuring the variation in shear strain: either plotting the variation of $\theta'$ or plotting the variation in the strain ratio $X/Z$ of the deformed grain aggregates (Figure 10.21C).

## BRITTLE–DUCTILE SHEAR ZONES

Many shear zones exhibit a mixture of brittle and ductile structures. This may be due to their development in a region of the crust, at intermediate depths, where both brittle and ductile conditions obtain, depending on the rheology of the material traversed by the shear zone and on the strain rate. Another reason for the coexistence of brittle and ductile structures is the evolution of the shear zone

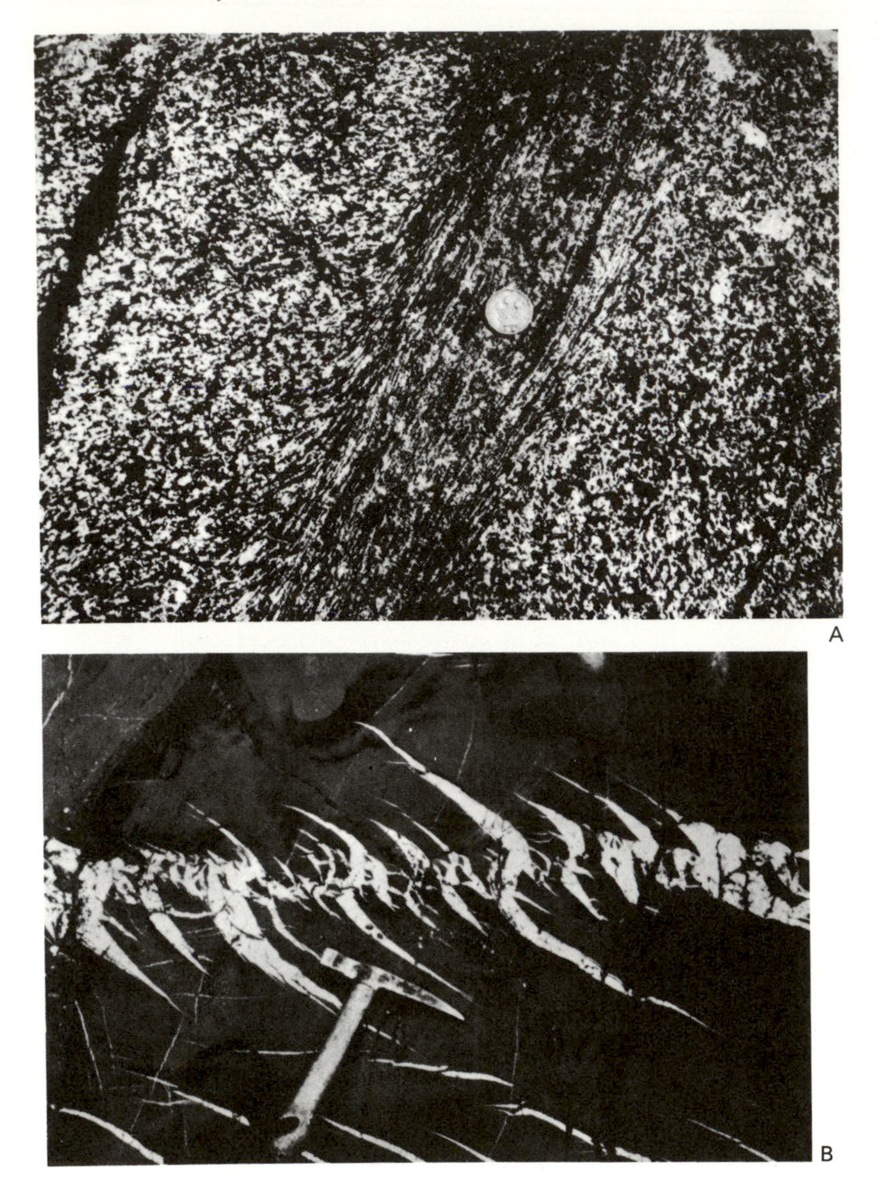


**Figure 10.22** A. Sigmoidal arrangement of fabric in a ductile shear zone in Lewisian metagabbro, north Uist, northwestern Scotland. Note that the fabric commences at an angle of 45° to the orientation of the shear zone and bends into near parallelism with the shear direction in the central part of the shear zone, where it is most intensely developed. (From Ramsay, 1980, figure 2C.) B. Dilational quartz-filled vein fissures in sandstone, north Cornwall. The en-echelon veins mark a dextral shear zone which has subsequently rotated the veins into a sigmoidal shape. (From Ramsay, 1980, figure 2B.)

with time from ductile to brittle as the area is gradually exhumed, so that earlier-formed structures within the zone are ductile and later-formed structures are brittle.

Brittle–ductile shear zones often contain vein arrays which can be used to determine the sense of shear on the zone (Figure 10.22B), since the extension direction (Z in Figure 10.21A) is indicated by the dilation of the veins. Earlier-formed veins may be folded by later movement on the shear zone to produce a sigmoidal pattern, which is also a good indicator of the sense of movement (Figure 10.22B).

Brittle–ductile shear zones are characterized by the combination of folds and faults, and in particular by the association of faults of different type. Figure 10.23 shows the orientations of normal faults, reverse faults and conjugate sets of strike-slip faults, as well as folds, within an idealized strike-slip shear zone. The orientation of these subsidiary structures

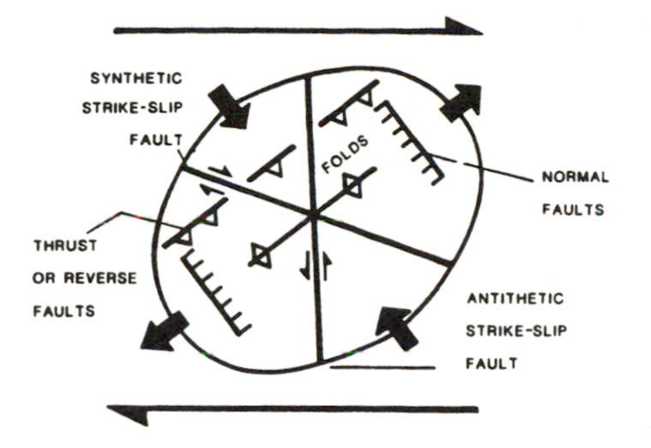


**Figure 10.23** Arrangement of folds and faults in an ideal dextral strike-slip brittle–ductile shear zone.

within such a zone conveys important information as to the sense of strike-slip motion.

## SHEAR-SENSE INDICATORS

Sense of shear (e.g. sinistral or dextral) can be deduced in ductile or semi-ductile shear zones using several different criteria (Figure 10.24). These include: (1) the vergence of asymmetric shear folds; (2) the directions of non-rotational ($\sigma$**-structure**) and rotational ($\delta$**-structure**) augen tails; (3) the orientation (sense of obliquity) of synthetic minor shear zones or shear planes (C), or extensional crenulation cleavages, in relation to the finite-strain planar fabric (S) – an arrangement that is termed **S-C structure**. All these structures are common in mylonites and often give the only method of determining the sense of shear on major ductile shear zones.

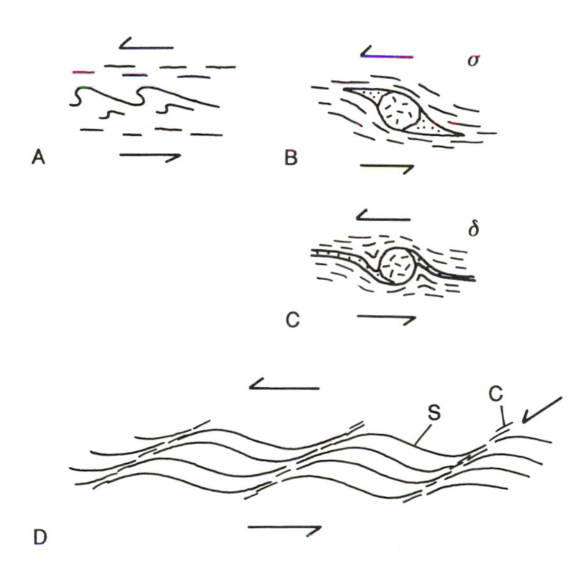

**Figure 10.24** Shear-sense indicators (see text).

## TRANSTENSION AND TRANSPRESSION

If a shear zone is thought of as a deformable sheet, the deformation may take the form of a combination of pure shear and simple shear, the pure shear component arising from net extension or compression across the zone, and the simple shear component from the strike-slip displacement of the boundaries of the zone. The addition of an extensional component across the zone produces **transtension** and a compressional component produces **transpression**. Transtension results in clockwise rotation of the extension axis (in dextral shear) and transpression in anticlockwise rotation. The reverse holds for sinistral shear.

## TERMINATION OF SHEAR ZONES

Like faults, shear zones do not continue indefinitely. A shear zone may end at another shear zone, in which case the displacement is transferred to the other shear zone e.g. Figure 2.13G shows how this can take place with faults. For example, major strike-slip zones may detach on subhorizontal detachment zones within or at the base of the crust. Alternatively, a shear zone may gradually die out into a wide zone of ductile deformation.

## FURTHER READING

Carreras, J., Cobbold, P.R., Ramsay, J.G. and White, S.H. (eds) (1980) Shear zones in rocks. *Journal of Structural Geology*, **2**, 1–287.

Cobbold, P.R. and Quinquis, H. (1980) Development of sheath folds in shear regimes. *Journal of Structural Geology*, **2**, 119–26.

Price, N.J. and Cosgrove, J.W. (1990). *Analysis of Geological Structures,* Cambridge University Press, Cambridge. [Good coverage of buckle folding.]

Ramsay, J.G. (1967) *Folding and Fracturing of Rocks,* McGraw-Hill, New York.

Ramsay, J.G. (1980) Shear zone geometry: a review. *Journal of Structural Geology*, **2**, 83–99.

Ramsay, J.G. and Huber, M.I. (1987) *The Techniques of Modern Structural Geology, Vol. 2: Folds and Fractures,* Academic Press, New York. [Contains a useful short list of further references.]

# STRUCTURAL GEOLOGY OF IGNEOUS INTRUSIONS 11

A detailed discussion and classification of igneous bodies is outside the scope of this book, but we should concern ourselves with certain aspects of igneous bodies that are directly relevant to structural geology. This relevance may arise in four main ways.

1. An igneous body may contain structures (particularly foliation and lineation) that are caused by deformation, either during or after intrusion.
2. The shape and orientation of many igneous bodies is a direct result of the pre-existing structure of the country rock into which they have been emplaced.
3. The forces within the Earth's crust that give rise to deformation in solid rocks also control the way that magma is emplaced and, to some extent, determine the geometry of the resulting igneous bodies.
4. Large igneous bodies may cause significant deformation of the country rock during emplacement.

## 11.1 STRUCTURES FOUND WITHIN IGNEOUS BODIES

It is often difficult in practice to distinguish between structures formed as a result of the movement of magma during emplacement (igneous structures) and those structures formed after emplacement as a result of deformation of the solid rock. The commonest type of igneous structure is **flow foliation**, which is caused by the alignment, during the flow of the magma, of tabular minerals that had formed before the final consolidation of the magma. A **flow lineation** may be formed in the same way by the alignment of elongate crystals, but is less common. Sometimes **flow banding** is formed in very viscous magmas such as rhyolite. This banding may locally be involved in intricate folding rather similar to the folds formed in solid rock under highly ductile conditions. Another type of banding or lamination is produced by gravitational settling within a magma chamber.

Many large igneous bodies, particularly granites, exhibit a foliation that is parallel to their margins and decreases in intensity towards the interior of the body. Studies of the shapes of xenoliths, for example, show that this decrease in intensity is related to a regular pattern of deformation attributed to the 'ballooning' effect of later pulses of magma on earlier, partly or wholly consolidated, material. However, other examples of this phenomenon have been attributed to deformation in the solid state brought about by upward diapiric flow under gravitational pressure (see section 12.4).

The most reliable method of distinguishing primary igneous structures from deformational structures is to look for evidence of undeformed crystals and igneous textures. An igneous rock deformed in the solid state, on the other hand, will usually show a penetrative fabric where individual crystals have been deformed.

## 11.2 STRUCTURAL CLASSIFICATION OF INTRUSIVE IGNEOUS BODIES

Intrusions are usually divided somewhat arbitrarily into major and minor. Major intrusions, termed **plutons,** are large bodies of various shapes and sizes comprising many cubic kilometres of igneous rock. Minor intrusions are normally sheet-like or

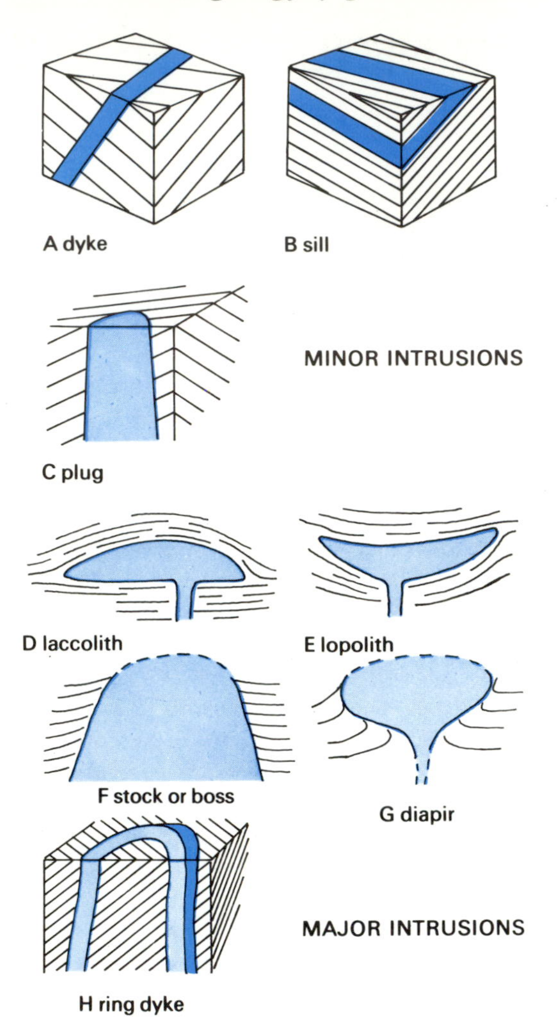


**Figure 11.1** Classification of igneous bodies based on form and structural relationships.

pipe-like in form, and at least one dimension is of the order of metres to tens of metres rather than kilometres.

SHEET-LIKE BODIES

The commonest types of minor intrusion are **dykes** and **sills**. These are essentially parallel-sided, sheet-like bodies, which differ only in that dykes are largely discordant with the country rock structure and normally steeply-dipping (Figure 11.1A), whereas sills are largely concordant and typically subhorizontal or gently-dipping (Figure 11.1B). Dykes often occur as a set of related bodies termed a **dyke swarm** which may have a subparallel, or sometimes radial, arrangement. **Cone-sheets** are a special type of dyke swarm where the dykes are arranged in a set of conical surfaces inclined towards a central point (Figure 11.7B, C).

Sheet-like minor intrusions are normally emplaced along fractures. The pattern of these fractures and how they relate to local and regional stress fields are of great interest to the structural geologist and are discussed below.

OTHER SMALL-SCALE BODIES

**Plugs** (Figure 11.1C) are pipe-like bodies of about 100–1000 m across that commonly fill the vent or neck of a volcano. Many eroded volcanoes are represented at depth by such bodies. The name **vein** is given to a small body (centimetres to metres in width) that may be either sheet-like or quite irregular in form. Veins of igneous rock are often apophyses or branches leading from a larger body, although, in certain metamorphic regions, granitic vein complexes are known that have no obvious source and therefore have been attributed to mobilization or fractional melting of country rock.

MAJOR INTRUSIONS

Large intrusive bodies, termed **plutons,** may be either broadly concordant or broadly discordant. The concordant bodies, which may exhibit considerable discordance locally, include **laccoliths** and **lopoliths**. Laccoliths make room for themselves by arching up the strata above the intrusion, to form a lensoid shape (Figure 11.1D). Lopoliths are less common. They are accommodated by a sagging or downbending of the strata beneath to form a saucer-shaped intrusion (Figure 11.1E). The type example of the latter is the Bushveld complex of South Africa, one of the largest igneous bodies in the world, with an outcrop area of more than 67 000 $km^2$.

Discordant plutons may be quite irregular in

form but very often they are roughly circular or elliptical in plan with steeply-dipping walls. The latter type are known as **stocks** or **bosses** (Figure 11.1F). The arched top of such bodies forms a **dome**. It is frequently difficult to establish the geometry of such large bodies at depth, and in the past many bodies were assumed to continue downwards indefinitely, until gravity surveys or more detailed mapping established their sheet-like or laccolithic form.

Many igneous bodies were once thought to have risen through the crust like very large blobs of liquid with an inverted tear-drop or pear shape. Such bodies are termed **diapirs.** The shape of a diapir swells upwards and outwards from a relatively narrow neck' (Figure 11.1G). However, it is now thought more likely that diapirs are formed by a ballooning mechanism and were fed through relatively narrow channels. The mechanism of emplacement of plutons is discussed in more detail below.

A special class of pluton consists of one or more bodies with a ring-shaped cross-section and steeply dipping walls. These bodies are called **ring dykes** (Figure 11.1H) and are often associated with stocks, radial dyke swarms and cone-sheets in what is termed a **central igneous complex**. There are several well-known examples of such complexes in the Tertiary igneous province of northwestern Scotland, including those of Ardnamurchan, Mull and Skye.

The largest major intrusions are known as **batholiths**. Such bodies are usually elongate and may be many hundreds of kilometres long. Where they are well known, they have been shown to be composed of a number of individual plutons of various forms. Batholiths are typically granitic in composition, although they may include a wide range of other rock types as well.

## 11.3 METHODS OF EMPLACEMENT OF IGNEOUS INTRUSIONS

There are four principal methods whereby magmas are emplaced within the crust to form igneous intrusions.

1. **Forceful emplacement** (Figure 11.2A), in which the intrusion makes space for itself by forcing the country rocks aside by deforming them. This is 'active' emplacement, where the pressure of the magma plays a significant part in

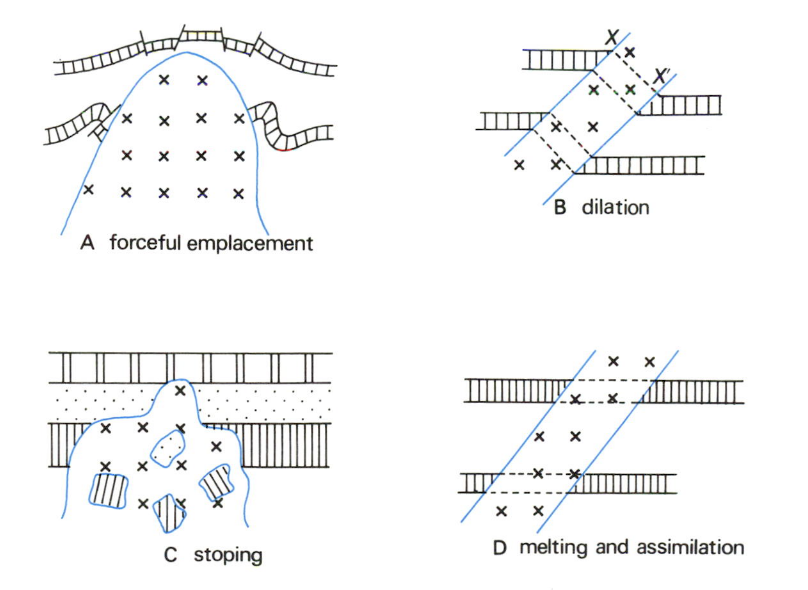


**Figure 11.2** Mechanisms of igneous intrusion.

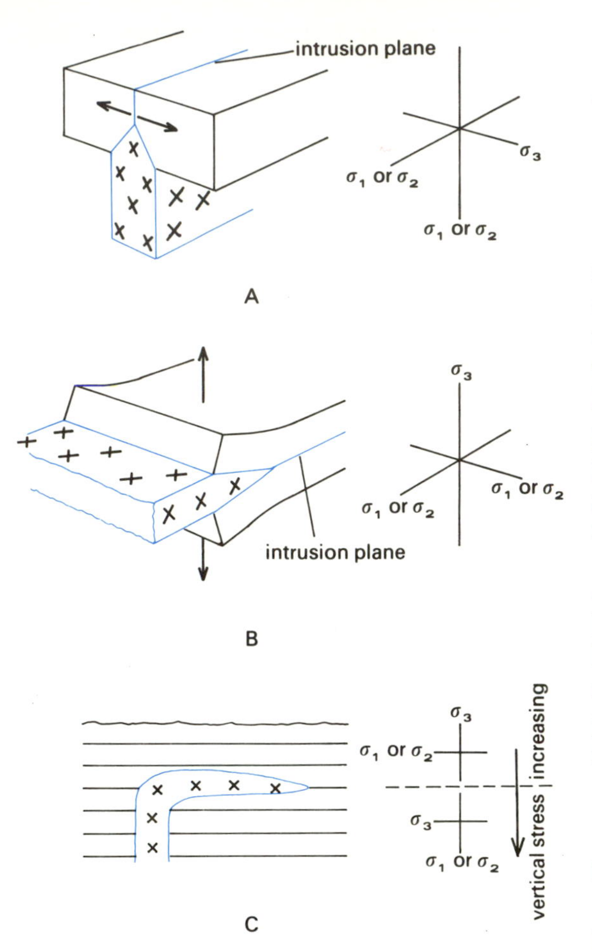


**Figure 11.3** Emplacement of dykes and sills. A. Dyke emplacement – intrusion plane ideally perpendicular to $\sigma_3$. B. Sill emplacement – implies a near-vertical orientation of $\sigma_3$. C. A dyke may feed a sill at a level determined by the change from $\sigma_1$ or $\sigma_2$ vertical to $\sigma_3$ vertical (see text).

creating the necessary space for the intrusion. The other three methods involve 'passive' or **permitted** intrusion.

2. **Dilational** emplacement (Figure 11.2B), in which the country rock moves aside, often under a tensional stress, to allow the magma to fill the space that has been created by the movement. The country rocks are not significantly deformed by this process.
3. **Stoping** (Figure 11.C), a mechanism in which the magma moves upwards by removing blocks of country rock which then sink downwards to create space.
4. **Melting and assimilation** (Figure 11.1D), a process in which the magma makes space for itself by melting and incorporating the country rock.

Melting does not appear to be an important mechanism on its own, although it probably plays a significant secondary role in many deep-seated intrusions. Stoping, as a method of permitted intrusion, is an important mechanism in certain high-level plutons, but in many other cases it probably plays only a minor part in the emplacement process. Neither melting nor stoping is of very great interest to the structural geologist since their role in deformation is negligible.

Both dilation and forceful emplacement, on the other hand, have important structural implications. Dilation is the main method of intrusion of sheet-like bodies and is controlled by the existing stress field. Forceful emplacement is important in large plutonic bodies which produce their own stress field.

## 11.4 DILATIONAL EMPLACEMENT OF DYKES AND SILLS

### DYKE EMPLACEMENT

The stress conditions governing dyke intrusion are shown in Figure 11.3A. The dyke may be considered to propagate itself upwards by a wedging effect, where the magma pressure (which is hydrostatic) acts perpendicular to the intrusion plane of the dyke. In a homogeneous body, the intrusion plane will correspond to the plane of $\sigma_1\sigma_2$, normal to $\sigma_3$, and emplacement will take place on condition that:

$$p \geqslant |\sigma_t| + \sigma_3 \qquad (11.1)$$

where $p$ is the magma pressure and $\sigma_t$ the tensile strength of the wall-rock.

Thus from the orientation of a regional dyke swarm, if we assume the country rock to be structurally homogeneous, we may deduce the orientation of $\sigma_3$. If structural heterogeneities exist in the country rock, however, the relationship is not so simple. If a previous fracture plane exists in the rock, emplacement of magma will take place along it, provided only that the magma pressure exceeds the compressive stress across the plane. However, in practice, only fracture planes making a relatively large angle with $\sigma_3$ are likely to be chosen.

## SILL FORMATION

Sills represent a special case of sheet emplacement where the intrusion plane is normally subhorizontal (Figure 11.3B). For emplacement to take place, the magma pressure must exceed the load pressure caused by the overlying strata. This is more likely to occur at high levels in the crust. We might expect, therefore, that a dyke may become a sill at a level determined by a minimum value of the vertical stress (which increases with depth). If the ratio of the two horizontal principal stresses remains the same, the decrease in the vertical stress will cause the stress axes to swap over, and a dyke may then change intrusion direction to become a sill (Figure 11.3C).

## CONTROL BY PRE-EXISTING STRUCTURE

A good example of the use of faults as well as bedding in the emplacement of a sill is provided by the Stirling Castle sill (Figure 11.4). Here the beds and the faults dip in opposite directions and the sill is able to keep to approximately the same favourable level by following the bedding dip down for a certain distance but periodically returning to a higher level along a fault.

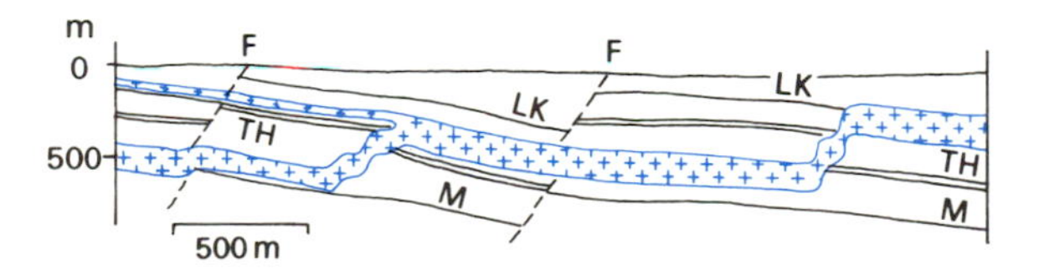


**Figure 11.4** The influence of pre-existing faults on the emplacement of the Stirling Castle sill in the Midland Valley of Scotland. The sill appears to have used both bedding and faults to maintain an approximately constant level. TH, Top Hosie Lst., M, Murrayshall Lst., LK, Lower Knott coal. (After McGregor, M.D. and McGregor, A.G. (1948) *British Regional Geology: the Midland Valley of Scotland*, HMSO, London, figure 15.)

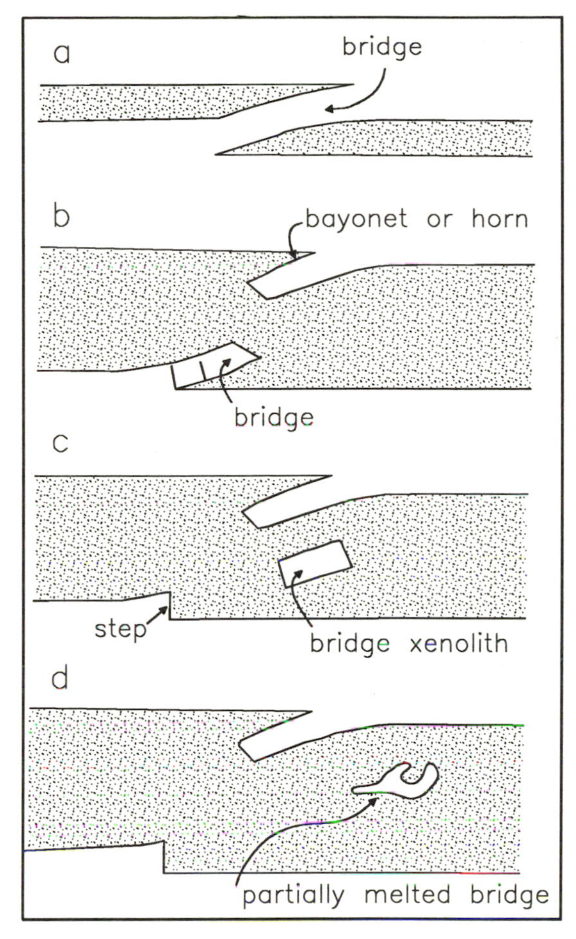


**Figure 11.5** Dyke enlargement by joining of adjacent overlapping dilational cracks. The bridge of wall rock between two adjacent overlapping cracks (a) is bent (b) and eventually broken (c) by magma pressure in the expanding cracks, leaving pointed bayonet structures as evidence of the process at the margins of the dyke. Ultimately, these too may be broken off, leaving only steps (d). (After Cadman, A., Tarney, J. and Park, R.G. (1990) Intrusion and crystallisation features in Proterozoic dyke swarms, in *Mafic Dykes and Emplacement Mechanisms* (eds Parker, A.J., Rickwood, P.C. and Tucker, D.H.), Balkema, Rotterdam.)

DYKE ENLARGEMENT BY JOINING OF DILATIONAL CRACKS

Dykes and sills are emplaced at the higher, brittle levels of the crust by the filling of dilational cracks ('tension gashes'), which are short extensional joints, often forming a set. Expansion of these dilated cracks to form a dyke takes place by the joining of adjacent overlapping segments as a result of the magma pressure in the expanding cracks bending and eventually breaking through the 'bridges' of country rock separating them (Figure 11.5). The resulting dyke margins often show evidence of this process in the form of **bayonet structures** and **steps**.

EN-ECHELON EMPLACEMENT

A rather different kind of structural control is responsible for **en-echelon emplacement**, where a set of intrusions occupies parallel planes which are consistently offset (Figure 11.6A). This arrangement may have several different causes. The source of the magma may have been an intrusion in a plane oblique to the individual intrusions but parallel to the zone defined by the set of intrusions (Figure 11.6B). This could be explained either by a change in the stress field between two successive levels in the crust, or, more likely, by the availability of a set of fracture planes at the higher level oriented obliquely to the intrusion plane at the lower level. Another common reason for en-echelon arrangement is intrusion under simple shear stress; thus, for example, a dextral shear stress would tend to open a set of oblique extensional fissures, as shown in Figure 11.6C. In some cases evidence of a simple shear component during emplacement is preserved in the form of a set of branching veins projecting obliquely in opposite directions on each side of the dyke (Figure 11.6D).

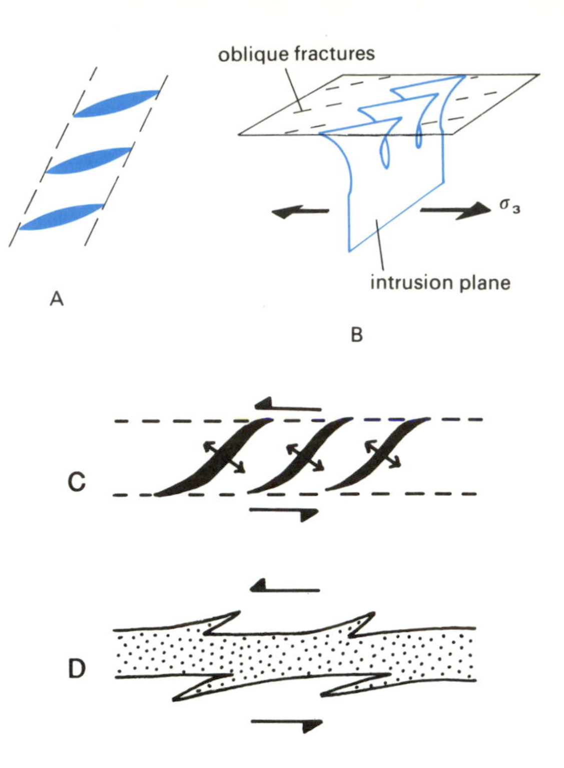


**Figure 11.6** Explanation of en-echelon dykes by the filling of oblique fractures. A. Plan view of en-echelon dykes. B. How the en-echelon dykes may be related to a single intrusion plane at depth. C. Formation of en-echelon dykes under simple shear. D. Formation of oblique branches under simple shear (see text).

## 11.5 EMPLACEMENT OF CONE-SHEETS AND RADIAL DYKES

Sets of sheet-like intrusions are commonly associated spatially with high-level plutons in central igneous complexes such as the well documented examples of Ardnamurchan, Mull and Skye, in northwestern Scotland. It is believed that the pattern and orientation of these intrusions are related to the local stress field generated by a pluton situated at depth below the centre of the complex (Figure 11.7). The magma pressure of the pluton exerts a compressive stress perpendicular to the margins of the body, resulting in curved sets of stress trajectories.

A possible arrangement of stress trajectories is shown in Figure 11.7A, assuming a simple dome-shaped pluton with circular cross-section. The arrangement has an umbrella shape with $\sigma_1$ corresponding to the radially arranged spokes, perpendicular to the surface of the pluton. The ribs of the umbrella are a set of parabolic curves which may correspond to either $\sigma_2$ or $\sigma_3$. The third set of trajectories forms a set of horizontal concentric circles

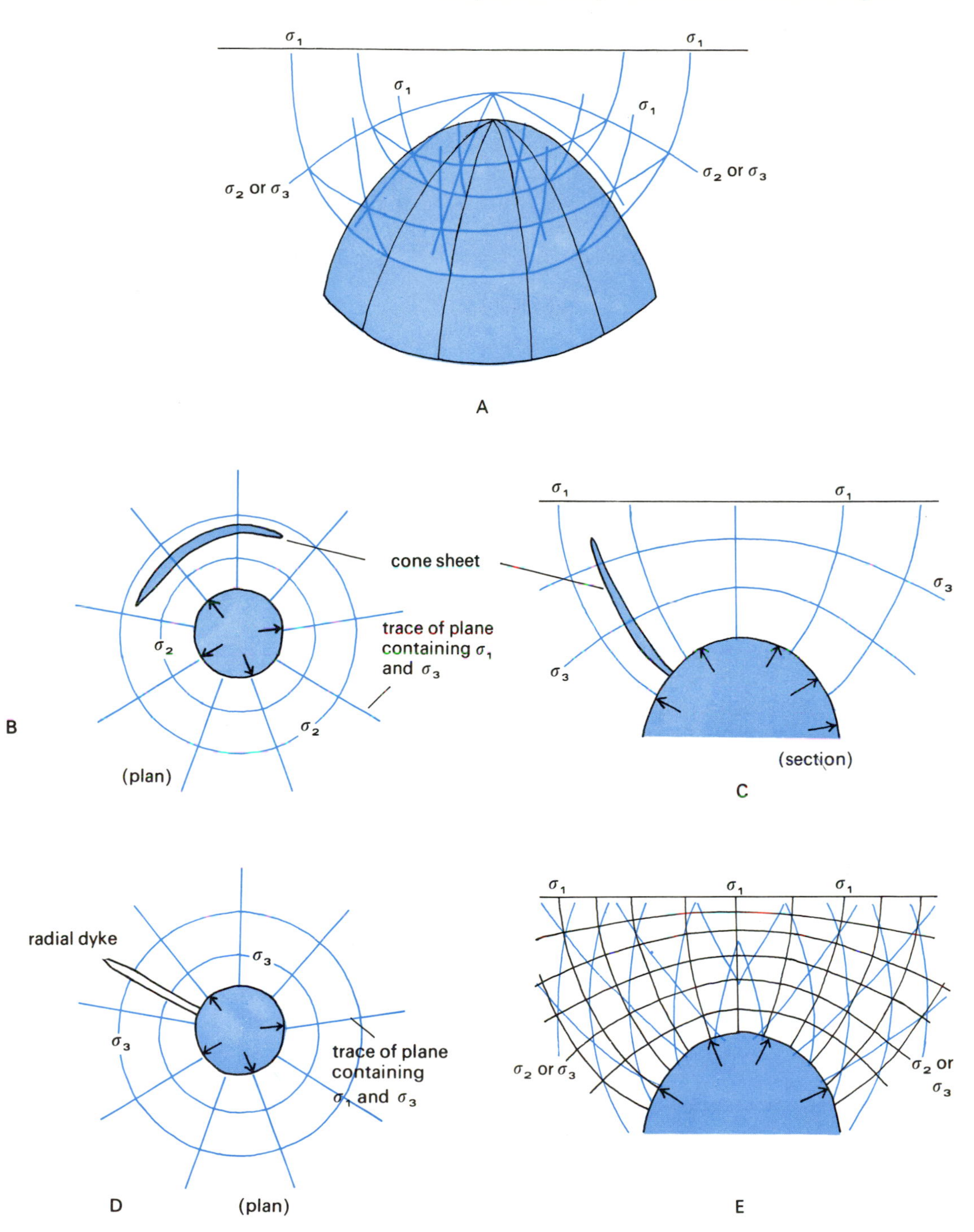


**Figure 11.7** Emplacement of cone sheets and radial dykes explained by stress fields generated by a pluton at depth. A. Stress trajectories generated by a dome-shaped pluton. Only $\sigma_1$ is uniquely determined. B, C. Orientation of stresses for the generation of cone sheets (B, plan; C, section). D. Orientation of stresses for the generation of radial dykes (plan view). E. Orientation of shear fractures generated by the stress system.

parallel to the rim of the umbrella. This arrangement can explain both the cone-sheets and the radial dykes that are associated with igneous complexes.

If the circular trajectories correspond to $\sigma_3$, then a set of vertical radial dykes may form (Figure 11.7D), whereas if the parabolic trajectories correspond to $\sigma_3$, then a set of cone-sheets may form (Figure 11.7B). It is possible that swapping of the $\sigma_2$ and $\sigma_3$ axes may result from an increase in pressure brought about by the intrusions themselves. This may be the explanation for the alternations of cone-sheets and radial dyke intrusions found in some central complexes.

It has been suggested that sets of conical shear fractures generated by this type of stress field (Figure 11.7E) might be used as intrusion planes for both cone-sheets and ring dykes. The former would follow inward-dipping shear surfaces and the latter outward-dipping surfaces. Such an explanation would relate all three types of minor intrusion found in central igneous complexes to the same basic mechanism.

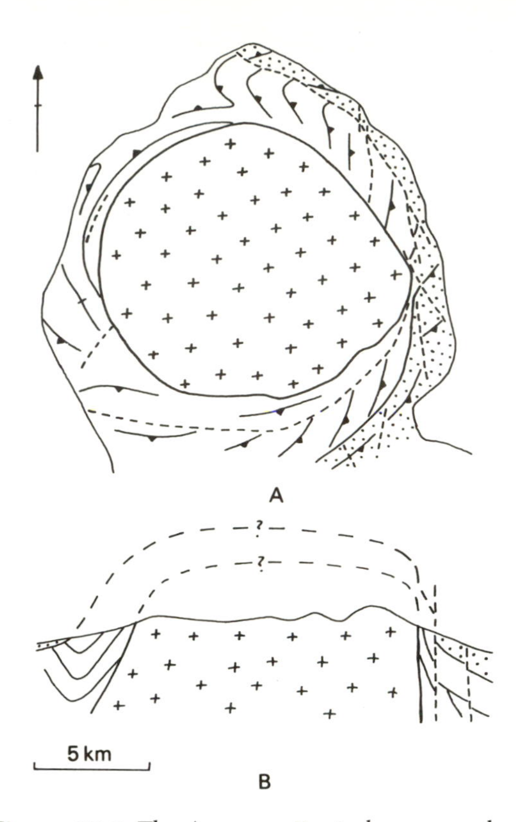


**Figure 11.9** The Arran granite stock: an example of forceful emplacement. The dotted ornament indicates Upper Devonian and younger strata. Form lines (continuous) and faults (broken lines) indicate the arrangement of folds and faults around the margin of the pluton. (After Read, H.H. and Watson, J. (1962) *Introduction to Geology*, Macmillan, London, figure 232.)

## 11.6 MODE OF EMPLACEMENT OF LARGE INTRUSIONS

From a structural point of view, the most interesting problem associated with the large intrusions is how they became emplaced within the crust and, in particular, how the space they occupy was created – the so-called 'space problem'. This problem becomes acute when we consider the size of a batholith. Has the country rock moved aside to accommodate this vast volume of rock; or has it become, in part at least, digested or assimilated by the magma; or has the country rock been somehow transformed ('granitized') to produce the igneous body? Each of these possibilities was argued for forcefully by its proponents during the 'granite controversy' of the 1950s. We need not concern ourselves with the detailed arguments as to the origin of granites, but there are important structural considerations which bear on the method of emplacement.

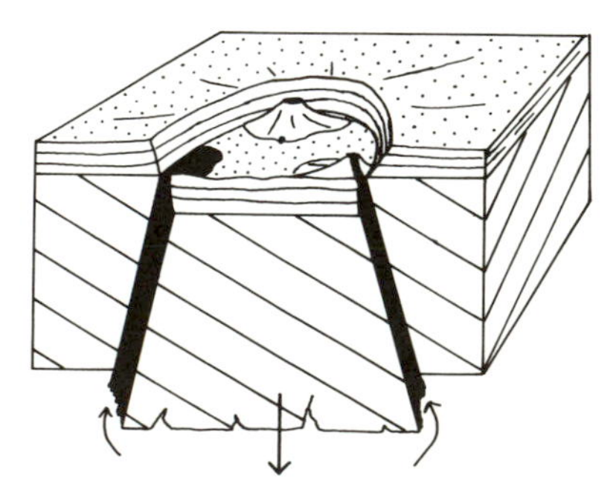

**Figure 11.8** Emplacement of a ring dyke by the subsidence of a central cylindrical block.

In the case of plutons emplaced as permitted intrusions, the structural relationships suggest a passive accommodation of the intrusive magma to the space left by the country rock as it moves aside, or subsides below an intrusion. Certain large ring-shaped intrusions have been regarded as permitted, created by the subsidence of a central cylindrical block (Figure 11.8). Other plutons appear to have been emplaced by the foundering of blocks of country rock detached from the roof of the intrusion in a form of large-scale stoping (see Figure 11.1C).

Plutons emplaced as forceful intrusions, in contrast, make space for themselves by actively pushing aside the surrounding country rock. Evidence of this is provided by folding and fracturing of the strata surrounding the intrusion. A good example of a forceful pluton is the Arran granite stock in southwestern Scotland (Figure 11.9). An arcuate fold follows the northern margin of this body, and there are also a number of arcuate faults parallel to the margin. Both the fold and the faults appear to be formed by the forceful emplacement of the granite. Most plutons show some evidence of forceful intrusion.

It seems likely that many major granite bodies (whatever the origin of the magma) are sheet-like or laccolithic intrusions formed as a result of the magma spreading outwards at a particular level of the crust, and do not necessarily extend to great depth. If this is the case, the space problem becomes much less severe than if we were to envisage large steep-sided batholiths extending to the base of the crust. It has been suggested that sheet-like bodies fed by relatively narrow dykes may, over a period of time, become inflated by successive pulses of magma to eventually assume a domed shape. This process is often referred to as **ballooning** and may account for the deformation seen in the marginal parts of certain large plutons, where earlier-formed solid or nearly solid portions of the magma chamber are compressed by the pressure exerted by the later magma batches. This type of mechanism combines active and passive processes, in that the feeder dykes may result from dilation in a regional extensional or transtensional stress field, whereas the resulting pluton may enlarge itself by essentially forceful means. The association of plutons with major shear zones in many parts of the world has led to suggestions that magma is channelled through the crust using these as pathways.

It has also been suggested that diapiric emplacement of igneous bodies may take place in the solid state under the influence of gravitational instability, and this possibility is discussed in the following chapter.

## FURTHER READING

Anderson, E.M. (1951) *The Dynamics of Faulting and Dyke Formation with Applications to Britain*, Oliver and Boyd, Edinburgh.

Duff, D. (1993) *Holmes' Principles of Physical Geology*, 4th edn, Chapman & Hall, London, pp. 162–87.

Price, N.J. and Cosgrove, J.W. (1990) *Analysis of Geological Structures*, chapter 3, Cambridge University Press, Cambridge.

Suppe, J. (1985) *Principles of Structural Geology*, Prentice-Hall, Englewood Cliffs, New Jersey. [Contains a useful chapter on intrusive and extrusive structures, with a short list of selected references.]

# GRAVITY-CONTROLLED STRUCTURES 12

The force of gravity affects all natural deformation processes and is an important component of all natural stress fields. The role of gravitational load pressure in controlling deformation has already been discussed. However, in this chapter we deal with structures that are primarily the result of the action of gravitational force rather than an applied external stress.

## 12.1 THE EFFECT OF TOPOGRAPHIC RELIEF

Differential erosion commonly leads to gravitational instability by exposing the ends of a set of beds along valley sides (Figure 12.1). The effects range from comparatively minor bending of the strata close to the ground surface, associated with **soil creep**, to gravitational collapse structures many hundreds of metres in length. The instability produced by a topographic slope is greatly accentuated if the beds dip towards the slope (Figure 12.1B). Competent layers resting on weak material may then slip towards the valley under gravity, particularly if their slip plane is weakened or lubricated by percolating ground water.

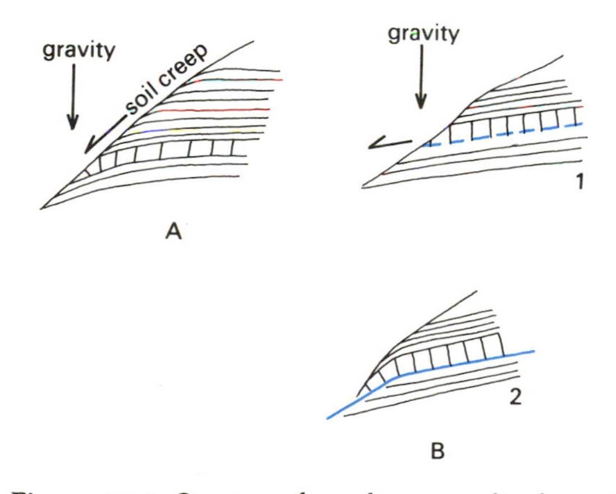


**Figure 12.1** Structures formed as a result of gravitational instability associated with a topographic slope. A. Bending of strata associated with soil creep. B. Slip of a competent layer over a weaker layer.

## 12.2 EFFECTS OF GRAVITY ON THRUST SHEETS AND NAPPES

Thrust sheets or nappes of the order of tens of kilometres in extent are important features of orogenic belts (see, for example, section 15.4). Most sheets are compressional in origin, but some have been attributed to gravitational sliding along low-angle normal faults. Because of subsequent changes in the attitude of the fault planes, it is not always possible to be certain whether a particular fault plane is a thrust or a plane of gravity sliding.

In the past there has been considerable controversy over the origin of the great thrust sheets (termed **nappes**) of the French and Swiss Alps, one school holding that the nappes were formed by gravitational gliding from an uplifted crystalline mass in the interior, and another claiming that the nappes are basically compressional in origin and formed by upwards and outwards flow of material squeezed from a highly compressed central root zone. Some Alpine geologists believe that an initial compression produces an uplift, which is followed by gravitational gliding down the slopes of the uplift.

A true gravity-gliding nappe or sheet should exhibit an extensional zone at the upper or proximal end of the sheet (i.e. nearest the uplift) and a compressional zone at the lower or distal end (i.e. furthest from the uplift). Low-angle normal faults on which the sheets rest are termed **detachment**

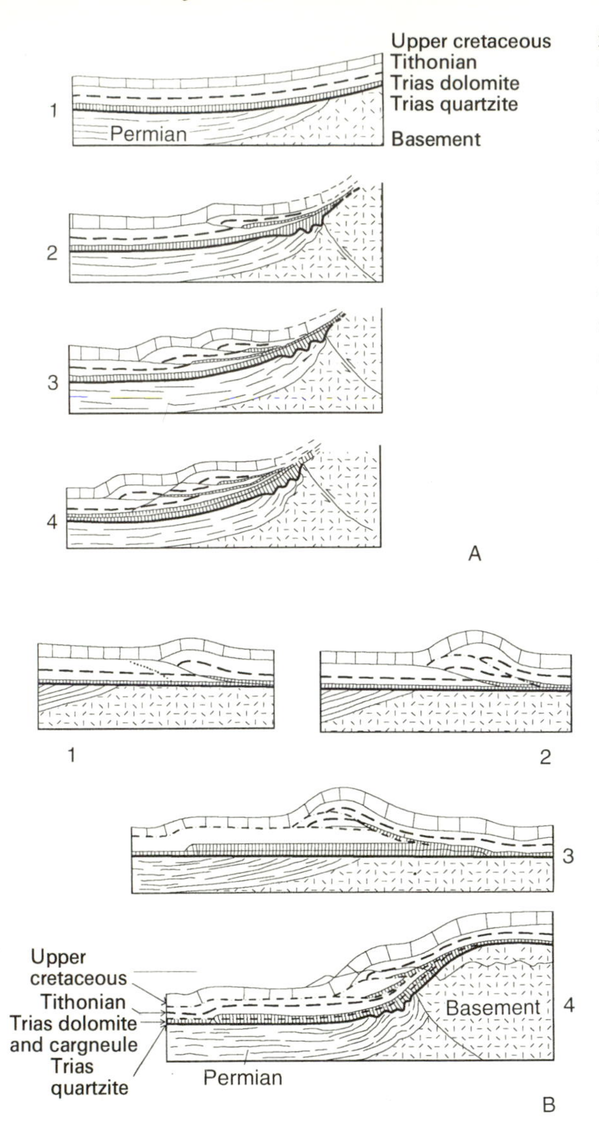


**Figure 12.2** Gravity sliding origin for the Tinée nappes. A. uplift followed by gravity sliding. B. thrusting followed by basement elevation followed by gravity sliding. (A and B after Graham, 1981, figures 14 and 15.)

**faults**. These faults cut up to the surface at the upper end of the sheet to allow the sheet to become detached. Graham (1981) has interpreted the Tinée nappes of the Maritime Alps of France in terms of a gravity sliding model (Figure 12.2). Here the 'flats' of the basal detachment faults dip away from the adjoining Argentera Massif, whereas the 'ramps' are subhorizontal. Unless the topographic elevation of the massif post-dates the fault movements, the attitude of the detachments clearly indicates normal faulting under a gravity-sliding mechanism rather than thrust faulting as an explanation for the fault structure (Figure 12.2A). However, it is recognized that the nappes may have originated as compressional structures and have subsequently become reactivated by gravity sliding once a sufficient elevation was achieved in the area of the basement high (Figure 12.2B).

## GRAVITY-INDUCED SPREADING IN LARGE CRUSTAL MASSES

Any mass of material forming a topographic high will possess gravitational potential energy; if the material is ductile, it may spread laterally under its own weight. This principle is well known in relation to the flow of glacier ice but may be applied with equal validity to any mass of sufficiently ductile material, such as the unlithified sediments in a delta. The gravitational spreading effect on a wedge of lower-density material overlying higher-density material can also cause the low-density material at the base of the wedge to flow 'uphill' (Figure 12.3), as occurs, for example, in the Greenland and Antarctic ice sheets.

## OROGEN COLLAPSE

The same principle has been applied on a large scale to orogenic belts, where it has been suggested that the crustal thickening induced by orogenic compression produces gravitational instability which can result in late-orogenic 'collapse' of the

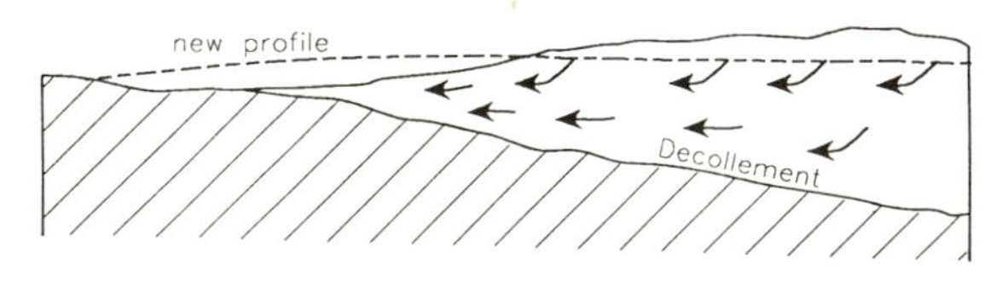


**Figure 12.3** Gravity spreading model for large crustal masses.

overthickened orogen to produce extensional spreading on a crustal scale. The elevated temperatures brought about by the crustal thickening, and the associated magmatism, are thought to cause the necessary increase in ductility.

## 12.3 SALT TECTONICS

The principle of gravitational instability also explains the mobility of salt, which commonly forms domes and other structures which 'intrude' the surrounding rock. These salt structures arise from the gravitational instability of a layer of low density (such as salt, or some other evaporite material) overlain by rock of higher density. If the salt layer and the strata above are perfectly regular and uniform there is no tendency for the salt to move. However, any irregularity in the system caused, for example, by a thickening of the salt layer, by folding or faulting in the beds above, or by local erosion, will lead to lateral pressure in the salt layer induced by the gravitational load. This pressure would lead ultimately, if able to continue to completion, to the flow of all the salt to the surface, where it would form a layer in a new gravitationally stable position.

In practice, of course, the above process is only partly completed. Structures caused by the movement of salt away from the source layer show a wide variety of forms, reflecting different stages in the upward migration of the salt, commencing with simple broad domes and proceeding to plug-like and mushroom-shaped forms (Figures 12.4 and 12.5).

Important advances in the understanding of salt tectonics were made in the 1980s as a result of seismic surveying during hydrocarbon exploration in the Gulf of Mexico, and led to the recognition of vast allochthonous salt sheets covering many hundreds of kilometres. These thin sheets, formerly thought to be irregular salt domes, are underlain by the overburden of the salt source layer and are thus important potential hydrocarbon reservoirs.

### TYPES OF SALT STRUCTURE

Salt bodies formed by movement of the salt away from its source layer form a hierarchy of types which may conveniently be described in order of increasing maturity (Figure 12.5). A **salt anticline** is an elongate upwelling of salt with a concordant overburden; a variant of this structure is the **salt pillow**, which is a subcircular upwelling. Another variant is the **salt roller**, which is a low-amplitude asymmetric structure where one (steeper) side is bounded by a normal fault. A **salt dome** is a dome-shaped upwelling with an envelope of deformed overburden. Where the mass of ductile salt has discordantly pierced the overburden, the dome becomes a **diapir**, which in turn can assume various

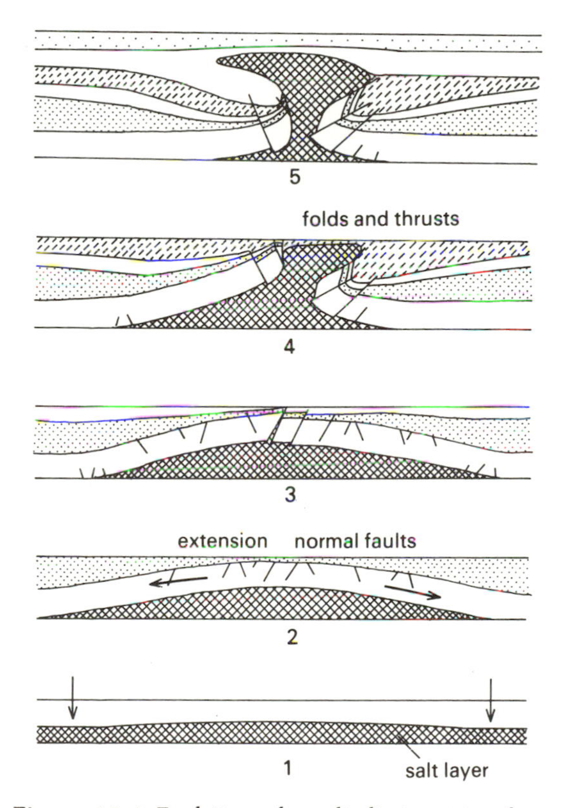


**Figure 12.4** Evolution of a salt diapir arising from the initial gravitational instability of a salt layer overlain by denser strata. As the diapir evolves, successive layers are deposited on the surface, each in turn becoming deformed as the diapir migrates upwards. Note the early extensional structures associated with the doming and later compressional structures associated with the 'neck' of the diapir. (After Trusheim F, 1960, *Bulletin of the American Association of Petroleum Geologists*, **44**,1519–41.)

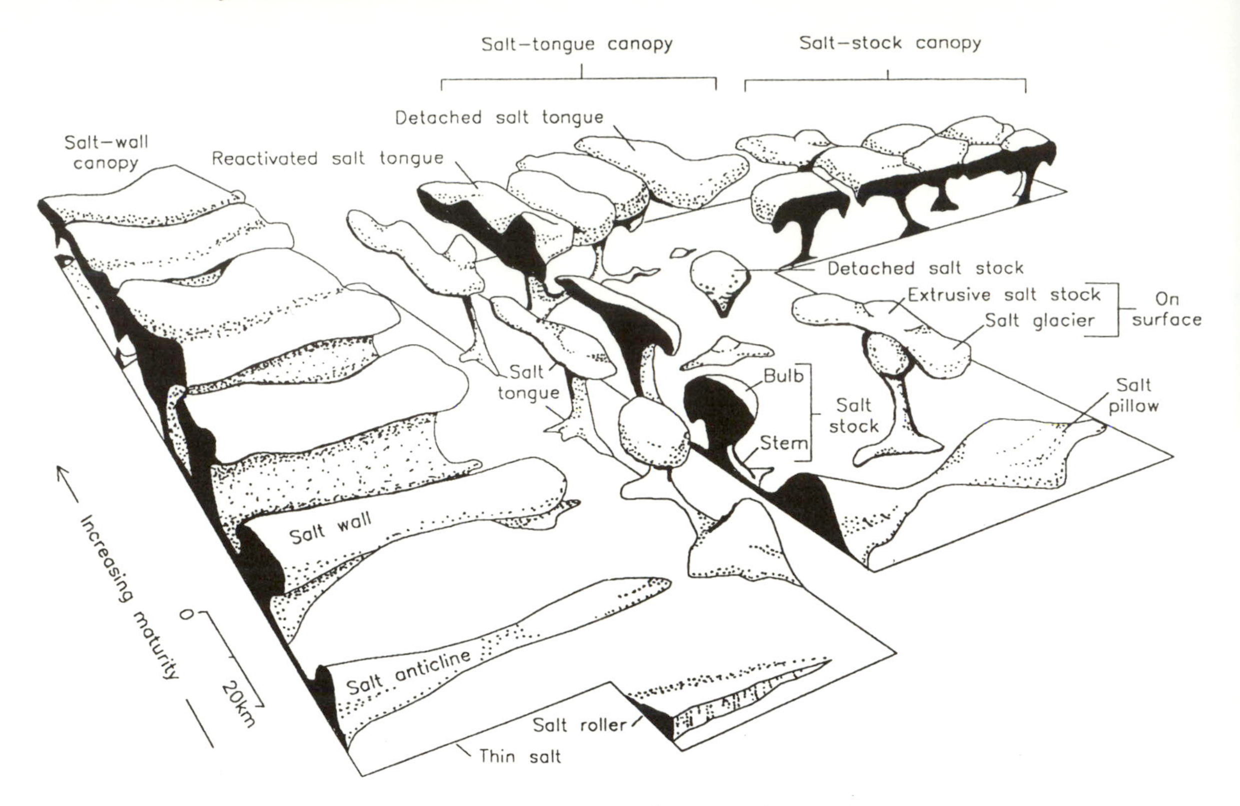


**Figure 12.5** Types of salt structure. (After Jackson and Talbot, 1994, figure 8.2.)

shapes. A cylindrical diapir is termed a **salt stock** or **salt plug**; where the top of the diapir begins to spread out, a **bulb** is formed, rising from a **stem**. Elongate diapiric structures are termed **salt walls**. Outwards spreading of a series of diapirs causes the bulbs or walls to coalesce to form a continuous sheet called a **canopy**. Sheet-like bodies of salt emplaced at a higher stratigraphic level than the source layer may themselves assume various forms, such as **salt tongues**, which are elongated bodies up to tens of kilometres in length, and may become completely detached from their roots. An exposed diapir will form extrusive sheets which flow over the surface in the form of **salt glaciers**.

## STRUCTURES ASSOCIATED WITH SALT DIAPIRISM

All salt diapirs show marked upwards bending of the surrounding layers against the walls of the body, often accompanied by reverse faulting. In certain cases tight compressional folding may occur in more ductile layers of the country rock. The strata above the diapir are affected by extensional tectonics, resulting in arching and thinning of the layers and in normal faulting. Intricate patterns of normal faults are commonly found above salt domes.

The margins of the diapir are typically marked by shear zones, which are both internal and external with respect to the salt body. Thus the essentially cylindrical body of a salt stock is surrounded by a ring-shaped shear zone, usually ductile within the salt but varying from brittle to ductile externally, depending on the rheology of the overburden (Figure 12.6). Within the salt body, the strain pattern is dictated by the direction and extent of flow. In a salt stock like that shown in Figure 12.6, the narrow 'neck' will exhibit very tight folding with a strongly developed vertical linear

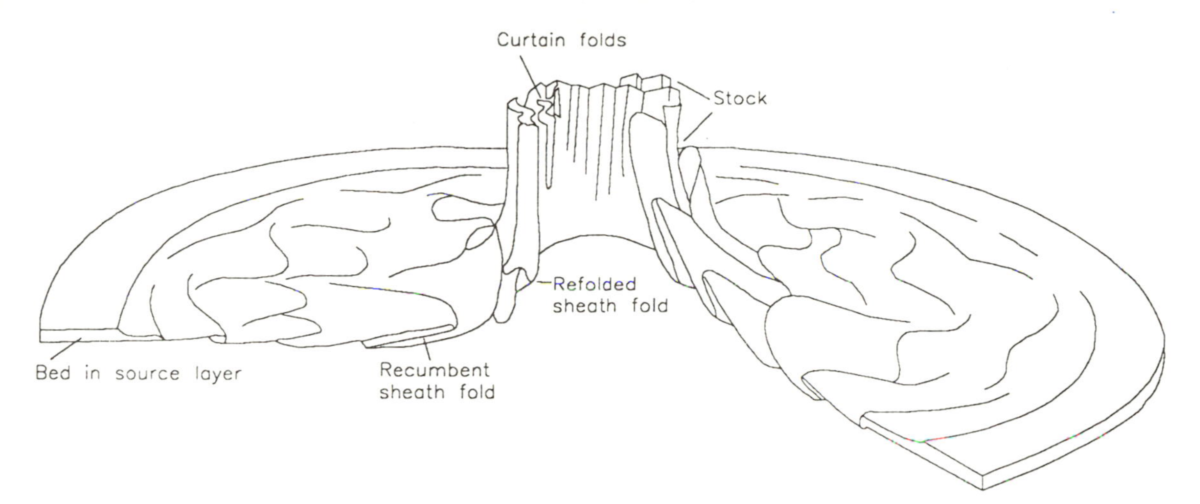


**Figure 12.6** Cutaway diagram illustrating the development of internal structures in a salt diapir. (After Jackson and Talbot, 1994, figure 8.5.)

elongation. Irregular flow of the salt from the source layer towards the growing diapir produces recumbent sheath folds which then rotate to form steeply plunging **curtain folds**.

REGIONAL EXTENSION IN SALT TECTONICS

Large-scale extensional structure promoted or aided by salt tectonics is common in continental slope regions such as the Gulf of Mexico. The salt provides a weak detachment layer in addition to initiating movement by breaking through the overburden. In such cases the extensional structures are entirely the result of gravitational spreading or gliding and are characterized by listric normal faults which detach on the salt layer. In regions of active crustal extension, such as the northwest German basin (Figure 12.7A), however, the base of the salt source layer is offset by normal faults which then initiate diapirism.

Extension, whether it be of gravitational or regional origin, is typically non-uniform, and causes local tectonic thinning, which in turn creates differential loading of the salt source layer (Figure 12.7B, C). This effect may create a set of parallel salt rollers, which form at the base of the footwalls of half-graben and are thus perpendicular to the extension direction. With increasing extension, the salt pierces the half-graben, forming a series of salt walls. Further movement of the salt may completely break through the overburden, creating a series of **rafts** which are then capable of lateral translation for many tens of kilometres, sliding down-slope on a detachment surface 'lubricated' by a thin layer of salt.

## 12.4 MANTLED GNEISS DOMES AND GRANITE DIAPIRISM

The crystalline regions in the cores of many orogenic belts, and in granitic or gneissose Precambrian shields, commonly exhibit dome-shaped areas of granitic material surrounded by a 'mantle' of metasedimentary or metavolcanic rocks. Structures of this type were first described from Finland and were given the name **mantled gneiss domes**. In many cases the basal mantle rock is a conglomerate, which indicates that the dome structure is a deformed unconformity between cover and basement. The material of the dome itself is often a complex mixture of rock types which, although predominantly granitic in composition, may include bands

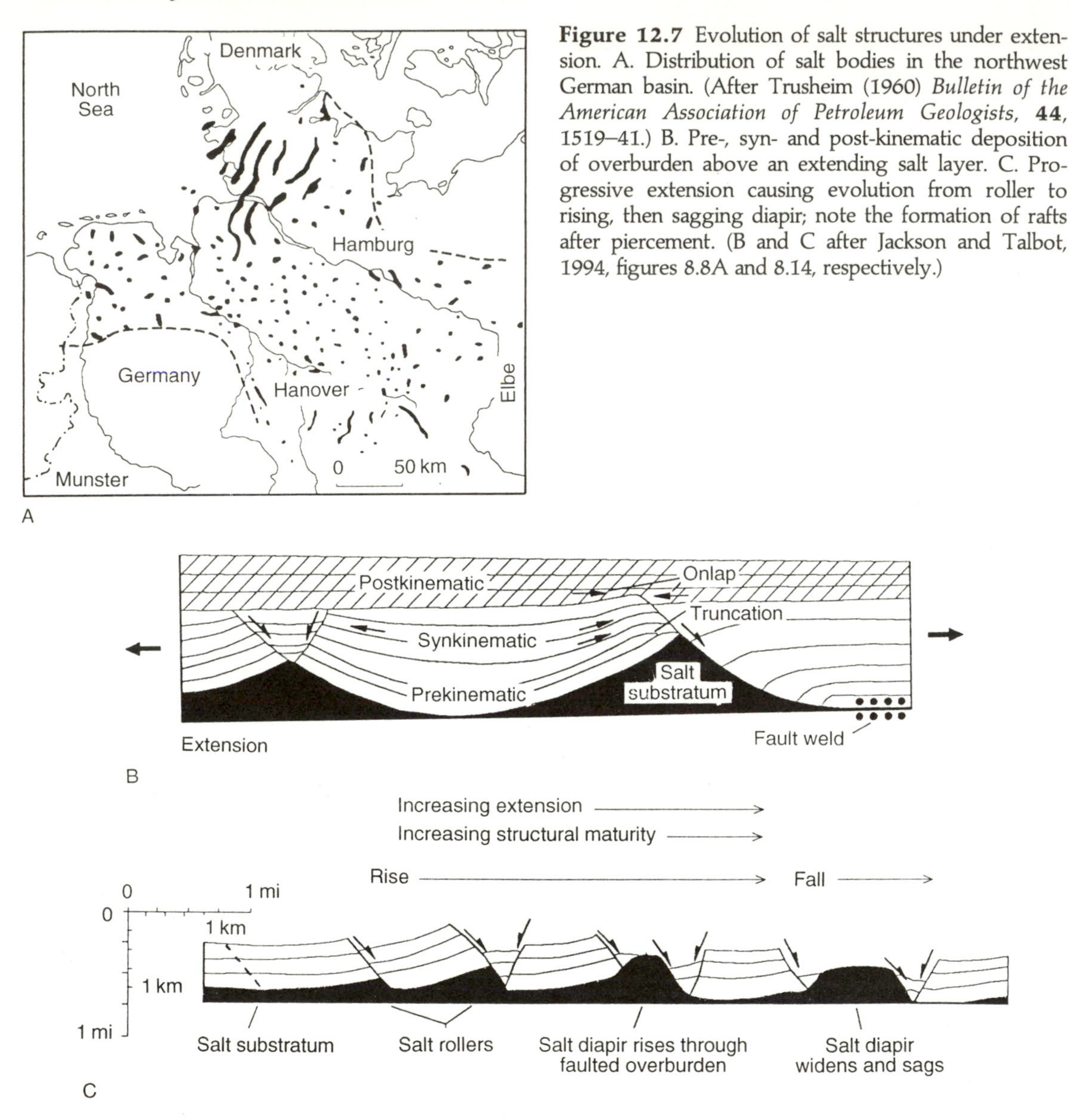


**Figure 12.7** Evolution of salt structures under extension. A. Distribution of salt bodies in the northwest German basin. (After Trusheim (1960) *Bulletin of the American Association of Petroleum Geologists*, **44**, 1519–41.) B. Pre-, syn- and post-kinematic deposition of overburden above an extending salt layer. C. Progressive extension causing evolution from roller to rising, then sagging diapir; note the formation of rafts after piercement. (B and C after Jackson and Talbot, 1994, figures 8.8A and 8.14, respectively.)

or small bodies of many other rock types. Some domes consist entirely of granite, which might suggest they are intrusive bodies. However, this is a likely explanation only for domes with discordant margins and fails to account for the mantling effect in many domes.

It is believed by some geologists that most gneiss domes originated in the same way as the salt domes described in the previous section, i.e. by diapiric flow of granitic material in the solid state, driven by the gravitational instability of a lower-density granitic layer overlain by a denser layer of, for example, basic volcanics. The mechanism has been tested experimentally using a centrifuge, and very similar structures have been produced (Figure 12.8). Other geologists have suggested that many domes may be the result of interference between two fold sets (Figure 3.18), the domes being produced at the culminations of crossing anticlines. Superimposed folding would tend to produce patterns of variable geometry, i.e. not equidimensional everywhere, and we should also expect there to be evidence of such large-scale re-folding within the layered supracrustal rocks.

Careful mapping of the strain patterns in certain Precambrian granite–greenstone terrains (see section 16.3) has revealed a number of features indicative of a diapiric process. For example, in the Bindura-Shamva greenstone belt and adjoining granitoid batholiths in Zimbabwe (Figure 12.9), a triangular area where the three granite domes meet is characterized by a triangular foliation pattern with vertical constrictional strains, contrasting with the flattening strains and subhorizontal stretching lineations around the margins of the domes in the regions of the inter-domal synclines. This type of pattern is consistent with the theoretical pattern produced by diapiric emplacement but is not compatible with a refolding origin.

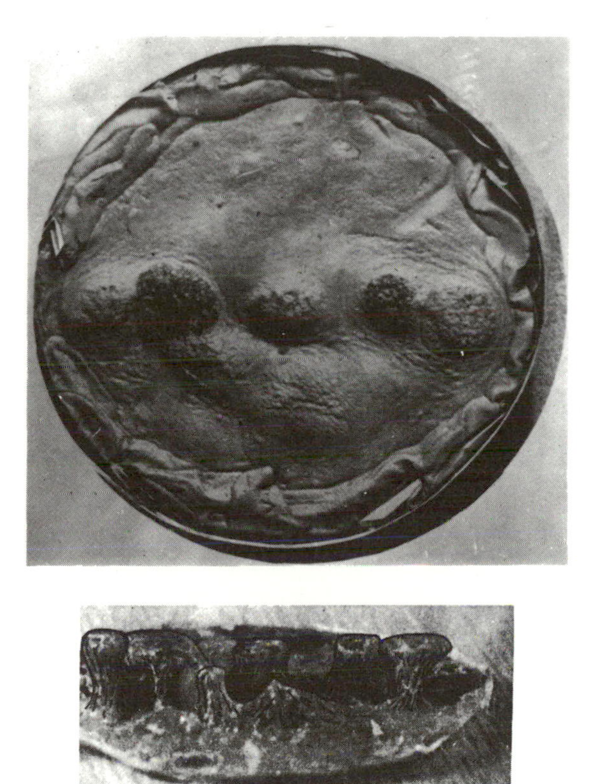

**Figure 12.8** Model of mantled gneiss dome structure produced in a centrifuge experiment. Domes of stitching wax in putty overburden are developed along the boundary of a deep-seated, less dense, source layer. Diameter of model 94 mm. The lower photograph shows a sideways view with the putty overburden removed. The domes are outlined in ink. This experiment simulates the effect of gravity in causing the diapiric uprise of less dense material into a denser overburden. (From Ramberg, H. (1962) *Bulletin of the Geological Institute of the University of Uppsala*, **42**, figures 27 and 28.)

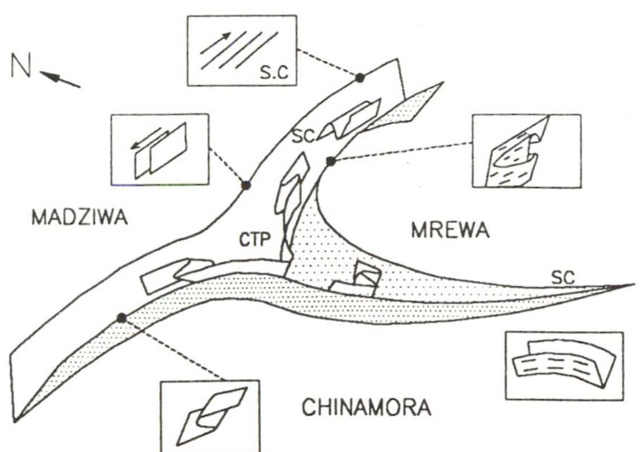


**Figure 12.9** Strain patterns and granite–greenstone relationships in the Bindura-Shamva greenstone belt of Zimbabwe. (After Jelsma, Van der Beek and Vinyu, 1993.)

It is thought that the structure of the Bindura-Shamva area supports a tectonic model in which deformation is caused by the remobilization of a buoyant granitic layer underlying the metasediments and metavolcanics of the greenstone belt; the granitic layer is assumed to have been originally intruded as a sheet-like or laccolithic body. It should be noted that some of the structural features exhibited particularly by the batholiths can equally well be explained by the ballooning mechanism described in section 11.6, and the relative importance of diapirism and ballooning in the emplacement of granite bodies is still unclear, as is the question of whether the granite flow responsible for the domes takes place in the solid state or is wholly or partly liquid.

## FURTHER READING

Graham, R.H. (1981). Gravity sliding in the Maritime Alps, in *Thrust and Nappe Tectonics* (eds K.R. McClay and N.J. Price), *Geological Society of London Special Publication*, **9**, pp. 335–52.

Jackson, M.P.A. and Talbot, C.J. (1994). Advances in salt tectonics, in *Continental Deformation* (ed. P.L. Hancock), Pergamon, Oxford, pp. 159–79.

Jelsma, H.A., Van der Beek, P.A. and Vinyu, M.L. (1993) Tectonic evolution of the Bindura-Shamva greenstone belt (northern Zimbabwe): progressive deformation around diapiric batholiths. *Journal of Structural Geology*, **15**, 163–76.

Ramberg, H. (1967) *Gravity, Deformation and the Earth's Crust*, Academic Press, London. [Describes experimental evidence for gravitational structures.]

PART TWO

# GEOTECTONICS

# MAJOR EARTH STRUCTURE 13

In the last four chapters of this book we discuss the significance of geological structures in the context of large-scale Earth processes. This branch of structural geology is often described as **geotectonics**. The development over the past three decades of plate tectonic theory gives us a way of integrating rock deformation into a model that attempts to explain the evolution of the crust as a whole. As a background to the plate tectonic hypothesis, which is dealt with in Chapter 14, we discuss major Earth structure and the distribution of tectonic activity. Chapter 15 relates deformation to the plate tectonic model, and, finally, Chapter 16 describes examples of this relationship in the geological past back to the Precambrian.

## 13.1 MAJOR TOPOGRAPHIC FEATURES OF THE EARTH

If we can imagine the Earth viewed from space with the water of the oceans removed, the major features of the structure of the crust would be readily apparent (Figure 13.1). A first-order division of the crust can be made into **continents** and **oceans**, and superimposed on these large features are rather linear elevations consisting of the **mountain ranges** on the continents and the great system of **ocean ridges**. Of a smaller order of magnitude are the deep **ocean trenches**. We shall now examine the nature and pattern of these large-scale structures, and their relationship to the tectonically active zones of the crust.

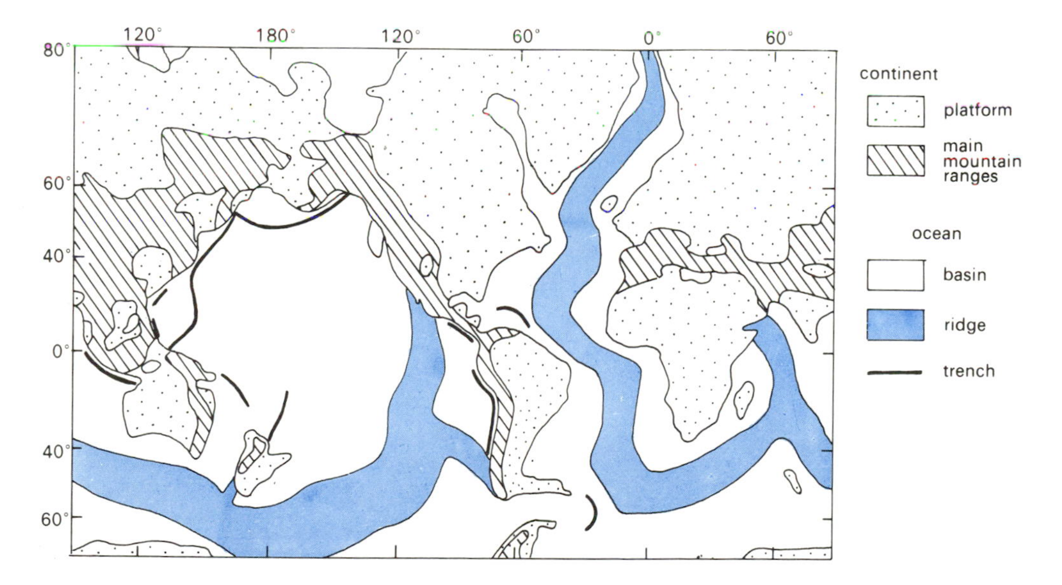


**Figure 13.1** The major topographic features of the Earth's surface as they would appear with the ocean waters removed. (After Wyllie, 1976, figure 3.7.)

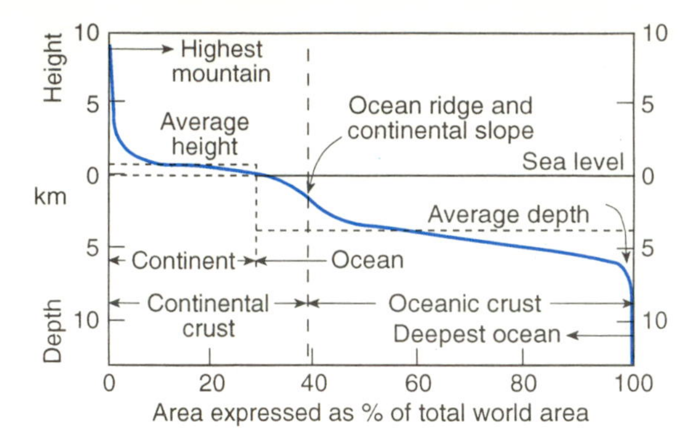


**Figure 13.2** Distribution of the topographic level on the Earth's surface expressed as a proportion of total surface area (see text). (After Wyllie, 1976, figure 3.11.)

## CONTINENTS AND OCEANS

The distribution of the topographic level of the Earth's surface is shown in Figure 13.2. It is clear that there are two dominant levels corresponding to the continents (average height about 1 km) and the ocean basins (average depth about 4 km) respectively, and that the proportion of the total area occupied by the extremes of height and depth (mountain ranges and ocean trenches) is very small. Elevations greater than 3 km make up only 1.6%, and depressions deeper than 5 km only 1%, of the total area.

The continents cover 29% of the Earth's surface and are distributed in a rather uneven way with 65% of the total land area in the northern hemisphere. The distribution of land and sea has varied considerably through geological time, and there has been a continuous change both of the relative positions of the continents and of the shoreline position. If we add the continental shelf and slope to the area of the continents, the total continental surface area is 40%, compared with 60% for the ocean basins, which more exactly reflects the relative proportions of continental to oceanic crust.

The existence of such a large difference in level is explained primarily by the difference in thickness between continental and oceanic crust. The fact that the Earth is in a state of general gravitational balance (**isostasy**) means that every sector of the Earth has approximately the same weight. This in turn implies that sectors with a higher surface elevation must contain a greater proportion of lower-density material to keep the total sector weight the same. The base of the crust, at the Mohorovičić discontinuity (Moho), marks a very significant change in composition and density (Figure 13.3). The mean density of crustal rocks is around 2.8 g/cm$^3$, whereas the peridotitic rocks of the uppermost mantle have a mean density of around 3.4 g/cm$^3$. The continents are situated on crust with an average thickness of 33 km and a mean composition close to that of granite, whereas the ocean basins are situated on crust with an average thickness of only 7 km and formed mainly of rocks with a gabbroic or basaltic composition (Figure 13.3). The ability of rocks to flow at depth means that material within the mantle can be transferred so as to maintain isostatic equilibrium and allow each part of the crust to sink or rise to the appropriate level. The difference of over 4 km in mean elevation between continents and oceans is therefore explained by the buoyancy of the thick continental crust.

## MOUNTAIN RANGES

The elevated and depressed features of the continents and oceans are generally linear and more or less continuous over long distances.

The high mountain ranges of the Earth form two main belts, one situated on the western side of North and South America, and the other (the 'Alpine–Himalayan belt') approximately at right angles to it, forming a sinuous belt from the Mediterranean through central Asia, and curving south-

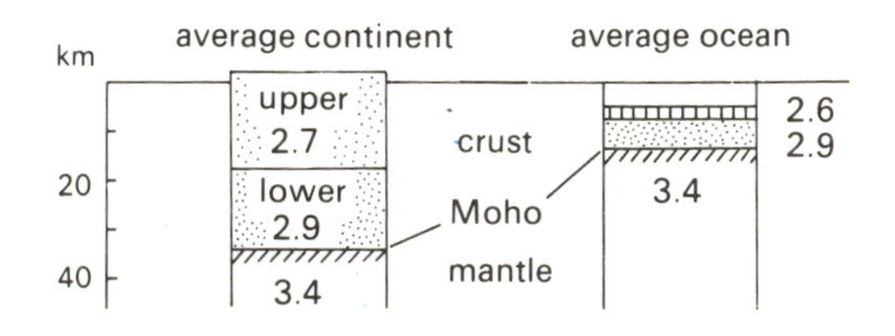


**Figure 13.3** Average cross-sections through continental and oceanic crust showing thicknesses and mean densities.

wards through Burma to Indonesia (Figure 13.1). Another type of mountain belt forms the series of island arcs which are found around the northern and western margins of the Pacific Ocean and in the northeastern Indian Ocean. The topographic relief of these chains is comparable to that of the continental ranges, but because they are partly submerged they appear less significant. These elevated features show extreme variation in dimensions, but, as a very rough approximation, we can regard their width as generally in the range 300–800 km, and their ratio of vertical to horizontal dimensions as in the range 1:100–1:200.

### OCEAN RIDGES

The great network of ocean ridges which figure so prominently on the Earth's surface (Figure 13.1) represents topographic relief of a rather greater order of magnitude volumetrically than the continental mountain ranges. They occupy about one-third of the surface area of the oceans and rise to between 2 km and 3 km from the ocean floor. Their width varies but is typically in the range 500–1000 km, and the ratio of their vertical to horizontal dimensions is thus around 1:500. These huge structures are isostatically compensated, the excess topographic relief being balanced by hotter, less dense mantle material beneath.

### OCEAN TRENCHES

The deep ocean trenches form a discontinuous system of arcuate features which either lie near the continental margins (as in the case of the South American trench) or border island arcs (as in the north and west Pacific). They differ in dimensions from both the other types of linear feature, being generally around 100–150 km in width and 2–3 km deep. The deepest trenches are over 11 km below sea level. They are thus very narrow, deep features with a depth/width ratio of around 1:50.

## 13.2 PRESENT-DAY TECTONIC ACTIVITY

The pattern of current tectonic activity is very closely related to the topographic pattern just described. This activity is of three types: (1) seismic movements (i.e. causing earthquakes), involving displacements of the crust with high strain rates; (2) aseismic crustal movements, with low strain rates; and (3) vulcanicity.

### SEISMICITY

The distribution of seismicity is shown in Figure 13.4. The vast majority of earthquakes, including all the severe ones, are concentrated in narrow belts which correspond with the linear topographic anomalies of Figure 13.1. More than 80% of the total earthquake energy is concentrated in the circum-Pacific belt alone. Thus the young mountain ranges, ocean ridges and ocean trenches, which represent extreme disturbances of the Earth's relief, are also the sites of severe tectonic activity. If we look at the distribution in terms of the depth of focus (point of origin) of the earthquakes, we find that the ocean ridges exhibit only shallow earthquakes, with focal depths down to 65 km, and that these are concentrated along a central rift zone or along faults which offset that zone. The deep earthquakes, with focal depths of over 300 km, are concentrated along the deep ocean trenches, especially around the Pacific.

A typical cross-section of a tectonically active segment of the northern or western Pacific shows that the earthquake foci lie on a plane outcropping in a trench and dipping below an adjacent island arc or continental margin. This inclined zone of earthquake activity is called a **Benioff zone** (Figure 13.5) and is a critical piece of evidence in favour of the process of subduction (see section 14.3).

Fault plane solutions (see section 9.3) generally indicate a compressive component acting across the Benioff zones, whereas ocean ridge earthquakes show tensional solutions. The great oceanic faults which offset the axes of the ridges (Figure 14.15) are generally strike-slip.

### ASEISMIC MOVEMENTS

Slow movements of the crust of the order of millimetres per year are detectable by precise

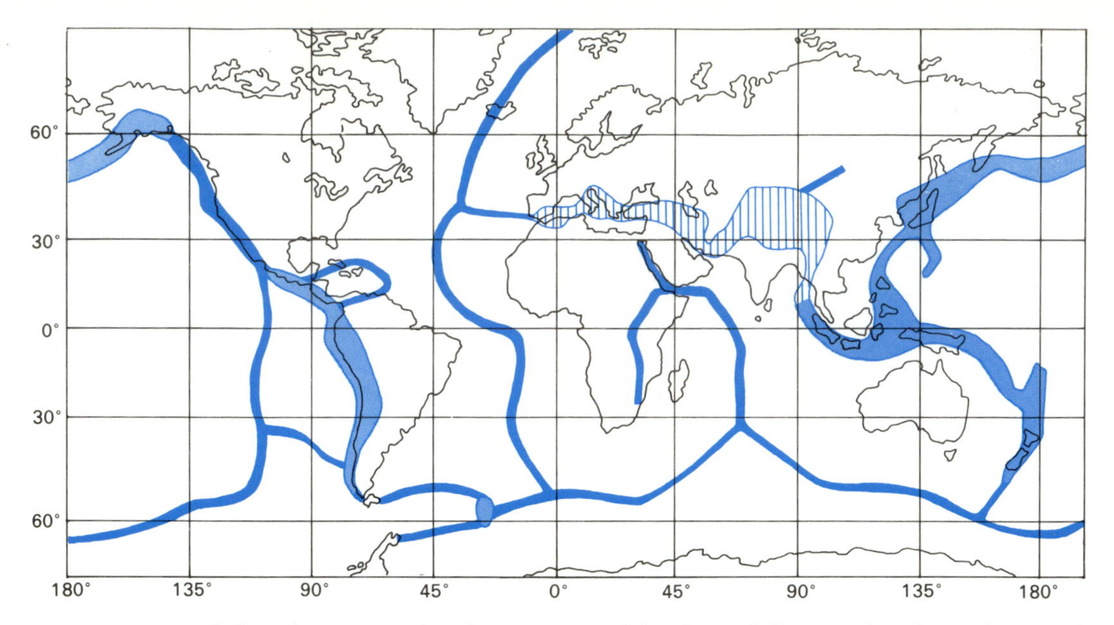


**Figure 13.4** Main belts of recent earthquake activity. Solid colour, shallow earthquakes only (> 65 km); vertical ruling, mainly shallow to intermediate earthquakes (65–300 km); stipple, shallow to deep earthquakes (> 300 km). (Based on Chadwick, P. (1962) in *Continental Drift* (ed. S.K. Runcorn), Academic Press, New York, figure 4, and Toksoz, M.N. (1975) *Scientific American*, November.)

measurements of height over a period of years. These rates of movement are comparable with, or even faster than, the rates determined from the geological evidence of uplift and sedimentation. All parts of the crust are subject to these aseismic movements, many of which are vertical, and are responsible for the creation of geological structures such as folds and basins, the uplift of mountain ranges and the lateral movement of the continents.

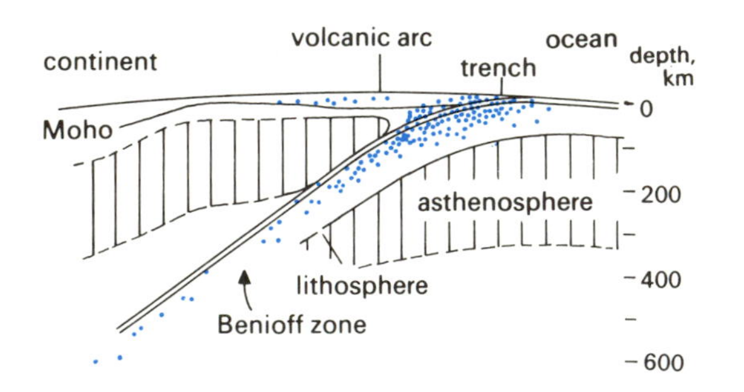


**Figure 13.5** Section across the Japanese island arc showing the concentration of seismic activity along the Benioff zone. Earthquake foci are shown as coloured dots. (After Uyeda, S. (1971 *The New View of the Earth: Moving Continents and Oceans*, Freeman, San Francisco, figure 5.18.)

## VULCANICITY

The distribution of present-day vulcanicity (Figure 13.6) bears a striking resemblance to the distribution of earthquakes (Figure 13.4). Clearly there is a very close relationship between vulcanicity and zones of tectonic instability. About 75% of the currently or historically active volcanoes are situated in the circum-Pacific belt, particularly along the volcanic island arcs. Many volcanoes are associated with the ocean ridge network but some are found along faults or lineaments within the ocean basins. The Alpine–Himalayan belt exhibits a rather sparse distribution of volcanoes, with the exception of the Mediterranean and Indonesian regions, where there is adjoining oceanic crust with sections of trench. The greatest concentration of vulcanicity within the continents is found in the African rift

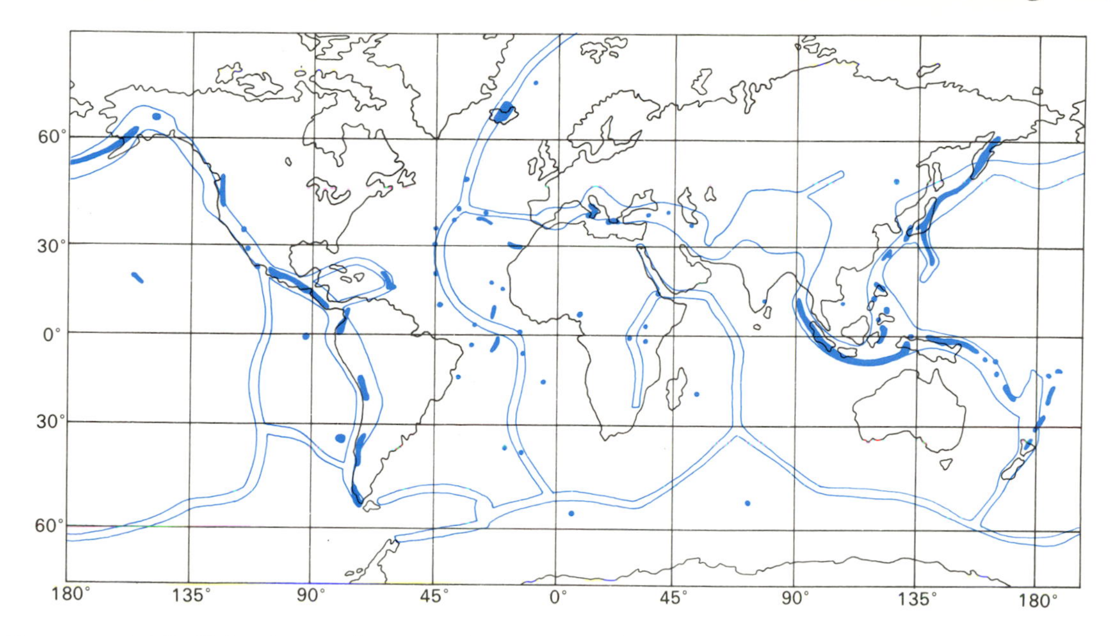


**Figure 13.6** World distribution of active volcanoes. Currently active volcanic areas are shown in solid colour. The seismic zones of figure 13.5 are shown for comparison.

system, which is a continental extension of the Indian Ocean ridge network.

### 13.3 STABLE AND UNSTABLE TECTONIC ZONES

The clear subdivision of the Earth into tectonically stable and tectonically active or unstable zones is one that characterizes most of the geological record. We have no direct information on the nature of the oceans before about 200 Ma ago, but the continental areas for any particular period, extending back to about the mid-Precambrian (c. 2500 Ma ago) show a marked contrast between stable regions, termed **cratons,** and unstable zones, termed **mobile belts**.

The cratons consist of undeformed flat-lying sedimentary cover on an older crystalline basement and exhibit only minor vulcanicity. Tectonic effects are confined to very slow vertical tectonic movements of the order of millimetres per year or less, resulting in broad sedimentary basins separated by uplifts (see section 15.6). **Orogenic belts** are a type of mobile belt characterized by highly deformed sediments, abundant and varied igneous rocks, and uplifted segments of deep crustal material. The tectonic processes that take place in these belts result from compression, and give rise to the formation of mountain belts. This is a result of crustal thickening, partly compressional and partly volcanic in origin, which leads inevitably to uplift due to isostatic forces. Another type of mobile zone is characterized by extension and includes rifts such as the African rift system (Figures 13.4, 13.6) and certain basins. Extensional rift systems are also typically associated with vulcanicity.

The continental crust of the present-day cratons is composed of the worn-down remnants of previous orogenic belts of different ages. The proportion of orogenic belts to cratons appears to increase generally as we go further back in time, until in the earliest Precambrian no evidence of cratons can be found, and there may have been a general state of mobility.

### FURTHER READING

See list at the end of Chapter 14.

# PLATE TECTONICS 14

## 14.1 HISTORICAL CONTEXT

In 1915 Alfred Wegener popularized the idea of continental drift in his book *The Origin of Continents and Oceans*. The relative movement of continents provided a convenient explanation for many otherwise puzzling geological phenomena – the geometric fit of the opposing coastlines of the Atlantic, peculiarities in faunal distribution and palaeoclimatological reconstructions, and awkward geological coincidences – that immediately made sense when continents were joined together in their presumed original positions to form a supercontinent, named **Pangaea**.

Many geologists found difficulty in accepting the proposed movements in the absence of a satisfactory mechanism, but most of the doubters were convinced by the results of palaeomagnetic work in the period 1950–60. Palaeomagnetic reconstructions for the Triassic period showed that the continents, when reassembled in the positions advocated by the believers in continental drift, showed a common Triassic magnetic pole (Figure 14.1A). The positions of the Triassic poles for the various continents at present are widely scattered.

An even more convincing test of the reassembly is provided by the **apparent polar wander** curves. These tracks of successive positions of the magnetic north pole are shown for the continents of Europe and North America from the Cambrian to the present in Figure 14.1B. The tracks are quite separate in the present positions of the continents. However, when the continents are reassembled, the tracks for the Carboniferous to Jurassic period, when these continents were thought to be joined together, are superimposed, though they diverge for the period after the Jurassic, when opening of the North Atlantic commenced.

The acceptance by most geologists of the idea of continental drift led to attempts to discover a mechanism, and in particular to attempts to understand the role of the oceanic areas in this process. The hypothesis of sea-floor spreading, put forward in 1962 by H.H. Hess and R.S. Dietz, linked continental and oceanic areas in the same general process, likened to a conveyor belt, in which slabs of crust were transported laterally carrying the continents with them. The ocean ridges were envisaged as the sites of creation of new oceanic material and the trenches as the sites of destruction (Figures 13.5 and 14.7).

Thus the two hypotheses of continental drift and sea-floor spreading became combined to form the nucleus of the new hypothesis of **plate tectonics**, which has evolved into a general theory to explain crustal movements and evolution. The critical evidence that led to the acceptance of the sea-floor spreading idea arose from the application of magnetic stratigraphy to the ocean floor, from which the direction and amount of movement of the ocean floor could be established.

Measurement of the magnetic field over the oceans had shown that the ocean floor exhibited a striped magnetic pattern with abrupt changes in magnetic intensity from one stripe to the next (Figure 14.2A). These stripes, which are mostly around 20–30 km in width, are parallel to the ocean ridges and, like them, are displaced by the oceanic fracture zones. A breakthrough came in 1963 with the proposal that these abrupt changes in magnetization could be correlated with reversals of polarity in the Earth's magnetic field. Since these reversals could be dated by comparison with continental rock sequences (Figure 14.2B), the magnetic stripe pattern effectively became a magnetic stratigraphy of the ocean floor, from

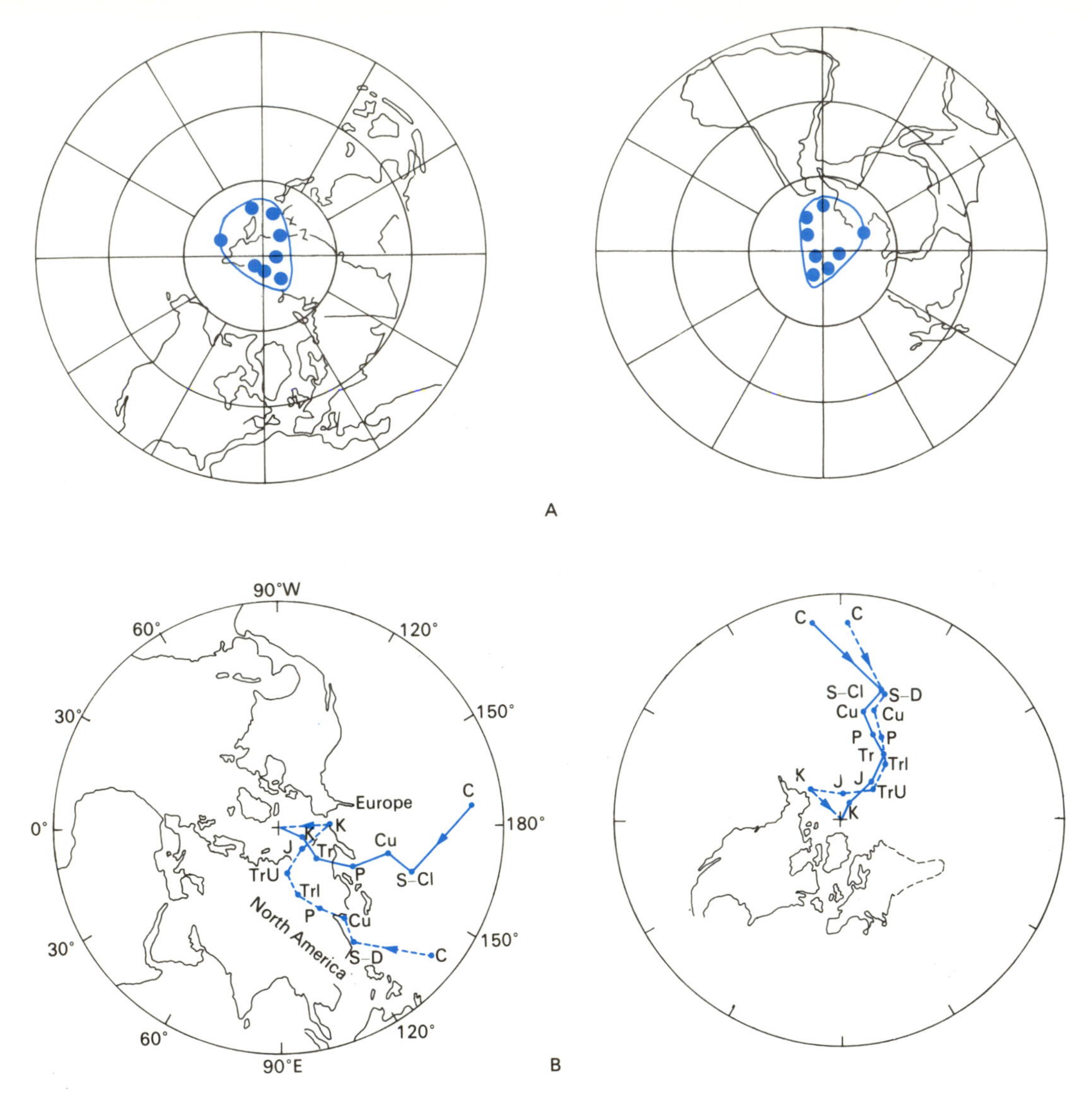


**Figure 14.1** Critical palaeomagnetic evidence for continental drift. A. North and south polar projections of the continents reassembled in their pre-drift positions, showing the positions of the Triassic magnetic pole for the various continents (coloured dots). B. The left-hand diagram shows the apparent polar wander curves for Europe and North America in their present relative positions. The right-hand diagram shows the superimposition of the curves between Silurian and Upper Triassic times with the continents in their pre-drift positions. C, Cambrian; S, Silurian; D, Devonian. Cl, Lower Carboniferous; Cu, Upper Carboniferous; P, Permian; Tr, Triassic; Trl, Lower Triassic; Tru, Upper Triassic; J, Jurassic; K, Cretaceous. (After McElhinny, N.W. (1973) *Palaeomagnetism and Plate Tectonics*, Cambridge University Press, Cambridge.)

which it is possible to establish the age of a large proportion of the oceanic crust (Figure 14.8) and thus to prove that the ocean spreading hypothesis was correct.

## 14.2 THE CONCEPT OF LITHOSPHERIC PLATES

The idea of plates arose from the observation that large areas of the crust have suffered very little

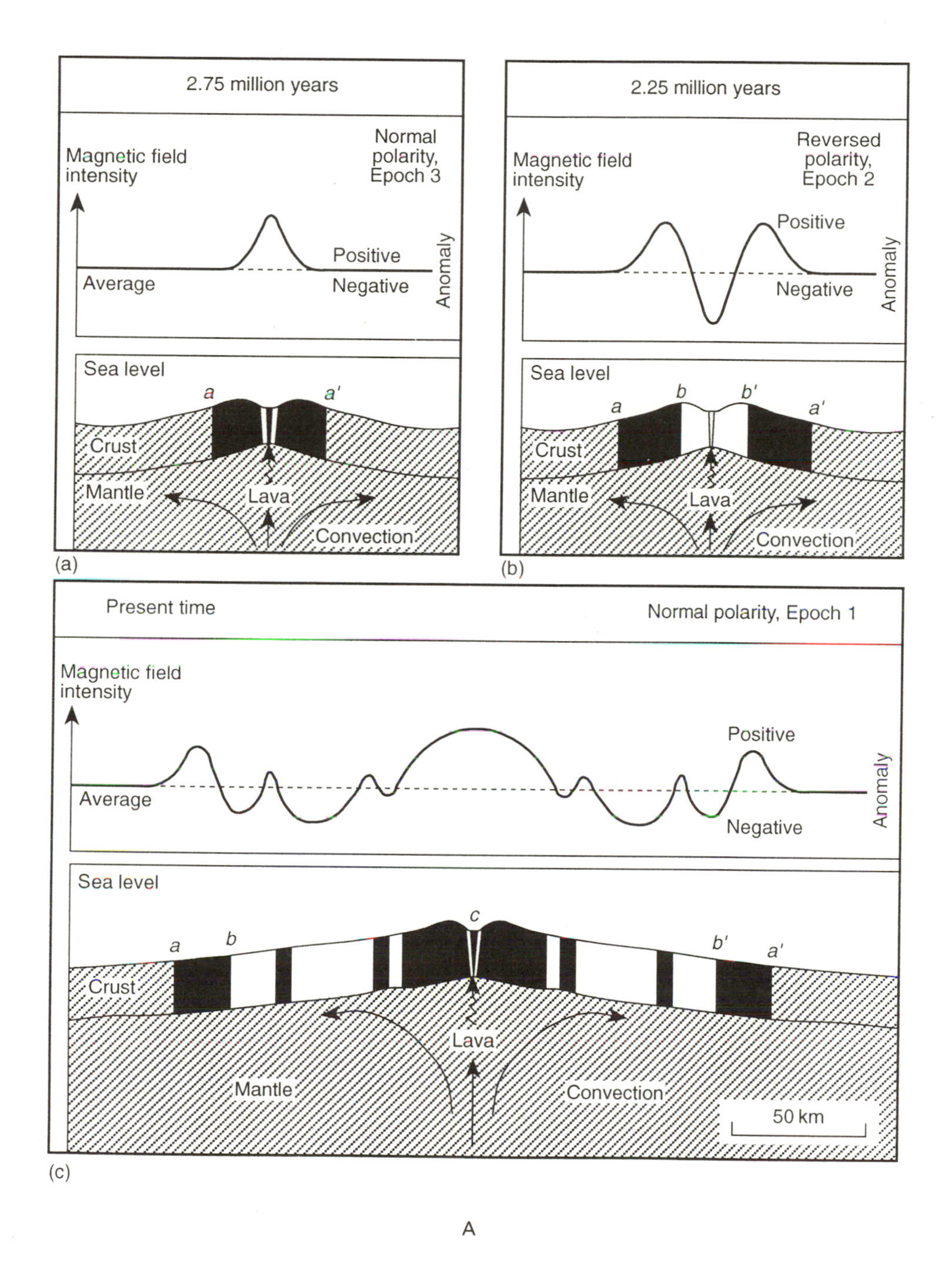


**Figure 14.2** Magnetic stratigraphy of the ocean floors. A. The magnetic 'tape recorder'. Diagrammatic representation of the process whereby sea-floor spreading and magnetic polarity reversals produce a series of differently magnetized stripes parallel to the ocean ridge crest. (a) at 2.75 Ma, (b) at 2.25 Ma, (c) at present. (After Wyllie, 1976, figure 10.5, with permission.)

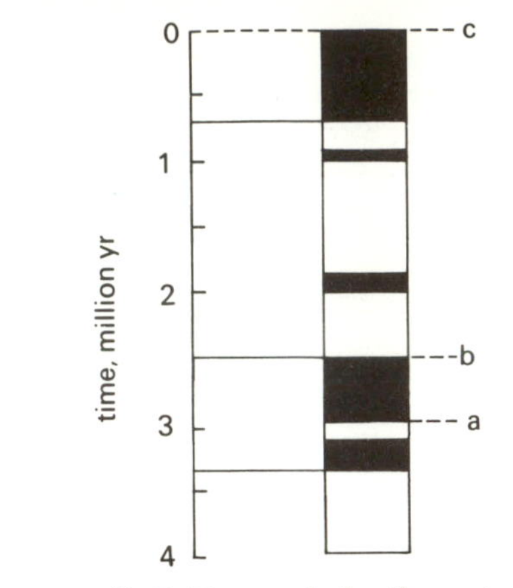


**Figure 14.2** (*contd.*) B. Time-scale for the past 4 Ma. (After Wyllie, 1976, figure 10.5, with permission.)

distortion although they have travelled laterally several thousand kilometres, if we accept the evidence for continental drift. The detailed and accurate 'jigsaw fit' of the opposing coastlines of America and Africa (Figure 14.3), after 200 Ma and 4000 km of drift, testifies to this lack of distortion. In the oceans also we find regular linear magnetic stripes and faults which have maintained their shape after tens of millions of years. This evidence reinforces the conclusions reached by studying the distribution of tectonic movements (see section 13.3) that there are large stable areas (cf. the continental cratons) that suffer little internal deformation and exhibit only slow vertical movements, while moving laterally as a coherent unit at rates 10–100 times faster.

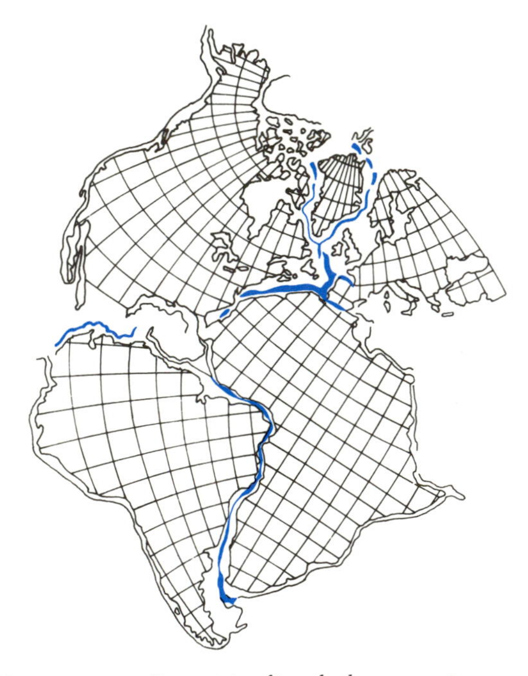

**Figure 14.3** Geometric fit of the opposing continental margins of the Atlantic. Solid colour represents areas of misfit. Matched at 1000 m below sea level. (After Bullard, E.C., Everett, J.E. and Smith, A.G. (1965) *Philosophical Transactions of the Royal Society of London, A*, **258**, 41–51.)

SEISMICITY AND PLATE BOUNDARIES

The obvious link between seismicity and present-day tectonic activity suggests that the seismic zones must represent the boundaries of these stable blocks of crust and that each block or plate can be delimited by a continuous belt of seismic activity. Taking the argument one step further, since the seismic activity represents fault movements with high strain rates, each plate must be in a state of relative motion with respect to each of its neighbours.

If we now examine the nature of these **plate boundaries** and their sense of movement, we can recognize three types (Figure 14.4).

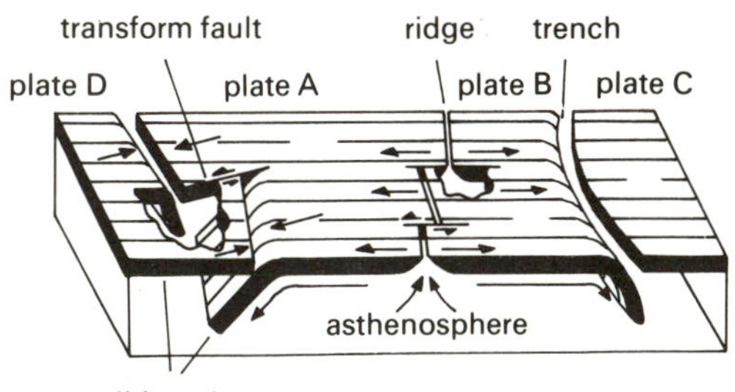

**Figure 14.4** Block diagram illustrating the plate tectonic model. (After Isacks, B., Oliver, J. and Sykes, L.R. (1968) *Journal of Geophysical Research*, **73**, 5855–99.)

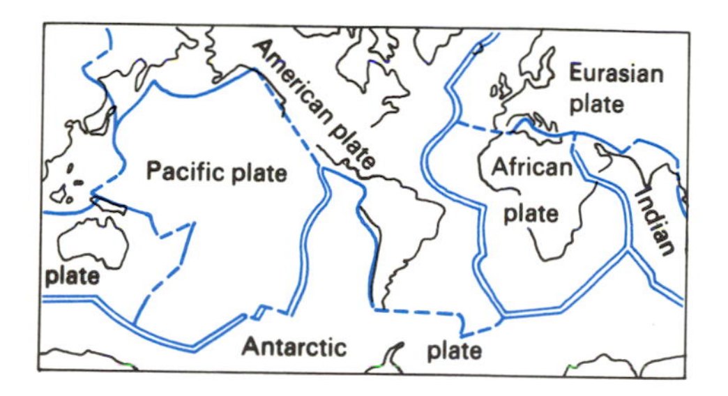


**Figure 14.5** The six major lithospheric plates. Ridges, double lines; trenches, single lines; transform faults, broken lines. (After Le Pichon, X. (1968) *Journal of Geophysical Research*, **73**, 3661–97.)

1. **Divergent boundaries**, where adjoining plates are moving apart (ocean ridges).
2. **Convergent boundaries**, where adjoining plates are moving together (ocean trenches and young mountain belts).
3. **Strike-slip boundaries**, where adjoining plates are moving laterally past each other with a horizontal strike-slip sense of displacement along steep faults.

The sense of displacement can be deduced from first-motion studies of individual earthquakes and in general confirms the relative movements inferred from other evidence such as palaeomagnetism and magnetic stratigraphy.

Following these principles, we can use the network of boundaries to divide the Earth's present surface into six major plates (Figure 14.5) the Eurasian, American, African, Indo–Australian, Antarctic and Pacific plates. There are also a number of smaller plates, associated especially with destructive boundaries around the margins of the Pacific Ocean. Note that plate boundaries may or may not correspond to continental margins. Continental margins that lie within plates, such as the Atlantic margins of Africa and America, are termed **passive margins**. Those that do correspond to plate boundaries (e.g. the western margin of the American continents) are termed **active margins**.

## LITHOSPHERE AND ASTHENOSPHERE

The vertical extent of a plate is much more difficult to determine than its horizontal extent. Alfred Wegener's original idea of pieces of continental crust moving across a plastic ocean was abandoned many years ago when it was realized that oceanic rocks could not behave in a sufficiently ductile manner near the surface. The possibility that the plates consist of pieces of crust sliding over the mantle must also be discarded. The oceanic crust is only about 7 km thick and could not remain undistorted or transmit horizontal stresses. Moreover, there is no evidence of a major change in physical properties that would suggest the existence of a zone of subhorizontal displacement at the base of the crust.

The evidence from earthquake waves suggests that the physical properties of the crust and uppermost mantle change gradually down to around 100–150 km in depth, where there is a more abrupt change in the seismic velocity profile. The rate of increase in seismic velocity drops through a zone about 100 km deep before rising again at greater depths. This layer of abnormally low seismic velocities is called the **low-velocity zone (LVZ)** and is thought to signify a decrease in density and in viscosity. We can therefore regard this zone as a more ductile layer where lateral flow of material could take place, or as a zone of ductile shear between plate movements above and the main part of the mantle beneath.

Isostasy theory demands a weak layer where lateral flow can take place, and this layer is called the **asthenosphere**. It has become convenient to identify the asthenosphere with the LVZ. The stronger layer above the asthenosphere is termed the **lithosphere** (Figure 14.6). The lithosphere thus includes the crust and the uppermost part of

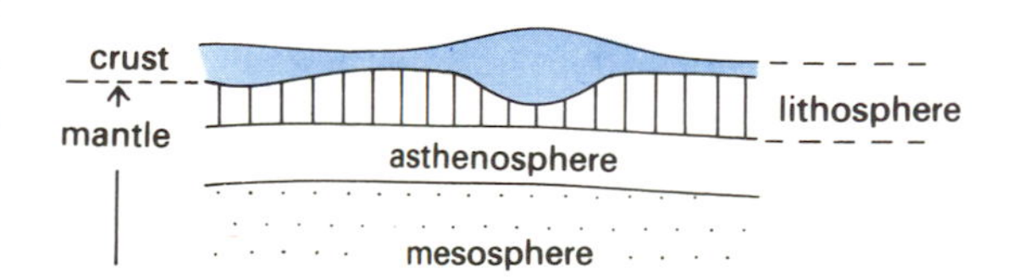


**Figure 14.6** Lithosphere and asthenosphere. The lithosphere includes the crust and the uppermost part of the upper mantle.

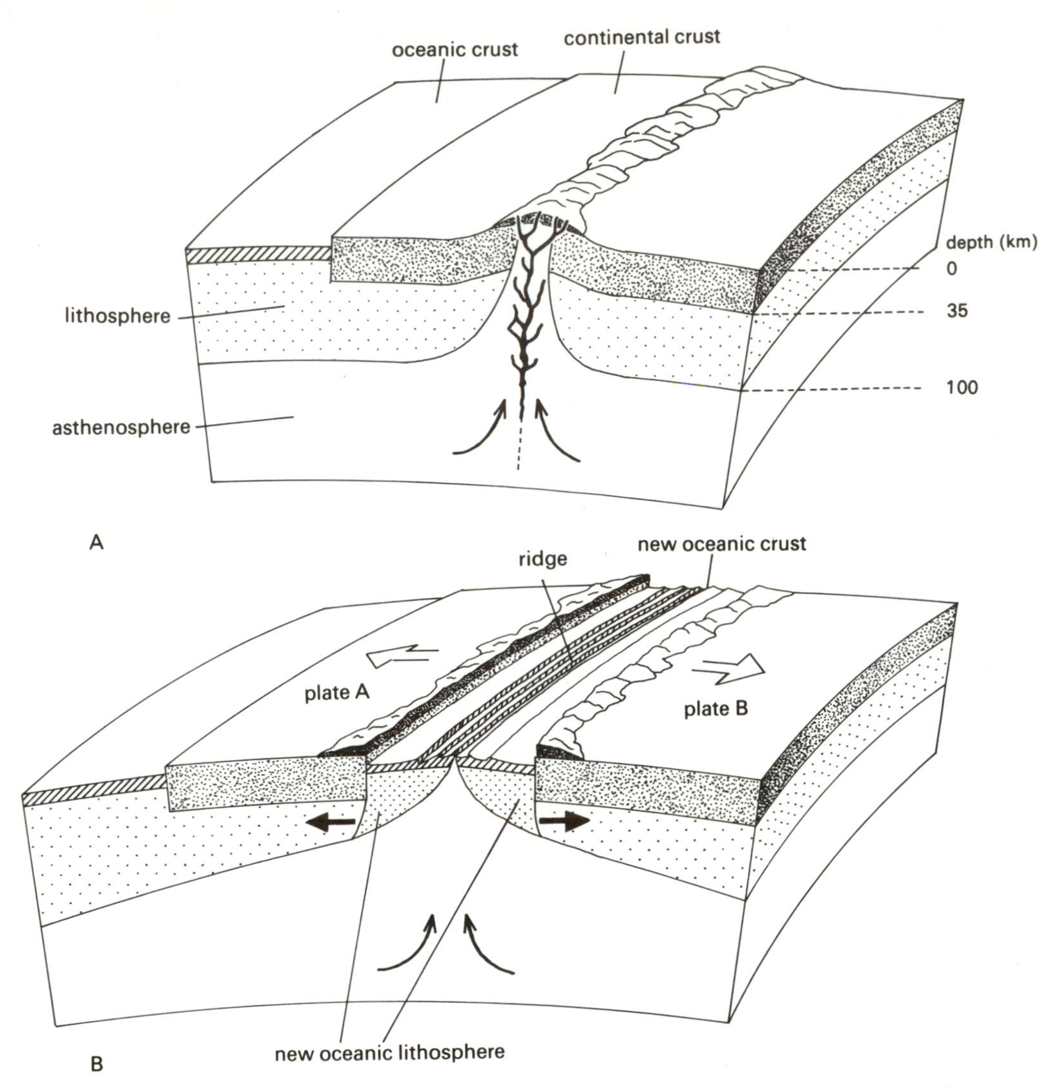


**Figure 14.7** Nature of constructive plate boundaries. A. Initiation of a constructive boundary as a continental swell as a result of the uprise of hot asthenosphere. B. Formation of new oceanic lithosphere along an ocean ridge between the two diverging plates. (After Dietz, R.S. and Holden, J.C. (1970) *Scientific American*, October.)

the upper mantle down to a variable depth of 80–120 km in the oceans and around 150 km or possibly deeper in the continents. Since the base of the lithosphere depends on a relatively gradual change in viscosity, which is strongly temperature-dependent, the base is not only gradational but varies both spatially and temporally in response to changes in temperature gradient. If we take as an example the oceanic lithosphere, this will be hottest near the site of formation at the ocean ridge crest and will cool gradually with time and distance from the ridge as it travels laterally away from it. This is reflected in the thickness of the lithosphere, which ranges from less than 50 km at the ridge crest to 120 km near the ocean margins (Figure 14.7).

## 14.3 NATURE OF PLATE BOUNDARIES

### CONSTRUCTIVE BOUNDARIES

Divergent plate boundaries along ocean ridges where adjoining plates are moving apart are called **constructive boundaries**, because new material is being added to them. The process is illustrated in Figure 14.7. Some of the new material is provided by upper mantle melts formed in the hot, low density region below the ridge. Part of this molten material is injected into the crust as basalt dykes or gabbro intrusions, and part is extruded on the ocean floor as basalt pillow lavas. At the same time, the mantle part of the lithosphere grows by the addition of ultrabasic intrusions and by ductile flow of material from the asthenosphere. The focus of activity at any given time is the central seismically active rift zone marked by dyke intrusions, extensional faulting and vulcanicity. This zone is only about 100 km wide, and the remainder of the ridge consists of warm lithosphere which gradually subsides to the level of the ocean basins as it cools and moves away from the central rift.

The **continental rift zones** represent incipient constructive boundaries. The African rift zone, the Red Sea rift and the Gulf of Aden rift meet in a triple junction in the Afar region of Ethiopia (Figure 15.1) which serves as a useful analogy for the way in which the major continents separated during the Triassic to Jurassic period. This example will be discussed in more detail in section 15.2. The characteristic features of the oceanic constructive boundaries are also found in the continental rifts but the nature of the vulcanicity is much more varied.

### DESTRUCTIVE BOUNDARIES

There are two types of convergent boundary at the present time. The first follows the deep ocean trenches and the second follows the belt of young mountain ranges of the Alpine–Himalayan chain (Figure 14.5). The evidence for the nature of the convergent movement along the trenches comes partly from earthquake first-motion study, which shows generally compressional solutions across the trench, and partly from the shape of the Benioff zone, which suggests that the oceanic plate dips down below the adjoining plate (Figures 13.5 and 14.9). This process is considered to be responsible for the convergent motion, and is known as **subduction**. Since subduction involves the destruction of plates, by returning old lithospheric material to the mantle, subduction zones are termed **destructive plate boundaries.**

Geological evidence for the subduction process is provided by examining the magnetic stratigraphy of the ocean floor adjacent to a destructive

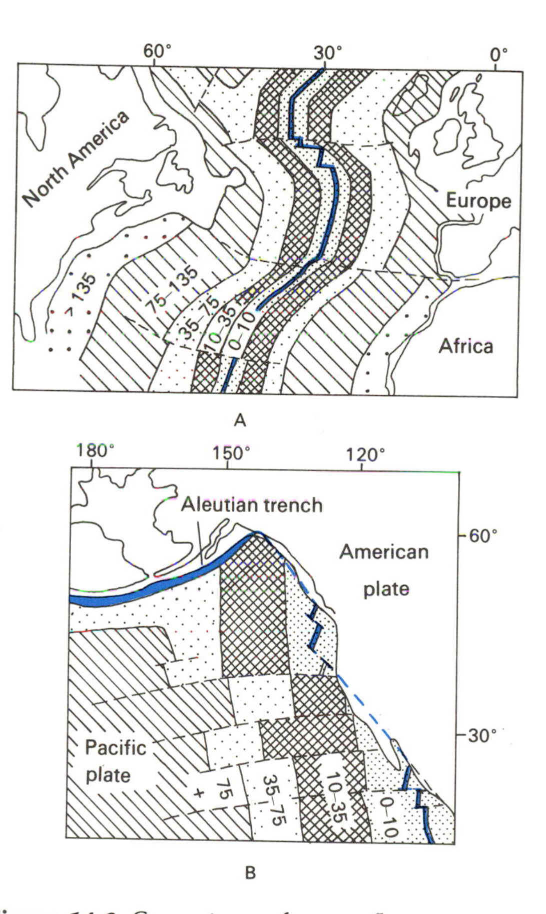


**Figure 14.8** Comparison of ocean-floor magnetic stratigraphy at constructive and destructive boundaries. A. Magnetic age pattern of the central Atlantic. Note concordance with continental margins. B. Magnetic age pattern of the northeast Pacific. Note discordance with the continental margin. Age in Ma. (After Larson, R.L. and Pitman, W.C. (1972) *Bulletin of the Geological Society of America*, **83**, 3645–61.)

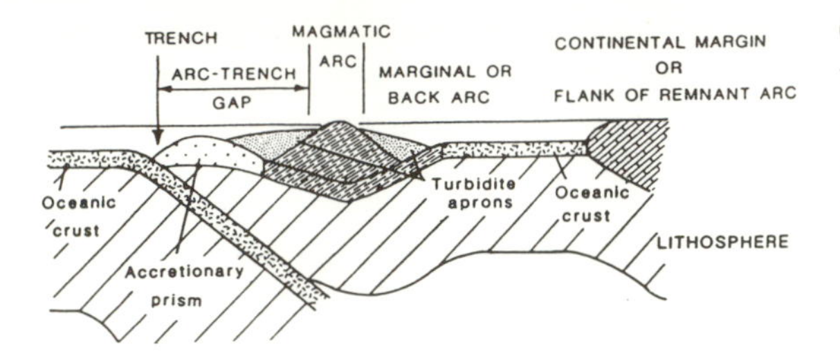


**Figure 14.9** Diagrammatic profile across an island arc/subduction zone, showing the main features. (From Windley, B.F. (1977) *The Evolving Continents*, Wiley, Chichester, figure 16.4.)

boundary and comparing it with the pattern at a constructive boundary (Figure 14.8). The magnetic stripes in the Atlantic (Figure 14.8A) are concordant with the coastlines; the oldest stripes adjoin the continents and reflect the time of separation. In the northern Pacific (Figure 14.8B), in contrast, the stripes are discordant with the Aleutian trench, and stripes of various ages occur along this plate boundary, demonstrating that their continuations have been subducted below the trench.

Certain subduction zones border continents, on the west side of South America for example. In this case there is a linear volcanic belt situated on the continent about 300 km from the Peru–Chile trench (Figures 13.1 and 13.6). Most subduction zones at the present time, however, are situated at island arcs within the oceans (Figures 14.9 and 14.15), so a section of oceanic crust intervenes between the subduction zone and the nearest continent.

A typical island arc (Figure 14.9) consists of a partially submerged volcanic mountain range 50–100 km wide (the magmatic arc) with a trench on its convex side between 50 and 250 km from the island arc. Between the arc and the trench is the **arc–trench gap** or **forearc**, which is a zone of sedimentary accumulation. A wedge of clastic material derived from the volcanic arc merges oceanwards with a zone termed the **accretionary prism**, where highly deformed arc-derived clastic material is intercalated with slices of oceanic material scraped off the descending slab. The arc–trench zone is not in isostatic balance. There is a mass deficiency along the trench and a smaller mass excess associated with the volcanic arc. This gravitational instability must be related to the subduction process, and the mass imbalance is thought to be supported by the lateral compressive stress associated with the convergent plates.

Certain island arcs are formed from pieces of continental crust that have perhaps become separated from a nearby continent. Others are built up by the addition of new volcanic material to oceanic

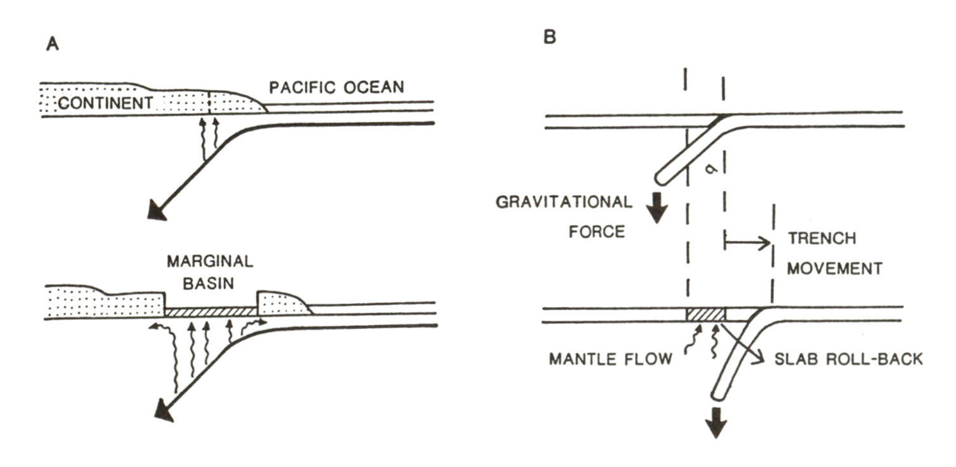


**Figure 14.10** Possible mechanisms of formation of a marginal back-arc basin. A. Secondary spreading due to heating of the upper mantle above a subducting slab. (After Uyeda, S. (1978) *The New View of the Earth: Moving Continents and Oceans*, Freeman, San Francisco, figure 5.22.) B. The trench (or slab) roll-back model. See text.

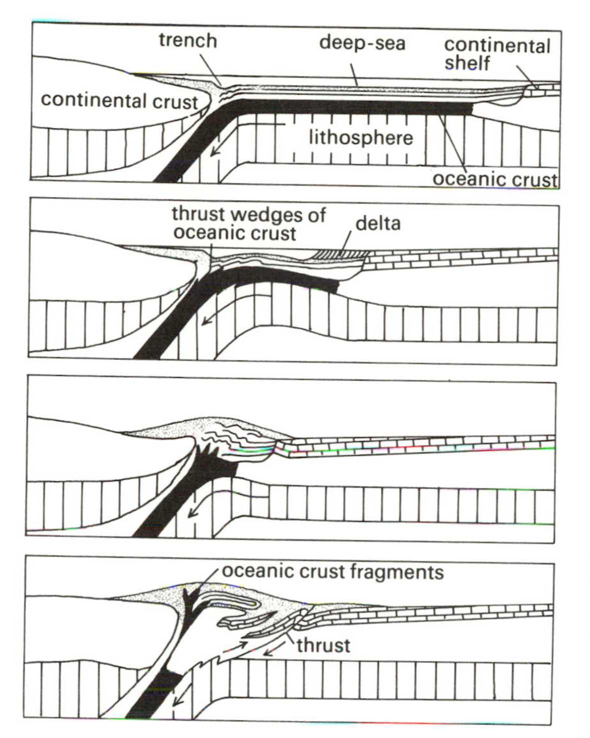


**Figure 14.11** Sequence of stages in the transformation of a subduction zone to a continental collision zone by the approach of two continents and the elimination of the intervening ocean. (After Dewey, J.F. and Bird, J.M. (1970) Mountain belts and the new global tectonics. *Journal of Geophysical Research*, **75**, 2625–47, figure 13.)

crust. The areas of oceanic crust between the island arc and the nearest continent form what are known as **back-arc basins**. These structures have been attributed to secondary ocean-floor spreading behind the island arc. Figure 14.10 illustrates two suggested mechanisms for producing the secondary spreading – the mantle diapir model and the trench (or slab) rollback model.

The other type of convergent plate boundary is often referred to as a **continental collision zone**. In the case of the Alpine–Himalayan belt, the plate boundary represents the collision of the Eurasian continent to the north with the African and Indian continents to the south (Figure 14.5). The collision can be deduced from the record of relative movements of these continents, which is well documented by palaeomagnetic data and ocean-floor stratigraphy. This record shows that the two continents gradually approached each other through the late Mesozoic and early Tertiary period as the intervening oceanic lithosphere was subducted beneath the southern margin of Asia. A plate margin of this type is transient in nature since continental crust, because of its low density, is not capable of subduction. Thus convergence of two pieces of continental lithosphere can only take place to a limited extent after the two continents come in contact with each other.

The line of collision is termed a **suture** and is important in recording older plate movements in the geological record of the continents (see Chapter 16). Along the Himalayan part of the suture there is geological evidence of the subduction zone that was responsible for destroying the large area of oceanic plate formerly separating Asia and India. The processes leading from a subduction zone to a collision zone are summarized in Figure 14.11.

In the case of the India–Asia collision zone, the convergent plate movements have resulted in a very wide belt of complex structures on the Asian side of the suture (see section 15.4 and Figure 15.7).

## CONSERVATIVE BOUNDARIES

Many sections of the boundaries of all the plates consist of steep faults with a lateral (strike-slip) sense of displacement. Because plate material is neither created nor destroyed along these sections but is conserved, they are termed **conservative boundaries**. Faults or fracture zones are very prominent features of the oceans, and the ocean ridge crests are repeatedly offset by them (Figure 14.15). These oceanic faults played a key role in the evolution of the plate concept. It was noted by J. Tuzo Wilson, in an influential paper in 1965, that parallel sets of such faults should be parallel to the spreading direction of the ocean ridges, and that divergent motion away from a ridge axis would be 'transformed' to a transcurrent motion along such a fault (Figure 14.12A), then perhaps transformed again to convergent motion at a trench (Figure 14.12B). He therefore called these

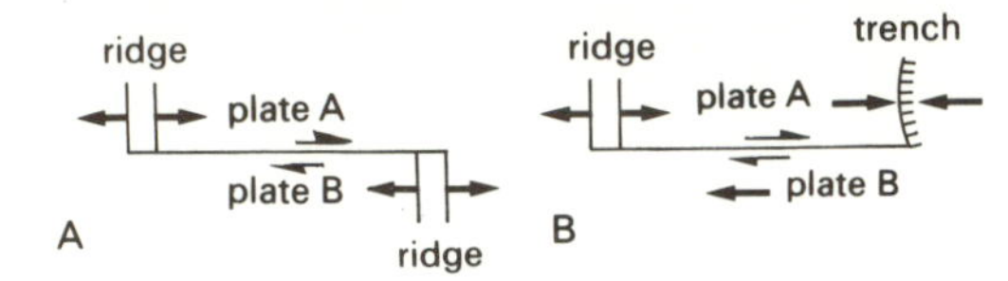


**Figure 14.12** Nature of a transform fault (see text).

faults **transform faults**, recognizing their fundamental difference from strike-slip faults on land.

A transform fault is part of a plate boundary, and must be parallel to the direction of relative motion of the plates on either side. It is therefore controlled by the relative velocity of the two plates, whereas a strike-slip fault, at least initially, is a response to stress (Figure 9.3C). However, in the case of major continental strike-slip faults, which are controlled by the relative movement of large crustal blocks, the distinction is less clear. Many transform faults appear to originate at abrupt changes in orientation of the severed continental margin (Figure 15.2), particularly where the margin is nearly parallel to the spreading direction.

One of the best-known examples of a transform fault is the San Andreas fault of California (Figure 14.13). This fault forms the plate boundary between the Pacific plate on its west side and the American plate on its east, transforming the divergent motion across the East Pacific ridge to the south to transcurrent motion over a distance of 2800 km until the boundary again becomes a spreading ridge west of Oregon. The direction of this fault thus tells us the direction of relative motion of the Pacific and American plates.

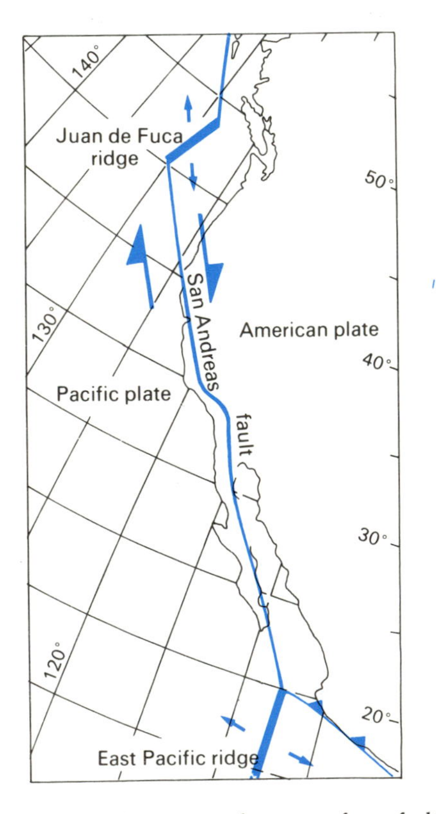


**Figure 14.13** The San Andreas transform fault (see text). (Based on Hallam, 1973, figure 24.)

## 14.4 GEOMETRY OF PLATE MOTION

Once it is accepted that plates behave as 'rigid' shells, their relative motion across the surface of the globe obeys the simple rules of motion on a sphere.

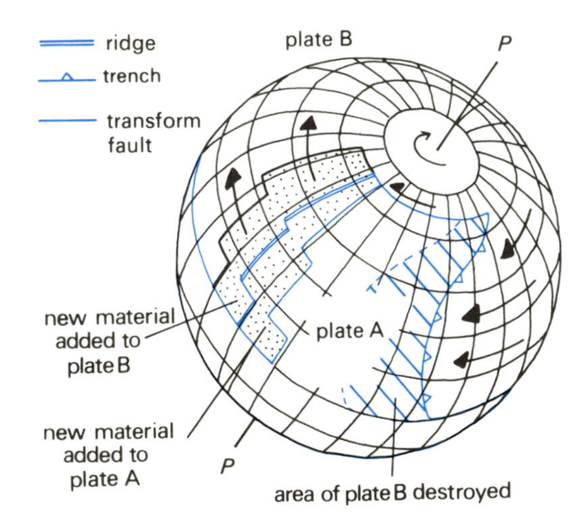


**Figure 14.14** Relative movement between two plates expressed as an angular rotation about a pole. The movement of plate B relative to plate A takes place parallel to the small circles about the pole of rotation along bounding transform faults. New material added to both plates at the ridge is balanced by material destroyed by subduction below plate A. (Based on Dewey, J.F. (1972) *Scientific American*, May.)

Any relative movement between two plates on the surface of a sphere can be described as an angular rotation about an axis that will intersect the surface of the Earth at two points called the **poles of rotation** for that movement. Figure 14.14 illustrates this principle. The displacement of plate B relative to plate A is an angular rotation about the pole *P*. The direction of movement is parallel to a set of small circles on the globe about the axis *PP*. If the displacement takes place by the opening of an ocean between two bounding transform faults, these faults will also be small circles about the pole of rotation, as they must be parallel to the direction of relative motion. The speed of relative motion may be described in terms of an **angular velocity,** which is the speed of rotation about the axis.

The transform fault method was first used to investigate the motion of the Pacific plate relative to the American plate. The pole for this motion is situated in the North Atlantic (Figure 14.15). This movement gives apparent velocities that increase southwards away from the pole of rotation. It is important to realize that, although the angular velocity is constant, the tangential velocity at the surface varies from a minimum at the pole to a maximum along the great circle at 90° from the pole (Figure 14.14). By using transform faults and spreading rates, the poles and relative angular velocities of several plate pairs were established. The relative velocities of the remaining plate pairs were then found using the 'triple junction' method (Figure 14.16).

Thus, working from plate boundaries with known relative motions, a complete picture can be built up of all plate velocities. Actual tangential or linear plate velocities are in the range 2–12 cm/yr and are illustrated in Figure 14.15 together with the poles of rotation for six major plate pairs. It must

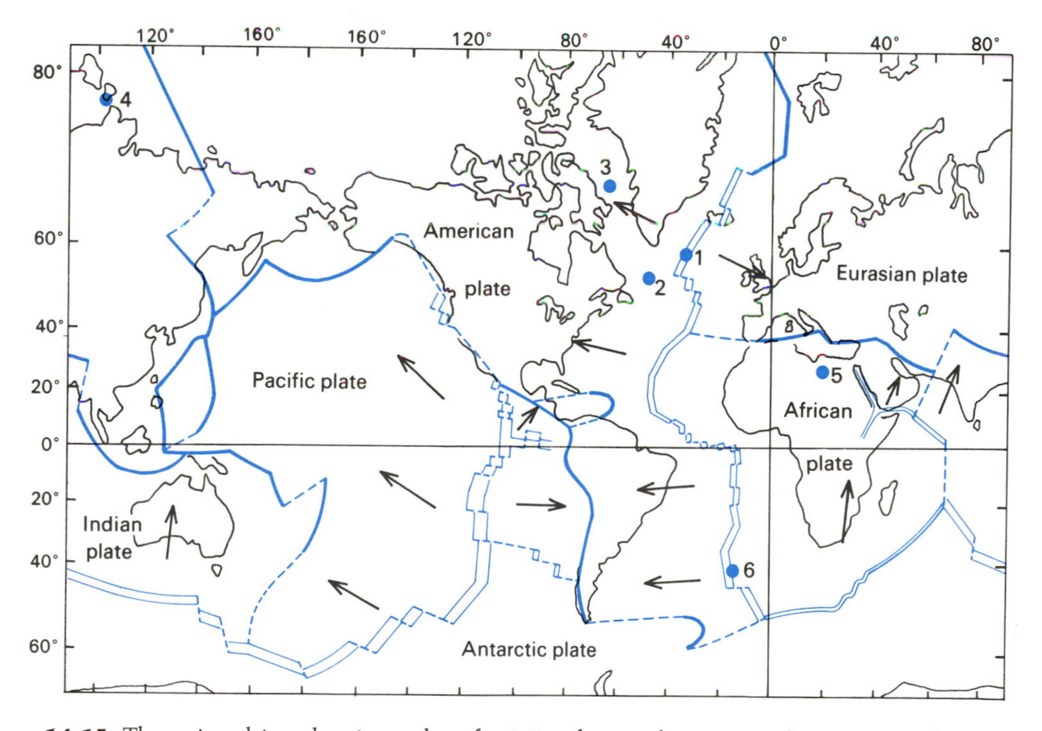


**Figure 14.15** The major plates, showing poles of rotation for six plate pairs and approximate linear velocity vectors relative to the Antarctic plate. 1, America–Africa; 2, America–Pacific; 3, Antarctica–Pacific; 4, America–Eurasia; 5, Africa–India; 6, Antarctica–Africa. (After Vine, F.J. and Hess, H.H. (1970) in *The Sea*, vol. 4, Wiley, New York.)

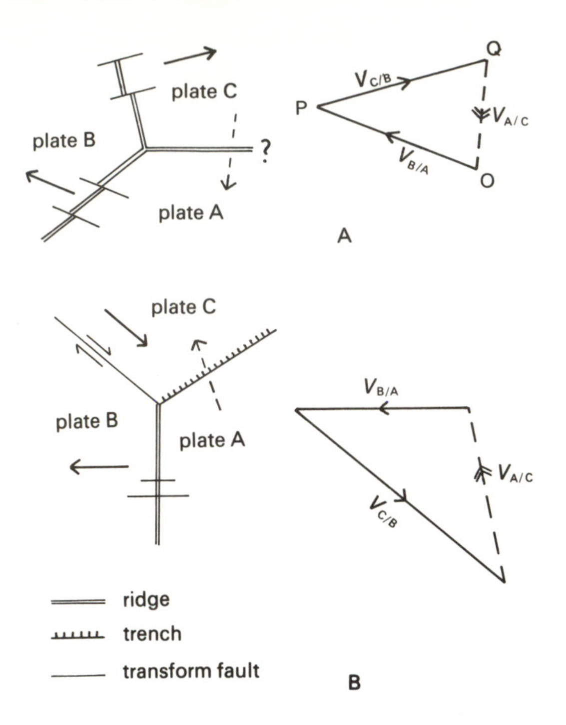


**Figure 14.16** Determination of the relative velocities of three plates meeting at a triple junction. When three plates meet in a **triple junction,** and the velocities of two of them are known, the velocity of the third can be calculated by drawing a **vector triangle**. In the example shown in A, three plates A, B and C, separated by three spreading ridges, meet at a triple point. The relative velocity of B with respect to A, $V_{B/A}$, is given by the spreading rate on the A/B ridge and the direction of the offsetting transform faults along that ridge. It can be represented by the vector OP, whose length is proportional to the velocity $V_{B/A}$. Similarly, the relative velocity of C with respect to B, $V_{C/B}$, can be represented by the vector PQ. If we suppose that the velocity of A relative to C, $V_{A/C}$, is unknown, it can be calculated by completing the vector triangle by joining QO. Let us assume for simplicity that plate A is stationary. The velocity of plate B is then given by $V_{B/A}$ or OP and the velocity of plate C by $V_{C/A}$ or OQ. B shows the method applied to a triple junction between a ridge, a trench and a transform fault. The spreading rate at the ridge can be used to determine $V_{B/A}$. The direction of relative motion between B and C is given by the transform fault. The direction of convergence across the trench may also be known from transform faults offsetting the trench. By completing the vector triangle, the rate of convergence between A and C, $V_{A/C}$ can also be found.

be remembered that linear velocities as shown on a map will vary in amount and direction depending on their position in relation to the pole of rotation.

Changes in relative plate motion can be recognized from discordances in the ocean stripe and transform pattern that indicate a change in the position of the pole of rotation. A good example is seen in the Indian Ocean (Figure 14.17), where the northward movement of India relative to the Antarctic plate to the south changed abruptly about 33 Ma ago to a northeasterly movement, causing a new ridge axis to be formed at an angle of 45° to the old direction.

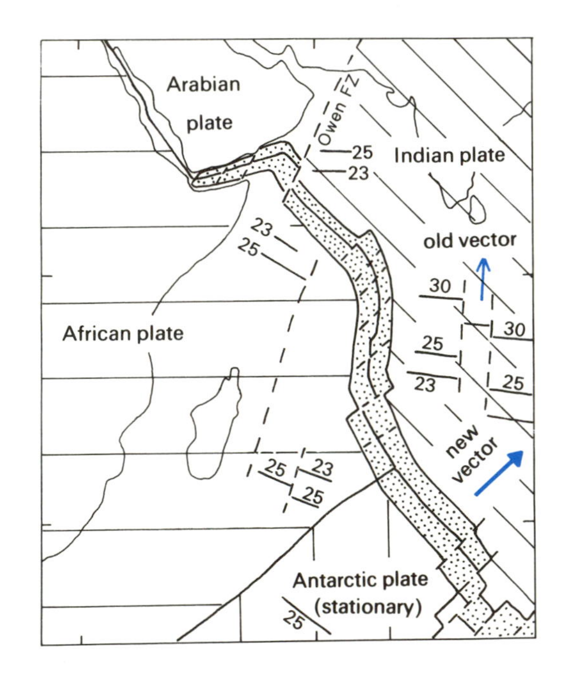


**Figure 14.17** The western Indian Ocean region showing the discordance in magnetic anomaly patterns and transform direction at anomaly 5 (33 Ma ago) caused by a change in spreading direction. Newer ocean floor (post-anomaly 5) stippled. Older anomalies (23, 25, 30) and transform faults give former relative velocity vectors. (Based on Laughton A.S., McKenzie, D.P. and Sclater, J.G. (1973) in *Implications of Continental Drift to the Earth Sciences* (eds D.H. Tarling and S.K. Runcorn), Academic Press, New York, figure 1.)

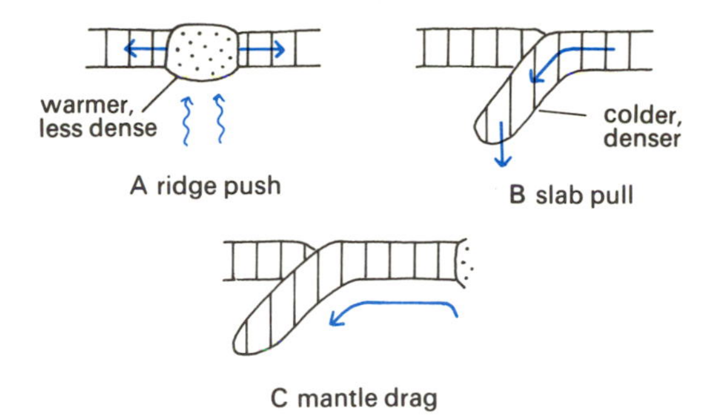


**Figure 14.18** Driving mechanism for plate motion (see text).

## 14.5 DRIVING MECHANISM FOR PLATE MOTION

Despite the explosion of research into plate tectonics that has taken place over the past three decades, there is still no general agreement on the fundamental mechanism that drives plate motion. As long ago as 1928, Arthur Holmes suggested convection currents in the solid mantle as a mechanism to explain crustal tectonics and continental drift, and it is now generally believed that some kind of convective flow pattern in the mantle provides the driving force for plate motion. However, there is still considerable debate about the nature and pattern of convective circulation, and whether it involves the whole or only part of the mantle.

The ultimate source of energy for tectonic processes is heat. The variation in distribution of the flow of heat leaving the Earth is converted into density imbalances which in turn provide gravitational energy. This can work in two main ways (Figure 14.18). First, the rise of hotter mantle material below an ocean ridge produces a large low-density bulge, which exerts a lateral gravitational pressure on the plates on either side. This is the **ridge-push** mechanism and operates in the same way as the gravitational spreading of orogens discussed in section 12.2. Secondly, the gravitational effect of cooler, denser material in and around the sinking slab creates a lateral force towards the trench on the subducting slab. This is the **slab-pull** mechanism.

Another suggested mechanism is **mantle drag**, in which lateral convective flow within the mantle effectively pulls the plates along. Although all three mechanisms play some part in driving plate motion, calculations of the likely magnitudes of the forces involved suggest that ridge-push and slab-pull are dominant and that mantle drag is much less important.

## FURTHER READING

Cox, A. and Hart, R.B. (1986) *Plate Tectonics: How it Works*, Blackwell Scientific Publications, Palo Alto, California. [Thorough and readable treatment of three-dimensional geometry and kinematics, with examples.]

Hallam, A. (1973) *A Revolution in the Earth Sciences*, Clarendon Press, Oxford. [A very readable account of the historical development of plate tectonic theory.]

Wilson, J.T. (ed.) *Continents Adrift and Continents Aground* (1972), Readings from *Scientific American*, Freeman, San Francisco. [Contains reprints of important articles relating to the development of plate tectonics.]

Wyllie, P.J. (1976) *The Way the Earth Works: An Introduction to the New Global Geology and its Revolutionary Development*, Wiley, New York. [An excellent and very readable account of plate tectonics at an introductory level.]

# GEOLOGICAL STRUCTURE AND PLATE TECTONICS

15

In Chapters 13 and 14 we discussed how the shape of the Earth's surface and the present pattern of tectonic activity can be explained in terms of plate tectonic theory. In this chapter, we examine the relationship between geological structure and plate movements in order to understand how geological structures may be explained by the plate tectonic model.

## 15.1 RECOGNITION OF INACTIVE PLATE BOUNDARIES

The recognition of presently active plate boundaries depends on seismic activity. In interpreting the geological record, we must use other criteria for the recognition of plate boundaries. To a great extent, the record over the past 200 Ma, since the break-up of the supercontinent Pangaea (Figure 14.1), can be reconstructed by extrapolation backwards in time, using particularly the oceanic record. However, before that period there is no oceanic record, since all older oceanic crust will have been destroyed, and we must rely exclusively on the interpretation of continental geology. By examining the way in which structures are related to present-day plate boundaries and plate movements, we can seek to interpret older structures by analogy.

## 15.2 STRUCTURE OF CONSTRUCTIVE BOUNDARIES

The continental record of structures associated with Mesozoic to present-day constructive boundaries is well documented. Evidence is available from the severed margins of the Atlantic and Indian Oceans and from continental rift systems, several of which may represent incipient constructive boundaries. Such boundaries are characterized by divergent plate movements and consequently are marked by zones of extensional faulting and commonly by vulcanicity.

A well-known example of an incipient constructive boundary is the Gulf of Aden–Red Sea rift, which meets the great African rift system in a triple junction in the Afar region of Ethiopia (Figure 15.1). The geological history of this area shows that a domal uplift around 1 km high and 1000 km

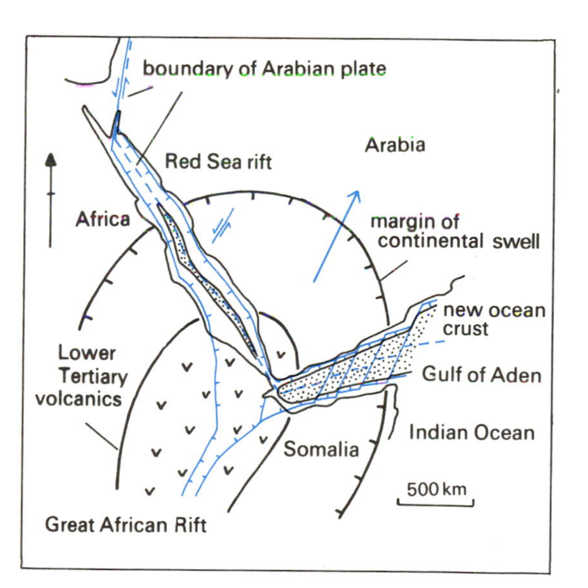


**Figure 15.1** Structure of the Red Sea–Gulf of Aden–African Rift system. (Based on Gass, I.G. (1970) The evolution of volcanism in the junction area of the Red Sea, Gulf of Aden and Ethiopian rifts. *Philosophical Transactions of the Royal Society of London, A*, **267**, 369–81, figure 1.)

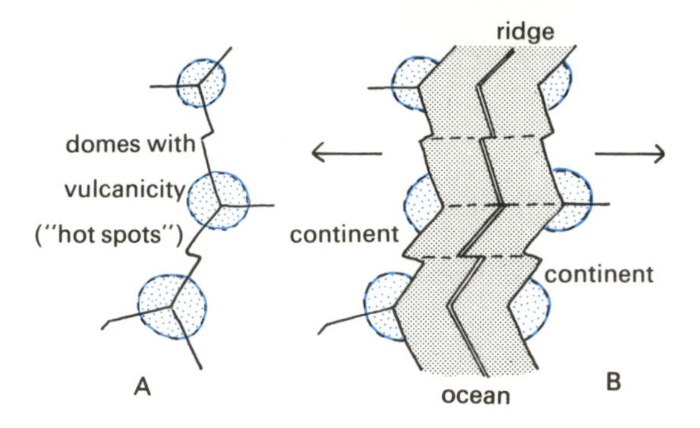


**Figure 15.2** Model for the formation of an ocean by continental rifting. (After Dewey, J.F. and Burke, K. (1974) Hot spots and continental break-up: implications for collisional orogeny. *Geology*, **2**, 57–60, figure 1.)

wide formed during Mesozoic times and was eventually broken by a three-pronged rift associated with deep-seated alkaline vulcanicity. Extensional normal faulting and crustal thinning produced the three rift valley systems, and was accompanied by rather higher-level basaltic volcanics. The final stage of separation is only seen in the Red Sea and Gulf of Aden rifts, where thin strips of oceanic crust have formed. The Gulf of Aden has been slowly opening over the past 20 Ma along a continuation of the Carlsberg ridge spreading axis in the northwestern Indian Ocean (Figure 14.17).

This triple rift system has been taken as a model for the process of continental break-up leading to the formation of an ocean (Figure 15.2). It has been suggested that the break-up of Pangaea took place by the linking together of a series of jagged rift-fractures of this type situated over mantle 'hot spots' marked by domal uplifts and volcanism. The third arms of the triple junctions have been called **failed arms**, since they never develop into oceans, but nevertheless they exhibit characteristic associations of structures, sediments and vulcanicity that enable them to be identified in the geological record.

An important feature of some of the rifted margins of Pangaea is the presence of coast-parallel dyke swarms associated with extensive outpourings of tholeiitic basalts, as can be seen for example in eastern Greenland and in the Deccan area of India.

In summary, therefore, we might expect to find the following tectonic features associated with past constructive boundaries: (1) a series of domal uplifts; (2) extensional faulting both parallel to the coastline and defining 'failed-arm' graben; (3) extensive vulcanicity, including extensional dyke swarms.

## STRUCTURES ASSOCIATED WITH CONTINENTAL RIFTS

Good examples of the types of extensional structures found in divergent tectonic regimes have been described from the Gulf of Suez, which forms the northwest segment of the Gulf of Aden rift. Here a set of parallel normal faults trends NW–SE

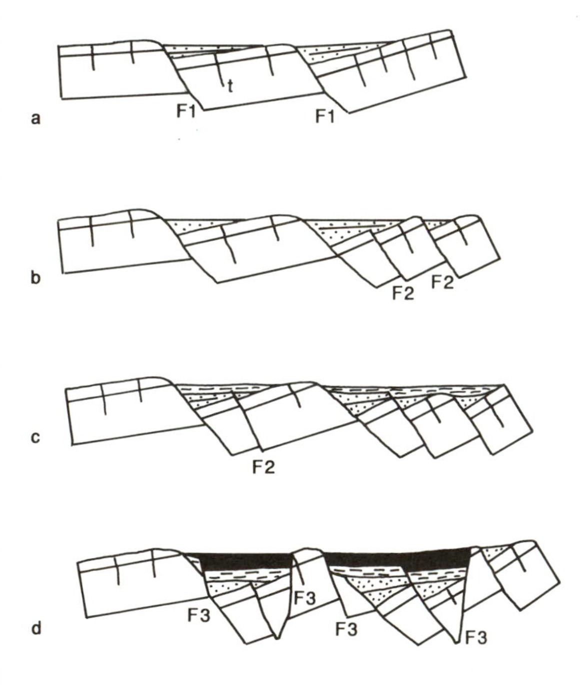


**Figure 15.3** Structure of the Gulf of Suez rift. Block diagrams a–d show the interpreted evolution of a system of tilted fault blocks accommodating to gradually increased extension from Miocene (a) to Present (d). F1–3 indicate successive generations of faults; t, early extensional fractures. (After Angelier, J. (1985) Extension and rifting: the Ziet region, Gulf of Suez. *Journal of Structural Geology*, **7**, 605–12, figure 5.)

parallel to the rift axis. The flanks of the rift consist of fault blocks tilted at 5–35° away from the rift axis. These are bounded by large normal faults marking the margins of the main graben (Figure 15.3). There is no evidence here of pre-rift doming on the plateaus bordering the rift. The inferred sequence of fault movements is illustrated in Figure 15.3. The initial fractures appear to have developed perpendicular to the bedding by pure extension, and they have been rotated during subsequent block tilting, when normal dip-slip movements took place. The rotated extensional fractures were then in a favourable orientation for secondary normal faulting to take place on them (Figure 15.3b). Fractures rotated to low dip angles were cut by younger normal faults and became inactive. The amount of extension estimated from the observed faulting is 20–30°, but this probably underestimates the real extension which may be much greater in the concealed part of the rift.

Similar extensional fault systems characterize the passive margins of the Atlantic and Indian Oceans, which represent constructive plate boundaries during Jurassic to Cretaceous times. Since these passive margin sedimentary sequences host important hydrocarbon reservoirs, their structure has been extensively investigated, mainly by seismic methods.

## STRUCTURES ASSOCIATED WITH BACK-ARC EXTENSION

Divergent tectonic regimes are also associated with the upper plates of certain subduction zones. The Basin and Range Province of the western USA is the type example of such a regime, situated on

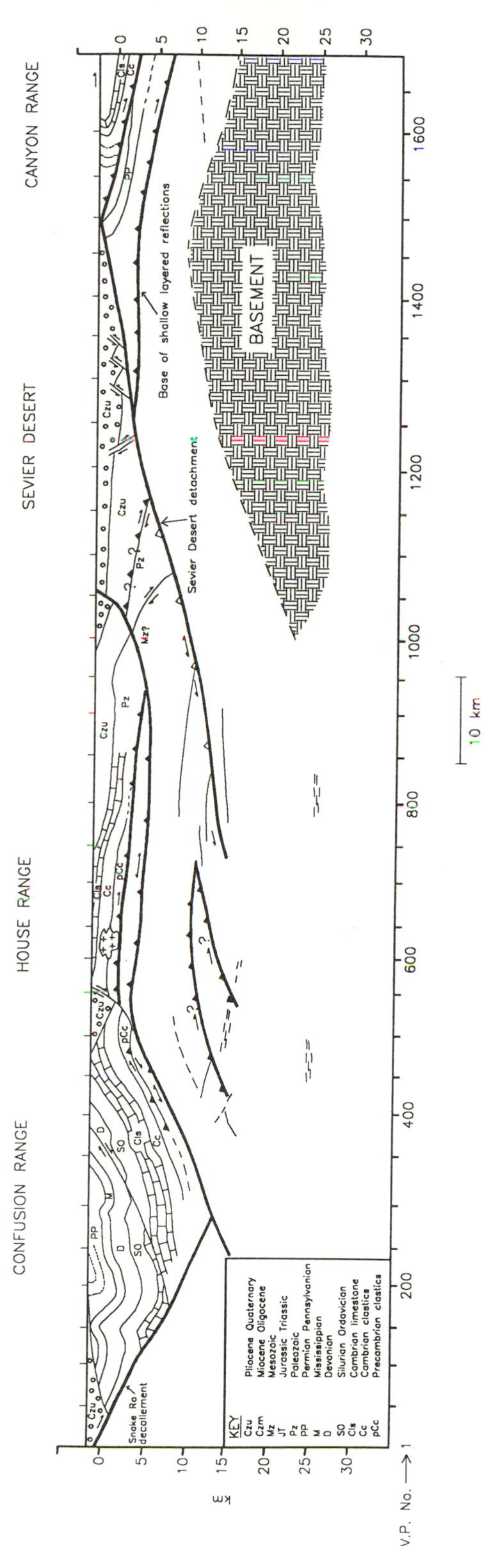


**Figure 15.4** Structural interpretation of COCORP deep seismic reflection lines across the eastern part of the Basin and Range Province. Solid toothed lines, thrust faults; ticked lines, normal faults; open-toothed lines, low-angle normal faults; solid-toothed ticked lines, thrusts reactivated as normal faults. (After Allmendinger, R.W., Sharp, J.W., Von Tish, D., Serpa, L., Kauffman, S. and Oliver, J. (1983) Cenozoic and Mesozoic structure of the eastern Basin and Range province, Utah, from COCORP seismic-reflection data. *Geology*, **11**, 532–6, figure 3.)

continental crust above a subduction zone which lay along the west coast of North America during the Mesozoic and early Tertiary (Figure 15.8B(1)). The province is about 1000 km wide at its maximum and consists of a series of linear ranges (horsts) separated by basins (graben); both horsts and graben are about 10–20 km in width. The present-day structure is controlled by normal faulting in response to WNW–ESE extension, which is confirmed by earthquake fault-plane solutions. Most active faults are steep and extension appears to be accomplished by a tilted fault-block mechanism. The amount of recent extension has been estimated at 17–20%.

Much larger extensions took place on now-inactive parts of the province during the Miocene. COCORP seismic reflection data from the eastern part of the province reveal a series of low-angle reflectors, one of which can be traced to the surface as the Sevier Desert detachment (Figure 15.4). This structure is one of several low-angle mylonite belts bounding metamorphic core complexes (see section 2.7 and Figure 2.13) and indicates extensions of the order of 30–60 km, much larger than those associated with the recent extension. The reduction in the amount of extension in the later Tertiary period has been attributed to the cessation of subduction on the southern part of the zone and its replacement by the San Andreas transform fault (Figure 15.8B).

## 15.3 STRUCTURE OF SUBDUCTION ZONES

Seismic reflection surveys have shown that modern ocean trenches typically have a V-shaped form with the steeper side towards the direction of dip of the subduction zone. Although the original floor of the trench is 2–3 km deep, there may be up to 2 km of sediment fill. These sediments are, at least initially, flat and undeformed along the trench bottom, but show complex folding and thrusting on the inner trench wall, where the Benioff zone comes to the surface.

The main processes and structures associated with the region between the trench and the volcanic arc of a subduction zone are shown schematically in Figure 15.5. This region is divided into a

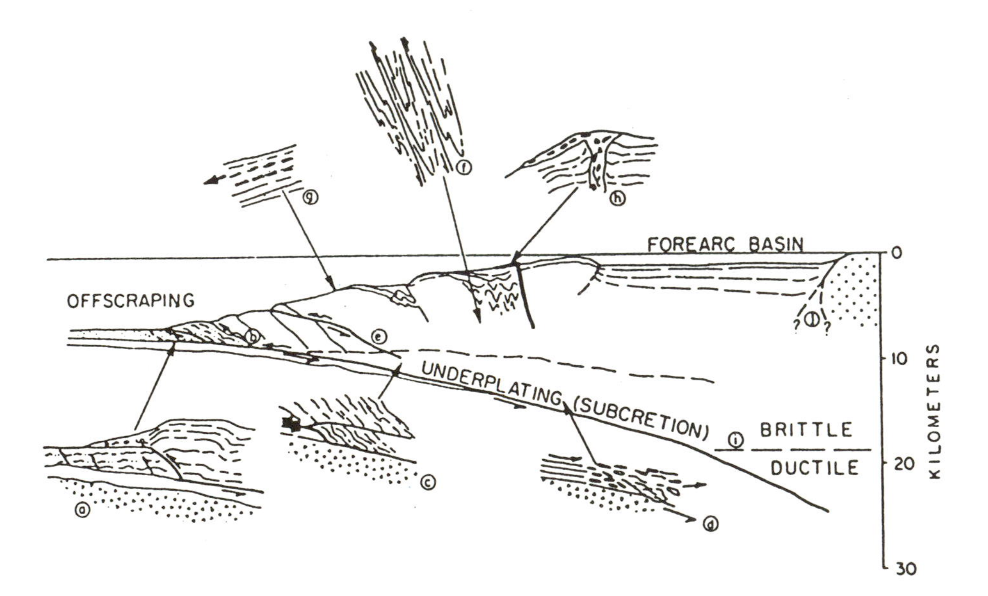


**Figure 15.5** Structure and processes in an idealized accretionary prism. a, frontal accretion by imbricate thrusting; b, décollement plane above subducting slab; c, d, underplating; e, later throughgoing fault; f, back-rotated steepened section; g, gravity sliding; h, diapir of disrupted water-charged sediment; i, brittle–ductile transition; j, basement defining edge of arc. (From Moore, J.C., Cowan, D.S. and Karig, D.E. (1985) Structural styles and deformation fabrics of accretionary complexes: Penrose Conference report. *Geology*, **13**, 77–9, figure 1.)

frontal deformed section, the **accretionary prism**, and an undeformed **forearc basin**. The leading edge of the accretionary prism is dominated by a process of accretion by offscraping of material from the ocean floor of the lower, subducting plate. This material forms a synthetic imbricate thrust complex. Further down the detachment plane above the subducting slab is a region where **underplating** ('subcretion') can take place, forming thrust duplexes. These effects result in thickening and raising of the accretionary complex, the more distal parts of which may exhibit steeply dipping, tightly folded strata which have been rotated backwards into a steep attitude by the continued emplacement of wedges of new material at the proximal end of the prism. Thickening and consequent instability causes gravitational sliding and slumping down the slope of the prism.

The structures of accretionary complexes can be studied more conveniently in zones of older trench sediments that have been uplifted above sea level between the present subduction zone and the volcanic arc. One such example is the Makran complex, discussed below. Other examples of uplifted Mesozoic–Tertiary trench assemblages include the Franciscan assemblage of the Californian coast ranges, parts of the thrust complex of the Banda Arc of Indonesia, and various sections of the Alpine–Himalayan orogenic belt. However, most of these examples are complicated by subsequent tectonic activity, particularly as a result of the collision of continental plates. An example of a Palaeozoic accretionary complex in the Caledonian orogenic belt of the British Isles is described in section 16.1.

### THE MAKRAN COMPLEX

The accretionary prism of the Makran lies along the continental margin of Iran and Pakistan on the north side of the Gulf of Oman (Figure 15.6). The complex is formed by the northward subduction of the oceanic part of the Arabian plate beneath the Eurasian plate. The subduction zone is terminated on its eastern side by the Owen–Murray transform fault that separates the Arabian and Indian plates (Figure 14.17) and continues northwards on the continent as the Quetta–Chaman fault system (Figure 15.7A). The subduction zone ends in the west at the Straits of Hormuz, where the Arabian and Eurasian continents are in contact. A sequence of sediments 6–7 km thick covers oceanic crust in the Gulf of Oman, which is thought to be between 70 and 120 Ma old.

The active volcanic arc consists of a chain of Cenozoic volcanoes situated 400–600 km north of the coast. There is no topographic trench, and the accretionary complex is unusually broad, about 300 km in width, more than half of which lies onshore. Seismic reflection profiles across the offshore part of the complex show a linear pattern of ridges with intervening troughs. Folding appears to have taken place initially at the southernmost or frontal part of the prism, which seems to have migrated southwards at a rate of 10 km/Ma. These frontal folds are then incorporated into the accretionary complex by uplift along a basal thrust. Little subsequent deformation appears to have occurred in this sector of the complex. However, 70 km to the north of the present front, a further uplift occurs which eventually rises above sea level 100 km north of the front to form the onshore Makran complex. Here a thick, faulted flysch sequence is exposed, extending about 200 km inland to the north.

The onshore structure, summarized in Figure 15.6B, affects a concordant sequence of marine sediments commencing with Oligocene to mid-Miocene abyssal plain deposits, followed by upper Miocene slope deposits, and by a late Miocene to Pliocene shallow-water shelf sequence, indicating rapid shoaling of the sedimentary prism in the mid-Miocene. There is apparently no field evidence for the progressive growth of structures during deposition, although the growth of gentle folds might be undetectable owing to the effects of the later deformation.

The main deformation, which caused 25–30% shortening, occurred after the early Pliocene (4 Ma ago) at a time when the accretionary front probably lay 70–100 km south of the present shore line, and has resulted in a series of E–W to ENE–WSW, asymmetric, south-verging folds and associated reverse faults (Figure 15.5C).

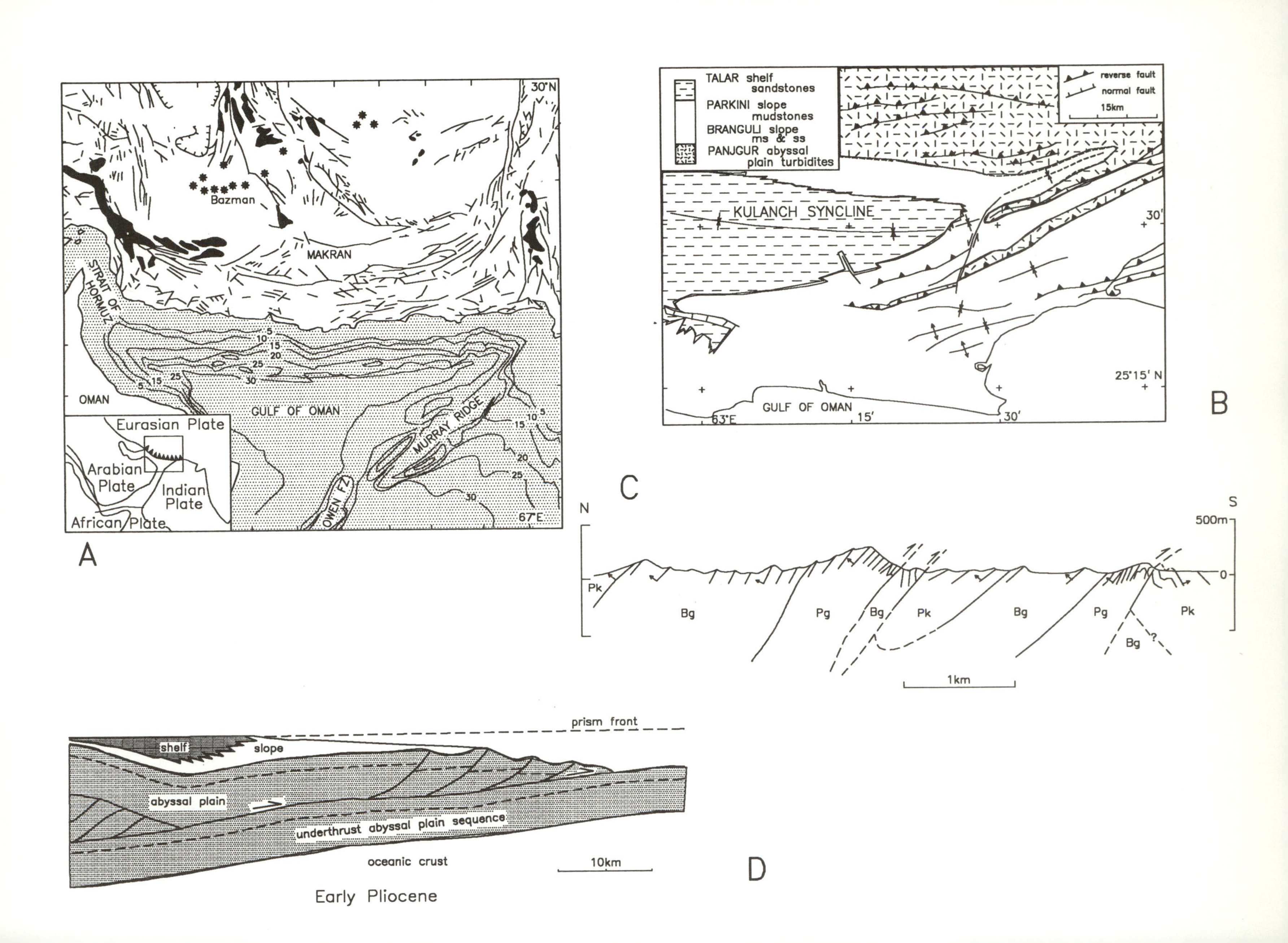
A
30°N
Bazman
MAKRAN
STRAIT OF HORMUZ
OMAN
GULF OF OMAN
MURRAY RIDGE
OWEN F.Z.
67°E
Eurasian Plate
Arabian Plate
Indian Plate
African Plate
B
TALAR shelf sandstones
PARKINI slope mudstones
BRANGULI slope ms & ss
PANJGUR abyssal plain turbidites
reverse fault
normal fault
15km
KULANCH SYNCLINE
30'
25°15' N
GULF OF OMAN
63°E
15'
C
N
S
500m
0
Pk
Bg
Pg
1km
D
prism front
shelf
slope
abyssal plain
underthrust abyssal plain sequence
oceanic crust
10km
Early Pliocene

The uplift of the onshore Makran and the accompanying deformation are thought to have been accomplished by underplating at depth (Figure 15.5D). It has been suggested that this process may have operated by the formation of a progressively widening duplex at a ramp in the basal thrust. Such a structure could have caused tilting of the upper part of the sequence, leading to shoaling and possibly to syn-sedimentary deformation, before the major folding and faulting.

## 15.4 STRUCTURE OF CONTINENTAL COLLISION ZONES

The collision of two continental plates produces a zone of very complex structure as illustrated by much of the Alpine–Himalayan orogenic belt. The reason for the complexity lies partly in the fact that the structures of two formerly separate continental margins have been brought together, often in a very intimate fashion. Figure 14.11 demonstrates schematically the effects of telescoping an active continental margin exhibiting subduction tectonics with an opposing passive continental margin. This is the simplest possible situation. The Alpine and Himalayan chains show evidence of several subduction zones and island arcs that have now coalesced.

**Figure 15.6** Structure of the Makran subduction zone. A. Location of the Makran subduction zone in the Gulf of Oman. The subduction zone marks the boundary of the Arabian and Eurasian plates (inset) and is truncated on its eastern side by a major transform fault zone. Stars mark volcanic centres along the active volcanic arc. The trench is obscured by a thick accretionary prism which is partly offshore and partly onshore. (After White, R.S. (1982) Deformation of the Makran accretionary prism in the Gulf of Oman (northwest Indian Ocean), in Leggett, J.K., *Trench-forearc Geology*, Special Publication of the Geological Society of London, **10**, 357–72, figure 1.) B. Simplified structural map of the onshore Makran accretionary prism showing major folds and reverse faults. Note lateral facies change between the Talar and Parkini formations. C. Section across the southern limb of the Kulanch syncline (see B) showing N-younging sequences bounded by reverse faults. Pg, Panjgur; Bg, Branguli; Pk, Parkini formations. D. Interpretative section across B showing suggested structure during the Pliocene. Note that the uplift of the northern limb of the syncline is attributed to a duplex structure. (B–D after Platt *et al.*,1985, figures 2–4.)

### THE CENTRAL ASIA REGION

The complexity of the tectonic pattern in Central Asia resulting from the collision between India and Asia is illustrated in Figure 15.7. We shall now discuss some of the reasons for this complexity.

The southern boundary of the Eurasian plate is marked by the Indus suture (now a steep fault), which lies on the north side of the Himalayas. At its western end, the suture is terminated by the large Quetta–Chaman transform fault zone. This fault zone connects to the south with the Makran subduction zone described above (Figure 14.15 and 15.6). At the eastern end of the suture, the plate boundary runs southwards again along a large dextral strike-slip fault which continues through Burma to connect with the Indonesian subduction zone. This fault also continues into the Indian Ocean as the now largely inactive Ninety-East ridge transform fault. The Indian plate, therefore, may be regarded as a two-pronged wedge that has driven northwards into the Eurasian plate between two large transform faults.

The initial collision appears to have taken place during the Eocene, when the leading edge of the Indian continent first made contact with Eurasian continental crust. Prior to this event, the intervening oceanic lithosphere, part of the 'Tethys Ocean', was being subducted below the Eurasian plate. Deformed remnants of Tethyan sequences are preserved immediately south of the suture (Figure 15.7C).

It has been estimated that the northward movement of India continued for at least a further 1500 km after the initial collision. Part of this movement must have been accommodated by crustal shortening within the Eurasian plate, the effects of which are to be seen in the broad zone of deformation that extends up to 3000 km north of the suture. Structures attributed to this deformation include compressional fold and thrust belts, strike-slip faults and extensional rift systems. Some of the convergent motion may also have been taken up by the closure of a small ocean basin or basins between the Tibetan block and the main Eurasian plate.

The compressional structures are confined to narrow belts separated by wide areas of comparatively undeformed crust. It is estimated that between 750 km and 1000 km of shortening has

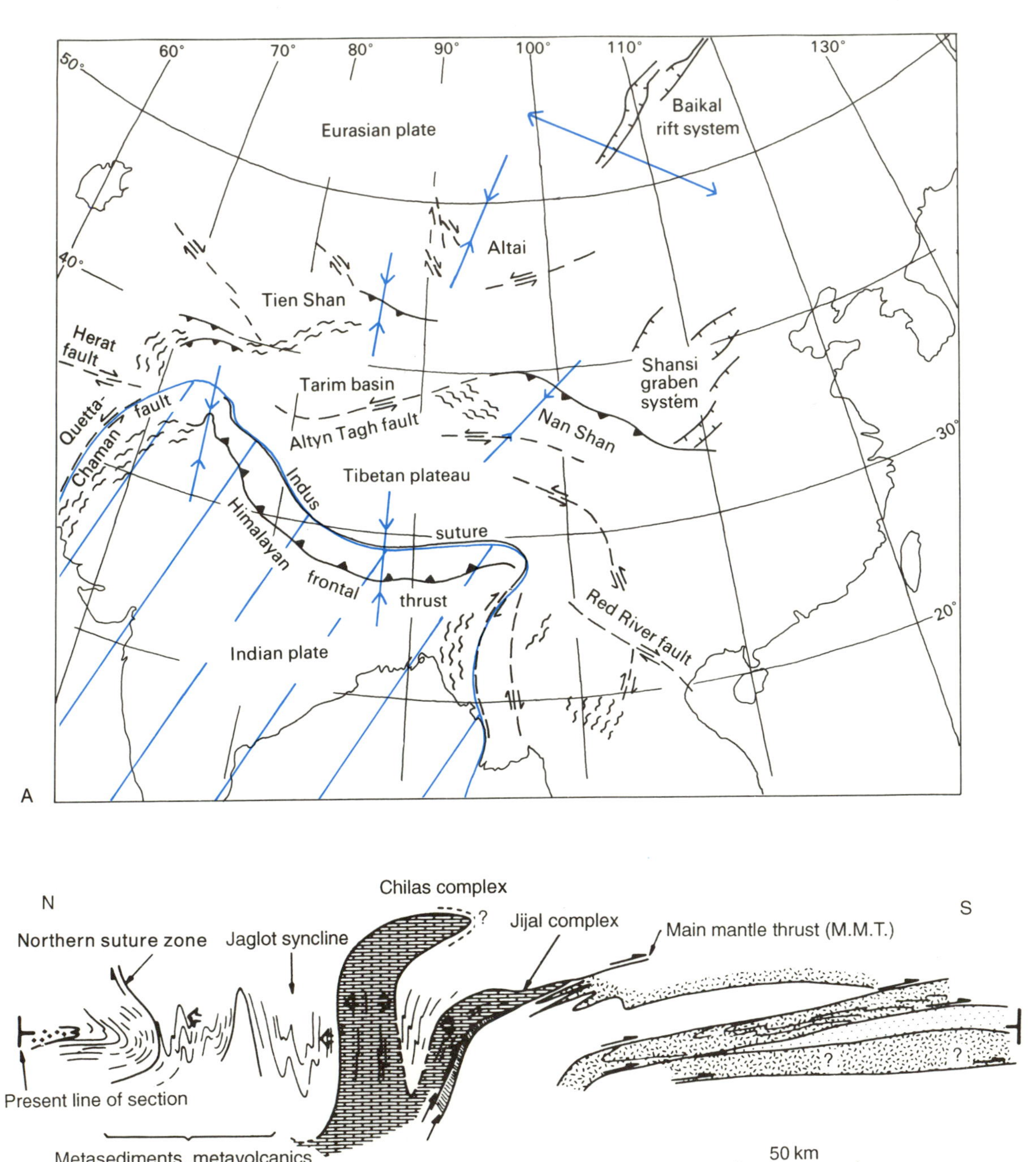

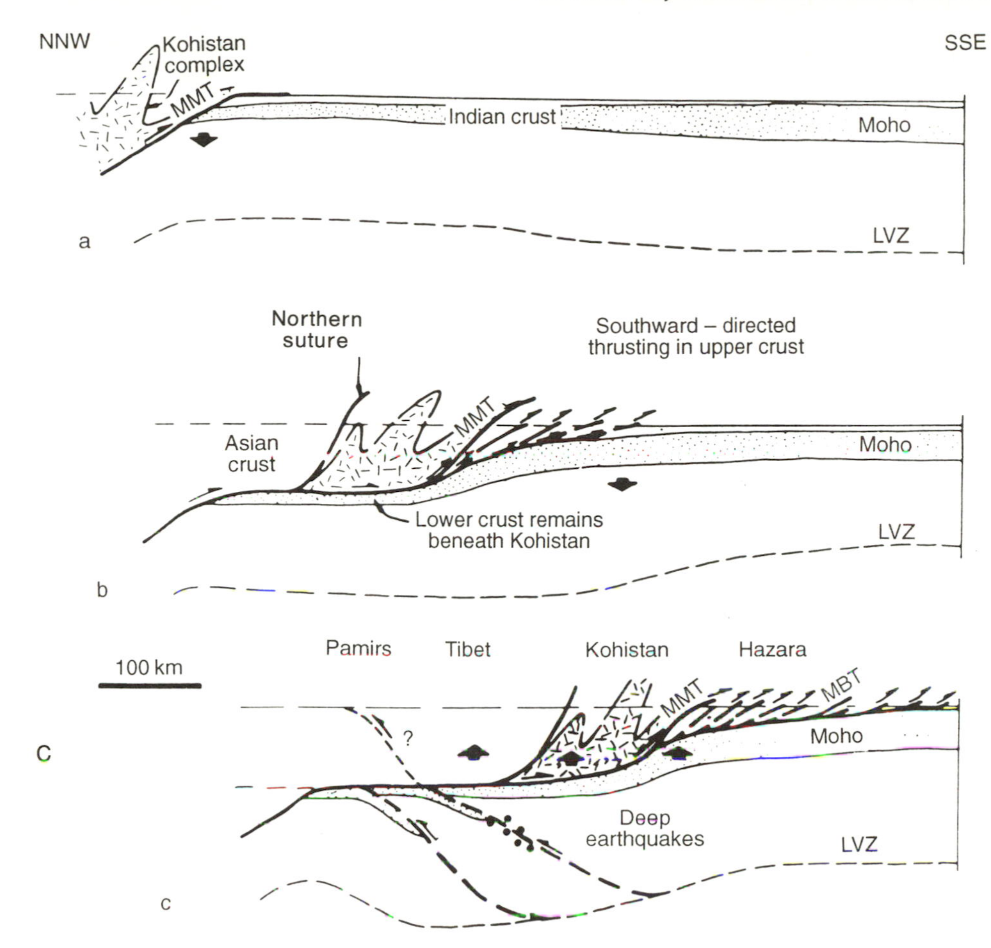


**Figure 15.7** Principal tectonic features of Central Asia thought to relate to collision with the Indian continent. A. Map showing principal compressional fold belts (wavy lines), thrusts (tooth-edged lines), extensional graben or rifts (lines with tick on downthrow side) and strike-slip faults (lines with arrows indicating sinistral or dextral motion). Coloured arrows indicate compressional or extensional fault plane solutions from recent earthquakes. (After Molnar and Tapponnier, 1975, figure 4.) B. Section across the Kohistan region in the Pakistan Himalayas, from the Indus suture to the frontal Himalayan thrust. (After Coward, M.P., Jan, M.Q., Rex, D., Tarney, J., Thirlwall, M. and Windley, B.F. (1982) *Journal of the Geological Society of London*, **139**, 299–308, figure 7.) C. Interpretative profiles showing how the structure of B can be explained by underthrusting of the Indian plate below itself along the main mantle thrust (MMT). (After Coward and Butler, 1985, figure 3.)

been achieved in these compressional belts and that the remaining convergence must have been taken up by lateral movements along the major strike-slip faults and by subduction of oceanic crust. In the west, northward movement of the western prong of India would tend to drive large blocks of crust westwards along several northwest–southeast dextral faults. In this way the Eurasian plate appears to have been shortened in a north–south direction and elongated in an east–west direction. Northeast of the main region of deformation, northwest–southeast extension is indicated by the Baikal rift and Shansi graben systems.

The most spectacular compressional effects of the collision are exhibited in the thrust and fold belt of the Himalayas, south of the Indus suture

(Figure 15.7B). Here a significant component of the crustal shortening, estimated at between 300 and 700 km, has taken place by the overlapping of thick crustal slices containing Precambrian basement belonging to the Indian plate. The crust in this area is estimated to be double its normal thickness, and Figure 15.7C shows in simplified form how this doubling of the crust could have been achieved, by underthrusting the main Indian crustal slab along a low-angle thrust plane (the Main Mantle thrust) outcropping about 130 km south of the suture. Above (north of) the Main Mantle thrust, deep crustal rocks belonging to the Kohistan complex have been upthrust and rotated into a steep attitude (Figure 15.7B). Because of their deep-seated origin, the structures of this complex are characterized by intense ductile folding rather than thrusting. The Kohistan complex is interpreted as an island arc and the Main Mantle thrust as a second suture separating it from the Indian plate. Both sutures represent subduction zones that originally dipped northwards beneath Asia.

Three stages in the post-collisional contraction process are shown in cartoon form in Figure 15.7C. After the intervening oceanic lithosphere had been subducted, the Asian and Indian continents made contact, sandwiching the Kohistan arc between them (profile a). Further convergence between the two plates took place by underthrusting Indian lithosphere beneath the Asian plate, accompanied at shallow levels by southward overthrusting, propagating from the Main Mantle thrust forwards into the cover of the Indian plate. This shallow overthrusting produced an imbricate thrust stack which detaches on the Himalayan boundary thrust. This is the frontal thrust, which outcrops 300 km south of the suture (see profile b). This imbricate thrust complex is an example of a 'thin-skinned' **foreland thrust belt.** At depth, steep backthrusts may have been developed (see profile c) as underthrusting proceeded.

## 15.5 STRUCTURE OF CONSERVATIVE BOUNDARIES: THE SAN ANDREAS FAULT

The best documented example of a conservative plate boundary on land is the large active continental transform fault zone in California known as the San Andreas fault (Figure 15.8; see also Figure 14.13). Here tectonic effects associated with the fault zone spanning a period of about 30 Ma have been studied in considerable detail.

A simplified sequence of steps illustrating the plate tectonic history of the region is shown in Figure 15.8B. Before about 30 Ma ago, the western border of the American plate was represented by a trench associated with the subduction of the northeast-moving Farallon plate (stage 1). The San Andreas transform fault seems to have commenced when a section of the northwest-moving Pacific plate met the American plate (stage 2). The sense of movement on the American plate boundary then changed from convergent to strike-slip along a particular section of the boundary between two transform faults on the Farallon plate. Subduction continued to the north and south. As time progressed, the length of the San Andreas fault grew as more of the ridge was subducted.

The effect on the geology of western California was complex (Figure 15.8A). A block of early, highly deformed trench sediments belonging to the original American plate boundary has now been carried northwards to become the Coast Ranges. Several other large blocks have also been displaced dextrally as part of a complex set of movements on a wide fault zone associated with the San Andreas fault.

An interesting feature is the bend in the main fault near Santa Barbara. The geometry demands that dextral strike-slip motion will cause compression across the region of the bend, and this accounts for the overthrusting found in this area. The bend is thought to be due to the effect on the American plate boundary of the intersection of the southern transform fault, which would at one time have been continuous with the major Garlock sinistral fault on land. When this transform fault reached the American plate boundary (stage 3 of Figure 15.8B) the direction of relative motion would have become convergent for a period until the next section of ridge collided with the continent. At this point, transform motion along the plate boundary would have been resumed, leading to the present situation (stage 4). During this movement the ridge

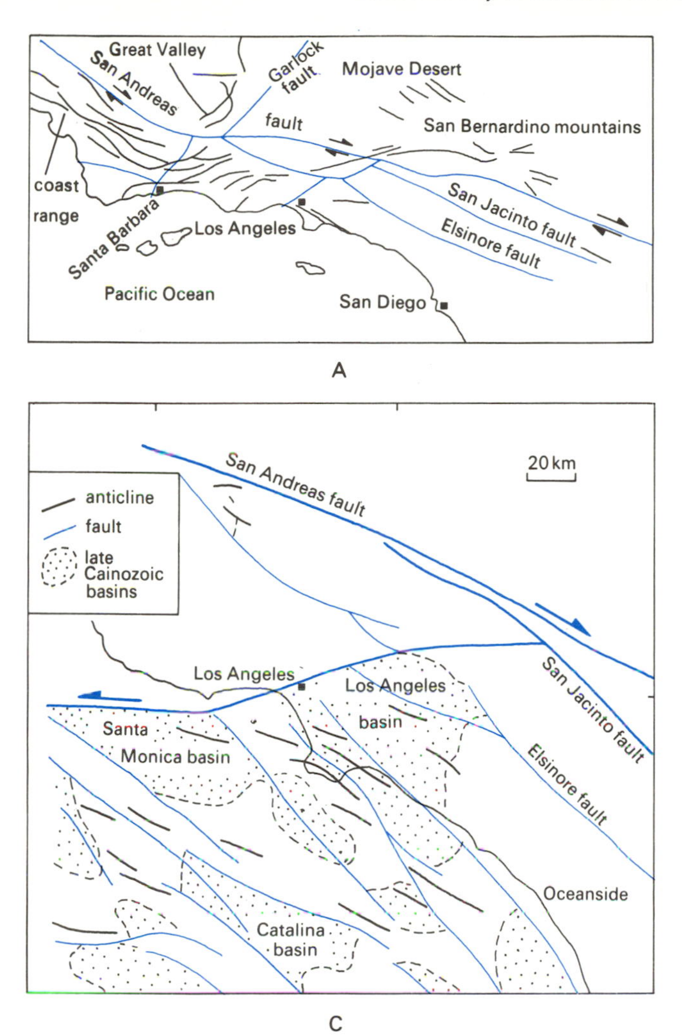


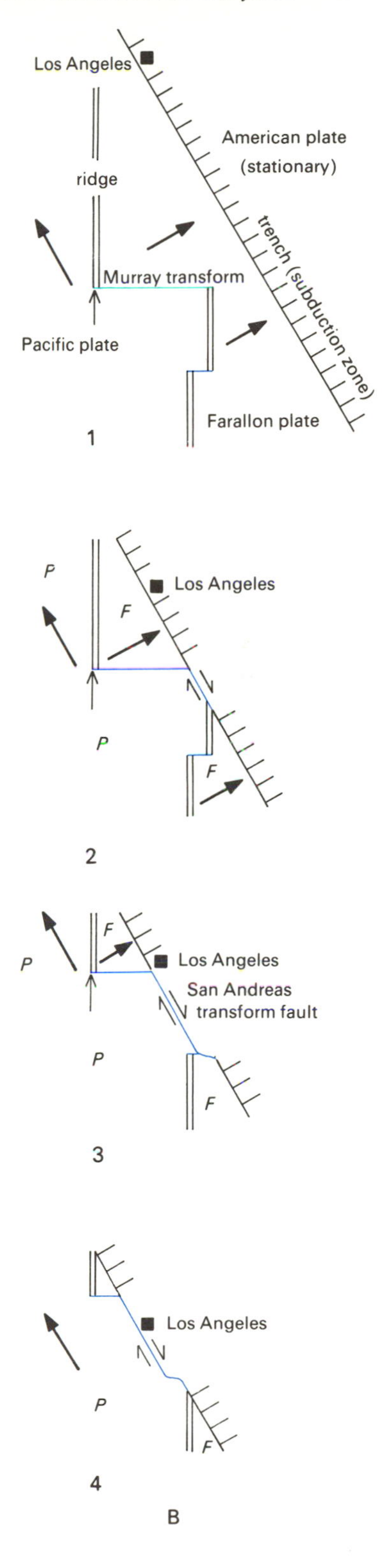


**Figure 15.8** Strike-slip fault tectonics illustrated by the San Andreas fault zone in Southern California. A. Fault distribution in Southern California with major strike-slip faults in colour. (After Anderson, D.L. (1972) The San Andreas fault, in *Continents Adrift and Continents Aground*, W.H. Freeman, San Francisco.) B. Evolution of plate movements in the area of the San Andreas fault, assuming the American plate to be stationary. 1, 53 Ma ago; 2, 30 Ma ago; 3, 10 Ma ago; 4, present. See text. (After Atwater, T. (1970) *Bulletin of the Geological Society of America*, **81**, 3513–36.) C. Detail of strike-slip fault tectonics in the Los Angeles sector of the San Andreas fault zone showing major faults, anticlinal fold trends, and sedimentary basins – see text. (After Howell, D.G., Crouch, J.K., Greene, D.S., McCulloch, D.D.S. and Vedder, J.G. (1980) in *Sedimentation in Oblique-Slip Mobile Zones*, (eds P.F. Ballance and H.G. Reading), *Special Publication of the International Association of Sedimentologists*, **4**, 43–62, figure 10.)

system itself was moving northwards relative to a fixed point (e.g. Los Angeles) on the American plate.

The transform fault is a zone about 100 km wide between two relatively undeformed blocks. The total strike-slip displacement of the plates on either side has been distributed through this zone, partly as movements on a number of smaller faults roughly parallel to the main fault, and partly in the form of secondary compressional and extensional structures caused by the redistribution of stresses arising out of these fault movements.

Part of the fault zone is shown in Figure 15.8C. In this area a number of roughly parallel dextral strike-slip faults branch from the sinistral Santa Monica fault. Many of these faults terminate or overlap with similar faults. The blocks between the faults contain extensional 'pull-apart' sedimentary basins and also compressional folds. Note that the fold axes are oblique to the direction of strike-slip displacement, as predicted by the simple shear model (see section 10.6 and Figure 10.23). The sense of obliquity is evidence for dextral motion.

This case history illustrates the complexities of structure that may be associated with transform faults. In contrast to divergent and convergent tectonic regimes, where large areas are characterized uniformly by either compressional or extensional structures, strike-slip regimes exhibit both compressional and extensional structures as well as strike-slip faults.

#### OBLIQUE CONVERGENCE AND DISPLACED TERRANES

The concept of **displaced terranes** arose from observations in the North American Cordilleran orogenic belt, where many large pieces of continental crust (terranes) were shown to have been derived from much more southerly latitudes and were therefore considered to have been transported to their present position by obliquely convergent plate motion and sutured to the North American continent. A piece of continental crust situated on a subducting oceanic plate is likely to cause a change from subduction to strike-slip motion along the suture after collision has occurred because of the buoyancy of the continental material. Although the term displaced terrane may be applied to any exotic piece of crust, it has more usually been applied to examples showing a component of strike-slip motion (see section 16.1).

### 15.6 STRUCTURE OF INTRAPLATE REGIONS

Although geological structures are concentrated along plate boundaries, and plate theory precludes significant lateral distortion within plates, deformation does occur in intraplate regions, albeit at generally much slower rates than apply to plate boundary regimes. Zones of compressional or extensional structures exist at considerable distances from the nearest plate boundary. Compressional belts are much less common than extensional belts, and may usually be explained as far-field effects of continental collision. A good example of such a belt is the Tien Shan compressional fold/thrust belt of Central Asia, which is situated several hundred kilometres north of the Indus suture across the relatively undeformed Tarim basin (Figure 15.7).

During periods of plate-wide extension, such as affected Pangaea in the Triassic before break-up, extensional rifts were widespread. Some developed into the newly created Atlantic and Indian Oceans, but in others active extension ceased and the structures continued as basins.

In contrast to these typically linear features, depressed or elevated areas representing large-scale warps or flexures with wavelengths of the order of hundreds of kilometres upwards are characteristic of the stable cratons, where they are usually referred to as **basins** and **uplifts** respectively (see section 13.3).

Many basins have an equidimensional shape with no pronounced elongation or alignment; others are more linear or are quite irregular. The intervening uplifts contribute the sedimentary fill to these basins over the period of their existence. Most basins evolve gradually by the successive addition of sedimentary formations, many of which may thicken towards the centre of the basin and thin towards its margins, the cumulative effect being to enhance the basin shape of the floor. Compared

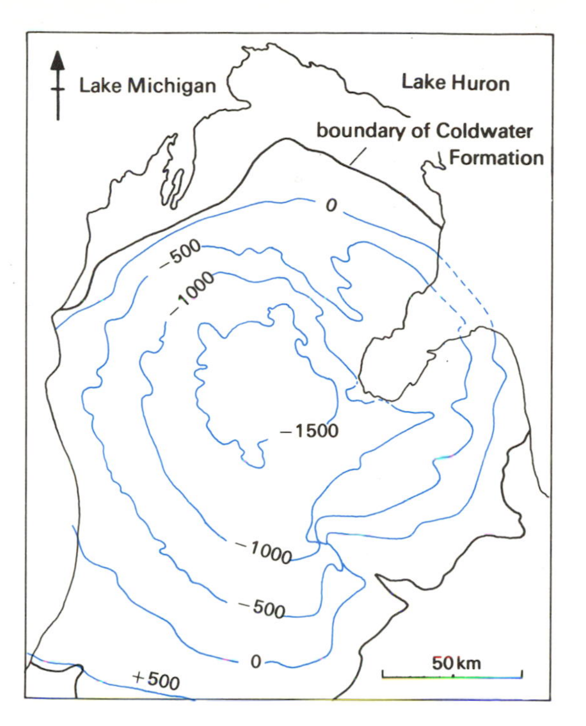


**Figure 15.9** Map showing the shape of the Michigan basin, northern USA. Contours are drawn on the position of the Coldwater formation at depth, at intervals of 500 feet. (Based on De Sitter, L.U. (1964) *Structural Geology*, McGraw-Hill, New York, figure 299.)

to compressional fold structures, the dips are very low – of the order of 1° or less. Figure 15.9 shows the shape of the Michigan basin in the northern USA by means of a set of structure contours drawn at successively deeper levels of the Coldwater Formation. These show that the basin has an approximately circular plan, about 380 km across and over 750 m deep.

Most intraplate basins and uplifts are considered to be primarily gravitational in origin. Horizontal tectonic compression can be ruled out in most cases by the large size and lack of orientation of the structures. The principle of gravitational equilibrium (isostasy) requires any large mass deficiency in the crust, such as a basin, to be compensated by a mass of greater density beneath, so as to make the weight of that column through the Earth balance that of the adjoining columns. This compensation could be achieved by a thinner crust, so that the denser upper mantle material is closer to the surface – analogous to but on a lesser scale than the ocean basins. Some large basins are thought to originate by extensional thinning, whereas others may be the result of density variations within the upper mantle. Uplifts may be formed by the reverse process. When any disturbance of gravitational equilibrium takes place, such as by extensional thinning of the crust, gravitational forces act to restore that equilibrium by flow of material at depth and compensating uplift or depression at the Earth's surface.

## FURTHER READING

Anderson, D.L. (1972) The San Andreas fault, in *Continents Adrift and Continents Aground*, Readings from *Scientific American* (ed. J.T. Wilson), Freeman, San Francisco, pp. 88–102.

Coward, M,P. and Butler, R.W.H. (1985) Thrust tectonics and the deep structure of the Pakistan Himalaya. *Geology*, **13**, 417–20.

Molnar. P. and Tapponnier, P. (1975) Cenozoic tectonics of Asia: effects of a continental collision. *Science*, **189**, 419–26.

Park, R.G. (1988) *Geological Structures and Moving Plates*, Blackie, Glasgow and London. [An account of how geological structures may be explained by plate tectonics, looking particularly at the structures of divergent, convergent and strike-slip plate boundaries.]

Platt, J.P., Leggett, J.K., Young, J., Raza, H. and Alam, S. (1985) Large-scale sediment underplating in the Makran accretionary prism, southwest Pakistan. *Geology*, **13**, 507–11.

# STRUCTURAL INTERPRETATION IN ANCIENT OROGENIC BELTS

# 16

The plate tectonic model has been successfully applied to the geological history of the past 200 Ma. A detailed knowledge of plate movements over this period has been built up from our knowledge of oceanic magnetic stratigraphy and continental palaeomagnetism, and can be checked by conventional stratigraphic methods. How far back into the geological past the plate tectonic model can be extended, however, is a matter of debate, and one of the important tasks of the structural geologist is to attempt to relate patterns of deformation in old rocks to some scheme of crustal movements – if not strictly to the plate tectonic model as we know it, then to some alternative kinematic model that can explain the geological evidence.

Because of certain differences in the preserved record in the older crustal segments, particularly those formed in the Archaean, some geologists have questioned whether plate tectonics can be applied to the oldest part of Earth history. It has been suggested also that the nature of the plate tectonic mechanism may have changed with time.

Others have maintained that the plate tectonic model in essentially its present form can be applied to the oldest preserved Archaean history and that only relatively minor quantitative changes have occurred since.

The purpose of this final chapter is to illustrate how the various types and patterns of geological structure are used to help to provide tectonic interpretations in the older geological record. We shall examine three quite different orogenic belts, spread across nearly 3000 Ma of geological time: the Caledonian orogenic belt in the British Isles, the Early Proterozoic Eastern Churchill Province of Canada and its extension into SW Greenland, and the Archaean Superior Province of North America. The structure of these regions has been described and explained in terms of plate kinematic processes.

## 16.1 THE CALEDONIAN OROGENIC BELT IN BRITAIN

The problems of applying plate tectonics of pre-Mesozoic orogenic belts are well illustrated by the British sector of the Caledonian orogenic belt. Despite (or perhaps because of) the fact that this is one of the most comprehensively studied belts in the world, there is no general agreement on a tectonic interpretation and many different models have been proposed.

The British sector of the Caledonian orogenic belt is part of a long orogenic belt of Palaeozoic age extending from the southern USA to northernmost Norway. It is divided into three parts, the South-Central Appalachian sector, the Northern Appalachian–Newfoundland–British Isles sector and the Scandinavian–East Greenland sector. The southern sector was active throughout the Palaeozoic and the main orogenic episode was caused by collision between Laurentia (North America plus Greenland) and Africa in the Permian. In the northern two sectors (Figure 16.1), Lower Palaeozoic activity ended with the main phase of orogeny during the Devonian. The Scandinavian–East Greenland belt resulted from collision between Laurentia and Baltica (the northern part of Europe), whereas the collisions in the British Isles, Newfoundland and the Northern Appalachians took place between Laurentia and two or more microcontinents, thought to have been previously detached from Africa.

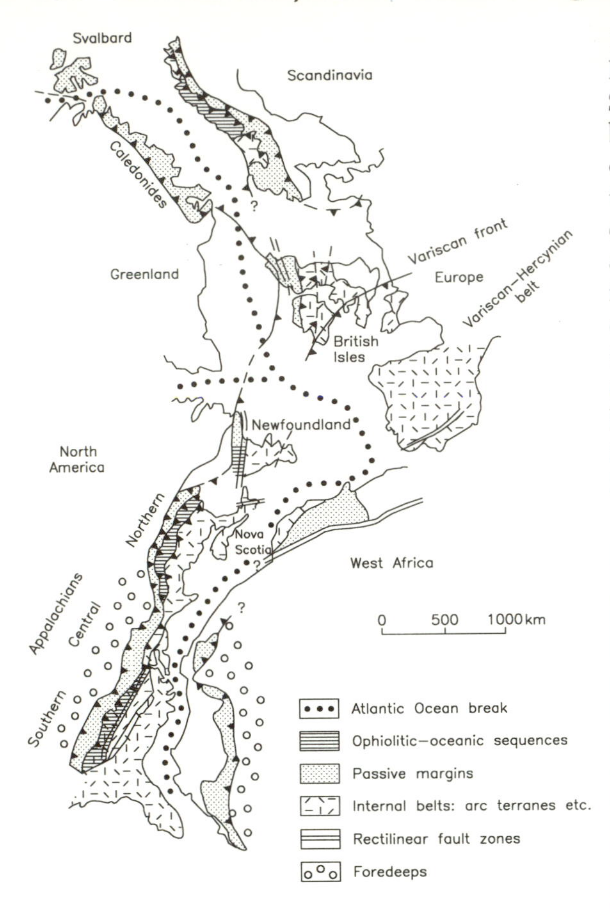


**Figure 16.1** The Appalachian–Caledonian orogenic belt.

In the British Isles (Figure 16.2), the suture marking the collision between Laurentia in the north and the southern microcontinent of Cadomia runs NE–SW across Ireland and through the Solway Firth dividing Scotland from England. We shall take as examples of the structural variation across this belt in the British Isles three important regions of Scotland: the Moine thrust zone at the northwestern margin, the Grampian Highlands metamorphic belt, and the Southern Uplands slate belt.

## THE MOINE THRUST ZONE

This zone marks the northwestern boundary of the Caledonian orogenic belt in Britain (Figure 16.3). It is up to 11 km wide and extends from Loch Eriboll, on the north coast of Scotland, to Skye, a distance of 190 km. The zone is one of the best known examples of a foreland thrust belt, and consists of several separate nappes resting on basal thrusts. These nappes die out laterally as the individual thrusts converge. The lowermost nappes are duplexes (see section 2.6), containing highly imbricated sequences of thin Cambro–Ordovician quartzites and limestones. These nappes have been thrust over the unfolded Cambro–Ordovician cover of the foreland, where it rests on Precambrian basement. The middle nappes contain Lewisian (early Precambrian) crystalline cores and exhibit large-scale recumbent folding. The uppermost

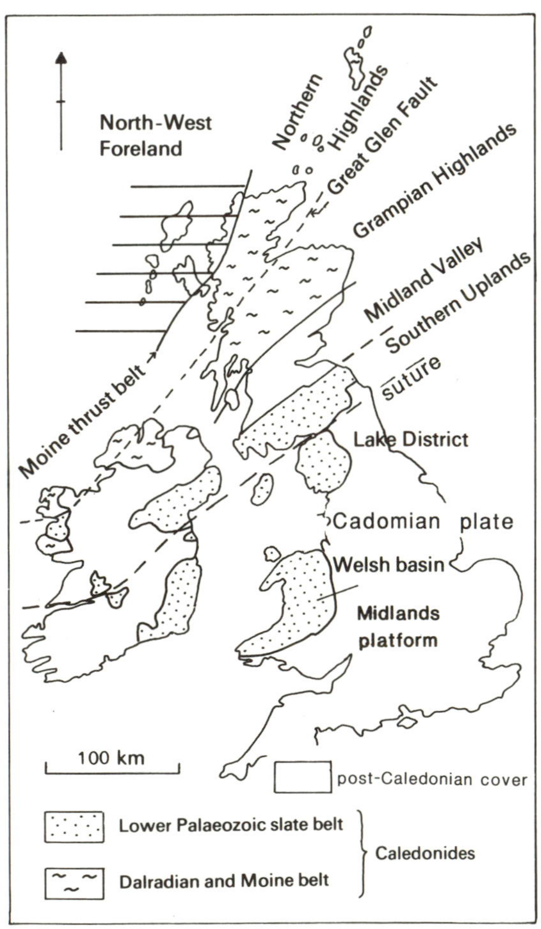


**Figure 16.2** Map of the British Isles showing the main tectonic zones of the Caledonian orogenic belt.

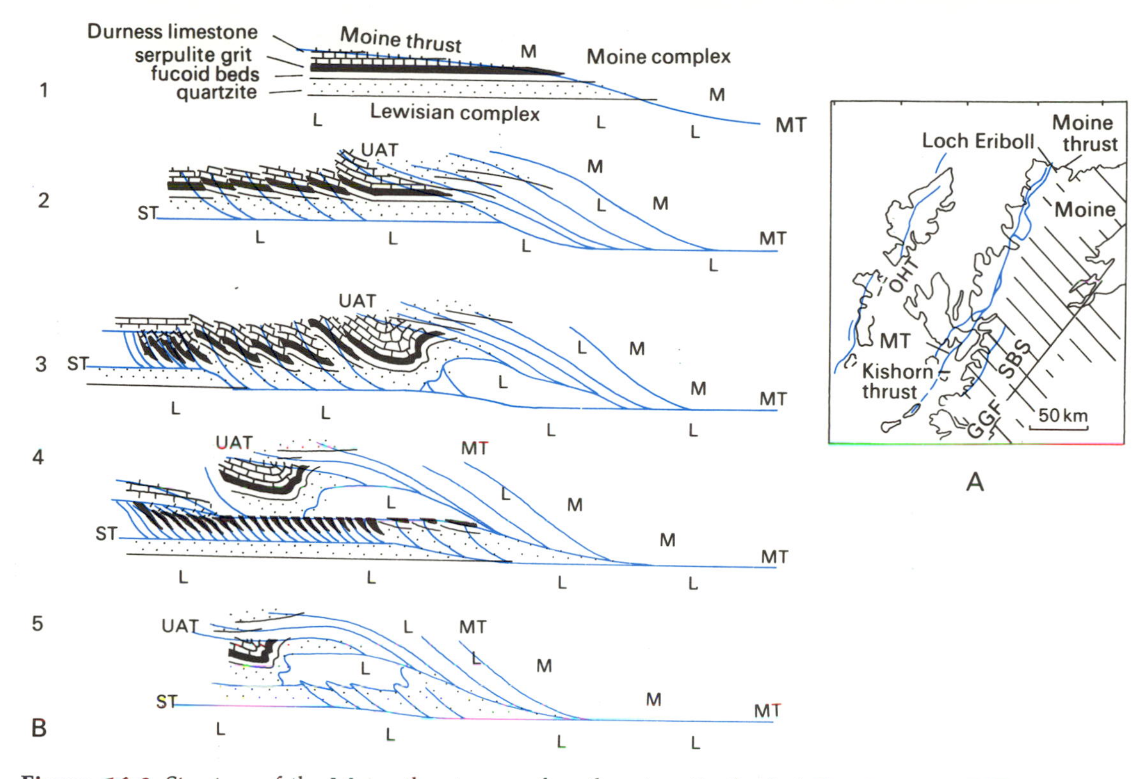


**Figure 16.3** Structure of the Moine thrust zone of northwestern Scotland. A. Location map. B. Diagrammatic sections across the north end of the thrust belt at Loch Eriboll showing stages in the evolution of the thrust zone. Stage 5 shows only part of the section. See text for explanation. MT, Moine thrust; OHT, Outer Hebrides thrust; ST, sole thrust; UAT, Upper Arnaboll thrust; SBS, Sgurr Beag slide; GGF, Great Glen fault. (After McClay and Coward, 1981, figure 7.)

nappe, which has a mylonitized base, is composed of the late Precambrian Moine complex. Lineations and strain markers on the thrusts indicate that the direction of movement on the thrusts has been towards the WNW, perpendicular to the Caledonian front. The total displacement on the thrust complex has been estimated at up to 100 km.

The evolution of the structure is summarized in Figure 16.3B. The thrusting is thought to have developed first in the east with the Moine thrust, which cut upwards and westwards from the basement into the Cambro–Ordovician cover. The sole (or basal) thrust developed next, in part following the base of the Cambrian. Continued movements caused an imbricate zone to form within the cover by the development of steep reverse faults which climb up from the sole thrust. At this stage, the sole thrust would form the floor thrust, and the Moine thrust the roof thrust, of a simple duplex structure containing the imbricate zone. Continued movements then caused the southeastern portion of the early duplex containing the Lewisian basement to climb up over the imbricated Cambro–Ordovician, forming overfolds as it did so to produce the middle nappes (e.g. the Arnaboll nappe – see stage 4 of Figure 16.3B). As the lower nappes moved westwards, they carried the upper nappes above them in piggyback fashion.

The Moine thrust zone formed in Devonian times during the closing stages of the Caledonian orogeny. It is younger than several other major thrusts and slides which formed during the early Ordovician Grampian orogeny of the Scottish Highlands, and has been attributed to the main

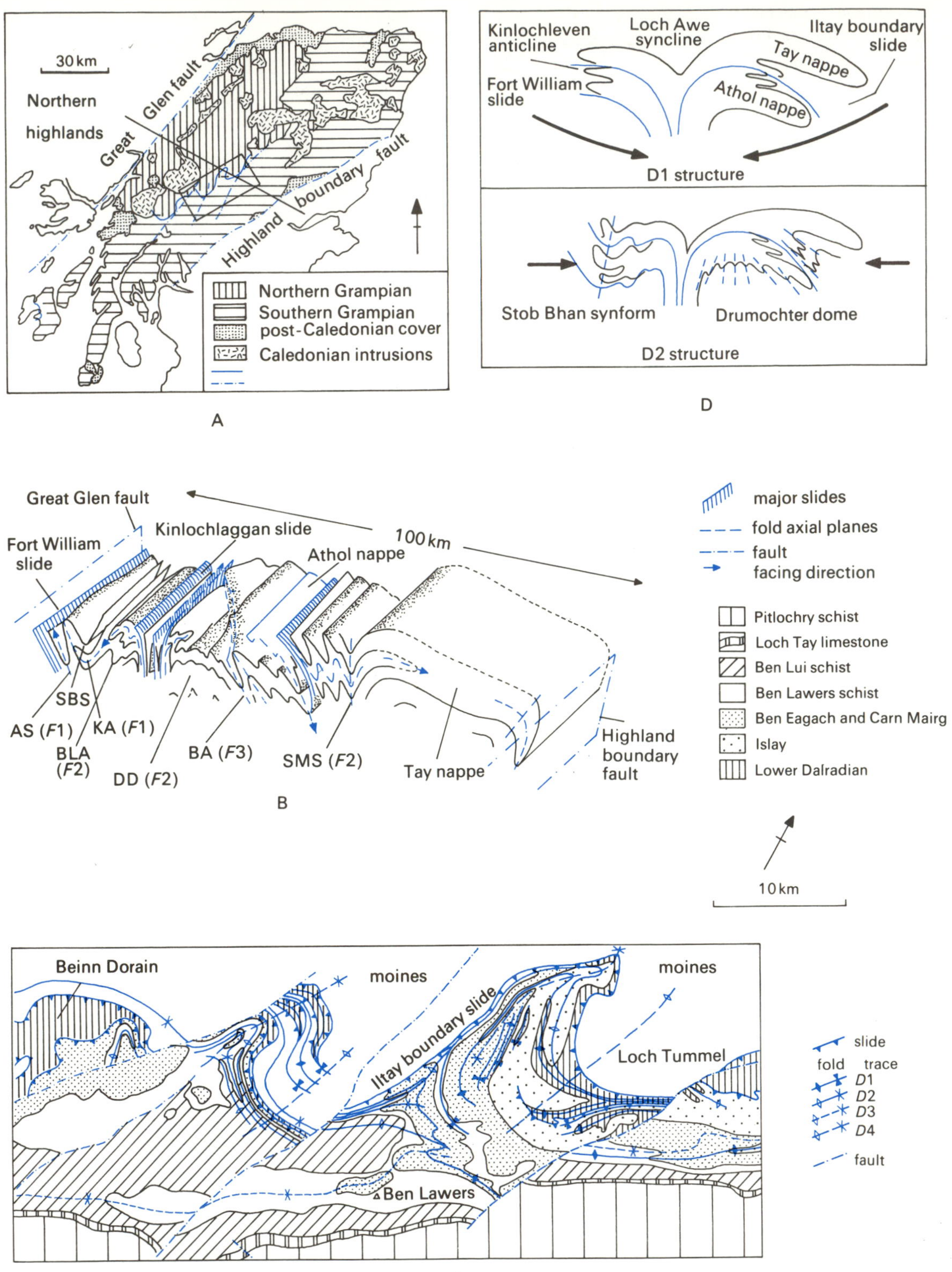

30 km
Northern highlands
Great Glen fault
Highland boundary fault
Northern Grampian
Southern Grampian
post-Caledonian cover
Caledonian intrusions
A
Kinlochleven anticline
Loch Awe syncline
Iltay boundary slide
Tay nappe
Fort William slide
Athol nappe
D1 structure
Stob Bhan synform
Drumochter dome
D2 structure
D
Great Glen fault
Kinlochlaggan slide
100 km
Fort William slide
Athol nappe
SBS
AS (F1)
KA (F1)
BLA (F2)
DD (F2)
BA (F3)
SMS (F2)
Tay nappe
Highland boundary fault
B
major slides
fold axial planes
fault
facing direction
Pitlochry schist
Loch Tay limestone
Ben Lui schist
Ben Lawers schist
Ben Eagach and Carn Mairg
Islay
Lower Dalradian
10 km
Beinn Dorain
moines
Iltay boundary slide
moines
Loch Tummel
Ben Lawers
slide
fold trace
D1
D2
D3
D4
fault
C

continental collision that occurred further south along the line of the Solway suture (Figure 16.2). There is an obvious analogy here with the Central Asia region discussed in section 15.4), in the sense that the most recent crustal shortening took place along the Moine thrust zone, which is situated nearly 300 km northwest of the collision suture.

## THE GRAMPIAN HIGHLANDS

The Scottish Highlands southeast of the Moine thrust zone contains the central metamorphic zone of the Caledonian orogenic belt, which exhibits the most complex structures. Many of the early investigations into structures and structural sequences took place there. The Grampian Highlands is that part which lies south of the Great Glen fault and is bounded in the southeast by the Highland Boundary fault (Figures 16.2 and 16.4A). Both these faults are considered to have acted as major sinistral strike-slip faults during the Caledonian orogeny.

The rocks of the Grampian Highlands consist of metasediments of the Grampian Group, which pass upwards into the late Precambrian to Cambrian Dalradian Supergroup. The sedimentary assemblage is thought to rest on stretched and thinned continental crust of the southern passive margin of the Laurentian continent. The major structure consists of large recumbent folds, tens of kilometres in amplitude, that appear to pass downwards into upright folds with much smaller amplitudes in a central steep belt (Figure 16.4B). These major folds are associated with slides of both thrust and lag type (see section 2.2). The major folds are accompanied by minor folds on all scales and by penetrative fabrics. The deformation took place under metamorphic conditions at depth and the structures are consequently ductile.

The main deformation, including possibly the first two phases of major folding, took place during the **Grampian orogeny** in Lower Ordovician times. This primary structure has been ascribed in one interpretation to severe compressive shortening at depth, causing an upwards and outwards flow of material squeezed from a 'root zone' (Figure 16.4D). It is suggested that this intense compression may have resulted from the collision of an island arc, situated in the region now occupied by the Midland Valley of Scotland (Figure 16.2), which moved towards the Highlands as a result of the subduction of a small intervening oceanic plate. Later strike-slip movements along the Highland Boundary fault have obscured the original relationships. It is thought that the extensional thinning of the continental crust beneath the thick pile of Dalradian sediments may explain the ease with which this zone subsequently became compressed. An alternative and more recent interpretation (Figure 16.5A) explains the structure by major crustal-scale ductile overthrusting from the south related to the obduction of a large-scale ophiolite nappe (i.e. one consisting of oceanic crust) (Figure 16.5B, stage c).

The major folds and slides are refolded by generally upright NW–SE major folds with wavelengths of several kilometres. There appear to be two or perhaps three generations of these folds, with varying orientations, that produce marked interference structures in the outcrop pattern. Each of these sets of major folds is associated with well-developed foliations and minor folds. Further deformation produced several sets of minor

**Figure 16.4** Structure of the Grampian Highlands of Scotland. A. Location map showing area of map C and line of profile B. (After Johnstone, G.S. (1966) British Regional Geology – The Grampian Highlands, HMSO, London, figure 3.) B. Structural block diagram across the Grampian Highlands to show the geometry of the major structures. AS, Appin syncline (F1); BA, Bohespic antiform (F3); BLA, Beinn na Lap antiform (F2); DD, Drumochter dome (F1/F2); KA, Kinlochleven anticline (F1); SBS, Stob Bhan synform (F2); SMS, Stob Mhor synform (F2). (After Thomas, 1979, figure 5.) C. Map showing complex interference structures between F1, F2 and F4 between Beinn Dorain and Loch Tummel (see A). (After Roberts, J.L. and Treagus, J.E. (1979) in *The Caledonides of the British Isles – Reviewed* (eds A.L. Harris, C.H. Holland and B.E. Leake), *Special Publication of the Geological Society of London*, **8**, 199–211, figure 1.) D. Simplified diagrams illustrating one interpretation of the primary deformation to explain: (a) generation of nappes and slides dunng D1, and (b) modification of D1 nappes by D2 major folds. (After Thomas, 1979, figure 6.)

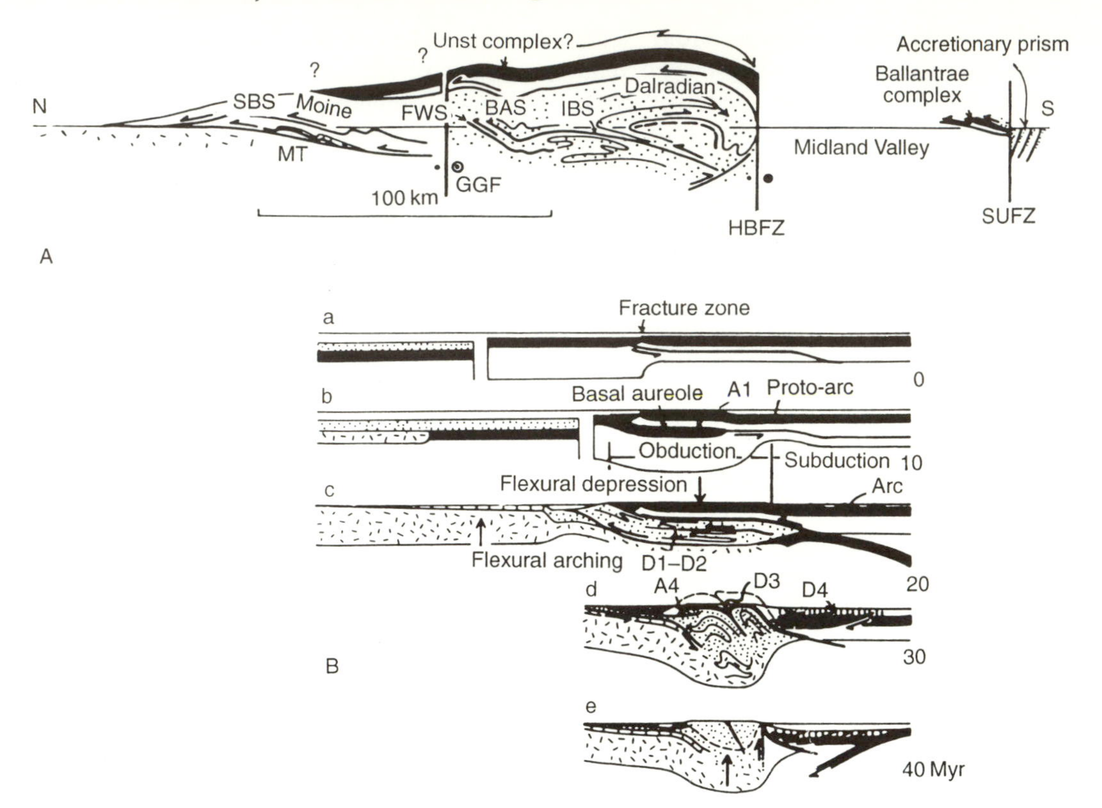


**Figure 16.5** A. Interpretative profile through the Scottish Highlands showing the main Grampian structures as NW-facing and the SE-facing folds (e.g. the Tay nappe) as backfolds (cf. Figure 16.4D). MT, Moine thrust; SBS, Sgurr Beag slide; FWS, Fort William slide; BAS, Ballachulish slide; IBS, Iltay boundary slide; GGF, Great Glen fault; HBFZ, Highland boundary fault zone; SUFZ, Southern Uplands fault zone. B. Sequential cartoons illustrating an evolutionary model for the Grampian orogeny: a–c, an ophiolite is overthrust on to oceanic lithosphere at a fracture zone and thereafter progressively obducted on to the continental rise and shelf, producing the D1–2 deformations in the Grampian Highlands; d, shortening and thickening of the sedimentary pile leads eventually to a reversal of subduction direction. (From Dewey and Shackleton, 1984, figures 2 and 3.)

structures, mainly crenulation cleavages and kink bands. Altogether eight separate phases of deformation have been recognized in the southwestern Highlands, but many of these are only of local significance. The complexity of the large-scale structure may be illustrated by a map of the Beinn Dorain–Schiehallion area, in the central part of the Grampian Highlands (Figure 16.4C). This map shows excellent examples of interference structures (see sections 3.7 and 10.3) attributed to the superimposition of four generations of major folds.

The later upright folds are ascribed to the compressional effects of collision, either in the Ordovician with the postulated volcanic arc terrane to the southeast (Figure 16.5B, stage d), or with the Cadomian plate in the late Silurian to Devonian period (Figure 16.5B, stage e). Probably structures representing both phases of movement are represented.

## THE SOUTHERN UPLANDS BELT

The Southern Uplands belt consists of a 60 km-wide zone of folded and faulted Lower Ordovician to Upper Silurian sediments and minor volcanics

(Figure 16.6A). The basal part of the sequence consists of oceanic basalts and cherts overlain by a very thick pile of greywackes and shales. Metamorphism is low-grade (sub-greenschist facies), but the mudstone lithologies are characterized by a slaty cleavage, which is well-developed in the south but less strong in the north. The beds strike uniformly NE–SW, parallel to the trend of the belt, and are generally steeply dipping to the northwest. Folds are intermediate in scale, with wavelengths in the range 5–50 m, and are typically asymmetric, verging towards the southeast (Figure 16.6B) (see section 3.6 for an explanation of vergence).

Major strike-parallel faults play an important role in determining the outcrop distribution. The belt consists of at least ten separate fault blocks, each containing sequences that generally young towards the northwest, but the belt as a whole becomes progressively younger towards the southeast.

This arrangement has led to the interpretation that the Southern Uplands represents a Lower Palaeozoic example of an accretionary prism or subduction complex. The combination of strike-parallel faults and asymmetric folds is thought to correspond to a steepened synthetic thrust belt, comprising a set of thrusts and related folds resulting from the underthrusting of an oceanic slab at a trench situated in the Solway Firth region in Ordovician times (Figure 16.6B, C).

Some of the faults show evidence of both dip-slip and strike-slip displacement, and it is thought that although many, if not all, of the faults originated as thrusts, some were re-activated as sinistral strike-slip faults after rotation into a steep attitude. Certain fault blocks are thought to have undergone considerable strike-slip displacement in relation to adjoining blocks, and the belt as a whole is regarded as a **displaced terrane** (see section 15.5) in relation to both the Midland Valley to the north and the English Lake District (i.e. the northern part of the Cadomian plate) south of the suture.

The slices with their bounding faults are believed to have been steepened by compression (Figure 16.6D), partly as a result of the continued subduction and partly due to subsequent continental collision. The folds within the slices are asymmetric, with long northwest-facing limbs and short southeast-facing limbs, often partly thrust out. The deformation is most intense in the ductile Moffat shale formation, which may have formed a detachment or sliding horizon (cf. section 2.6) for the underthrusting. The arrangement of the structures bears some similarity to the onshore Makran complex (see section 15.3 and Figure 15.6).

An interesting feature of the folding/cleavage relationship is that in the southern part of the belt, the cleavage strike is consistently clockwise of the fold axial planes (Figure 16.7). This arrangement is a strong indication that the cleavage developed under a transpressive regime, i.e. with a component of strike-slip displacement. Folds forming in such a regime would be oblique to the direction of strike-slip displacement (Figure 10.23) but with progressive compression would rotate towards it. In a sinistral regime, the later-formed cleavage would thus strike clockwise of the fold axial planes, i.e. at a larger angle to the strike-slip direction. This evidence adds support to the interpretation made from the faults that the later deformation of the Southern Uplands was significantly affected by sinistral strike-slip movements. The sinistral strike-slip component of the deformation is considered to reflect oblique convergence of the Cadomian and Laurentian plates, which resulted in partitioning of the deformation into compressional and strike-slip components.

Geophysical evidence shows that the deformed sedimentary sequence is underlain by continental basement and that a strong northwest-dipping reflector separates the two. This reflector comes to the surface along the line of the Solway suture, leading to the suggestion that the continental basement of the Southern Uplands consists of a wedge of the Cadomian continent which has underthrust the accretionary complex following the subduction of the intervening oceanic lithosphere.

A major problem with the tectonic interpretation of the Southern Uplands is deciding which structures are attributable to the late Silurian collision event and which to the earlier subduction process. It has been suggested that by Silurian times the two continents had come into contact and that Cadomia may have been underthrusting the Southern Uplands for much of the Silurian.

There is stratigraphic evidence that the central part of the belt was emergent during the Silurian (Figure 16.6C), which may be a response to this event. Thus the late-Silurian 'collision' may not have been the initial continent–continent collision but some other event which caused a steepening of the synthetic thrust belt. One possibility is that this second event was collision with the Lake District arc situated some distance south of the leading edge of the Cadomian plate. It is this second event that was presumably responsible for the development of the Moine thrust belt and for the later structures in the Grampian Highlands referred to earlier.

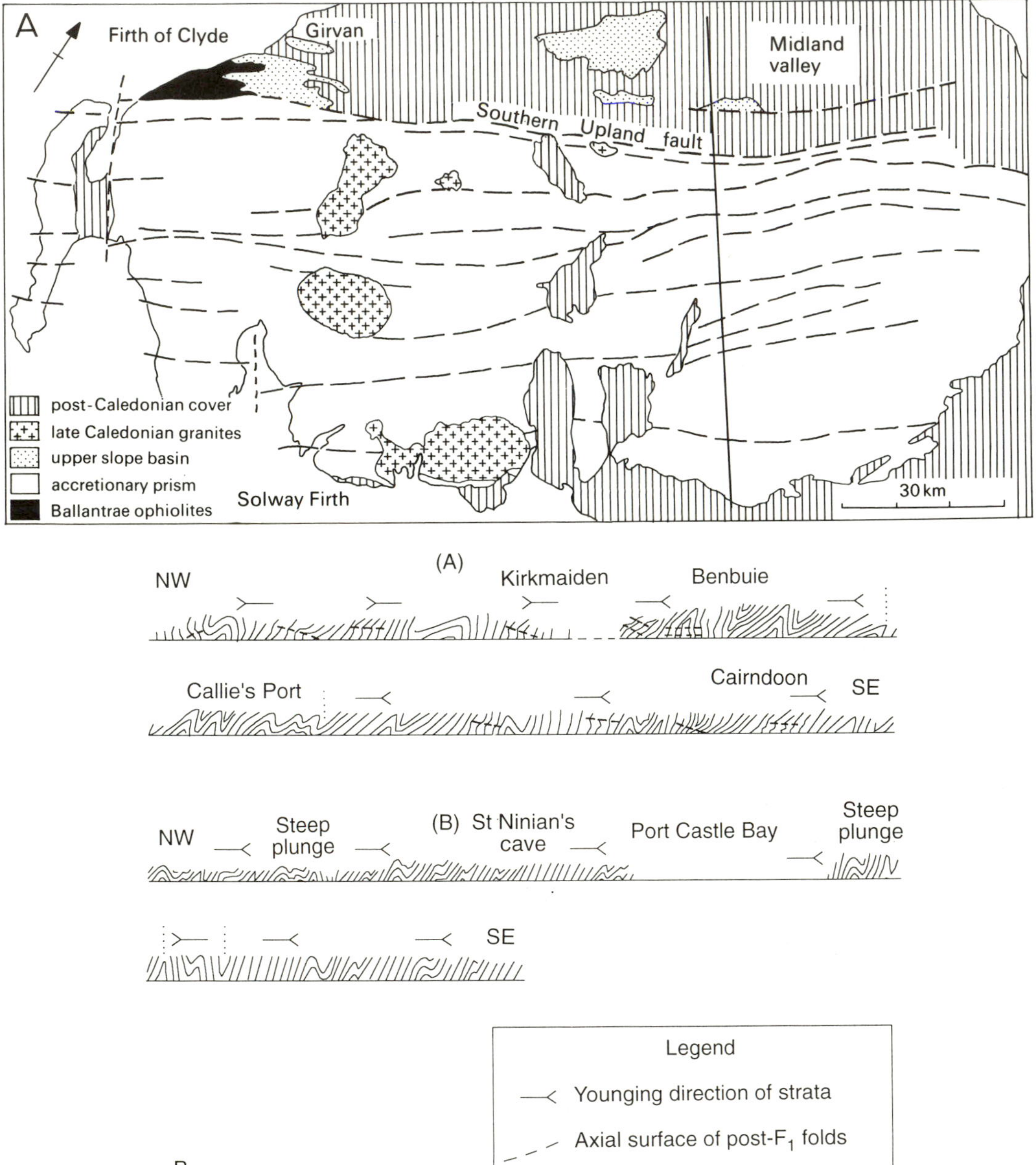

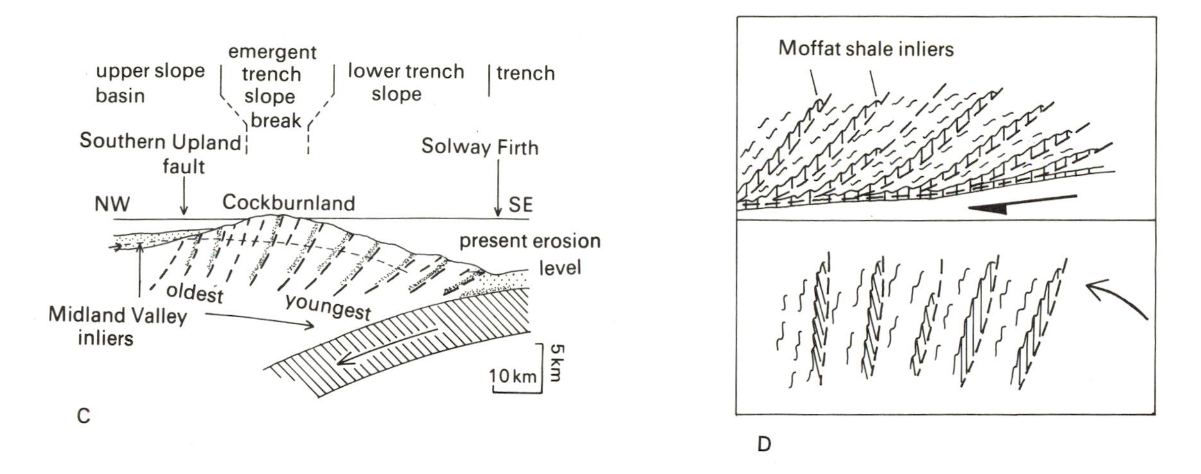


**Figure 16.6** Structure of the Southern Uplands of Scotland – a possible example of Lower Palaeozoic subduction tectonics. A. Simplified geological map of the Southern Uplands showing the principal faults. (After Leggett *et al.*, 1979, figure 1.) B. NW–SE cross-sections showing the pattern, scale and geometry of folds along the west side of Wigtown Bay, in the southern sector of the belt (see W on figure A). (After Stringer, P. and Treagus, J.E. (1981) Asymmetric folding in the Hawick rocks of the Galloway area, Southern Uplands. *Scottish Journal of Geology*, **17**, 129–48, figure 3.) C. Diagrammatic profile across the area to illustrate the underthrusting model. Each slice has been detached along the Moffat shale horizon (dotted). (After Leggett *et al.*, 1979, figure 6.) D. Diagram showing how the orientation of asymmetric folds and thrusts may be changed by steepening due to compression. Moffat shales, ruled. (After Eales, M.H. (1979) in *The Caledonides of the British Isles – Reviewed* (eds A.L. Harris, C.H. Holland and B.E. Leake), *Special Publication of the Geological Society of London*, **8**, 269–73, figure 2.)

Another problem concerns the palaeogeography of the Southern Uplands during the Ordovician. The tectonic model described above assumes northwestward subduction beneath an upper plate situated in the region of the present Midland Valley, which mainly contains younger rocks. The accretionary prism model assumes that an arc existed in this region, but no direct evidence of this arc exists. The Grampian orogeny, which affected the region to the north, occurred in the early Ordovician, while the earliest ocean-floor sediments of the Southern Uplands were being deposited. If the Grampian orogeny was caused by the obduction of oceanic plate from the south as suggested above and by collision of an arc situated on that plate, then the ancestral Midland Valley containing the arc must have been separated from Laurentia at this time by oceanic crust. It has been suggested that it was this collision that resulted in the initiation of the subduction that gave rise to the Southern Uplands (Figure 16.5B).

## 16.2 THE EARLY PROTEROZOIC EASTERN CHURCHILL AND NAGSSUGTOQIDIAN BELTS

The problems of reconstruction and interpretation become more acute when we consider the Precambrian orogenic belts. When we examine these belts, we find certain differences that set them apart from their more modern counterparts. While many belts show evidence of subduction-related magmatism, and ophiolite complexes marking collision sutures, as shown by the Alps or the Himalayas, for example, others reveal no evidence of plate collision by the removal of an intervening ocean, and matching of structures on either side of certain belts indicates that the cratons on either side may not have been displaced by more than a few hundred kilometres.

Some belts are interpreted as major shear zones where displacement has been strike-slip with little or no contraction across the belt. In the later

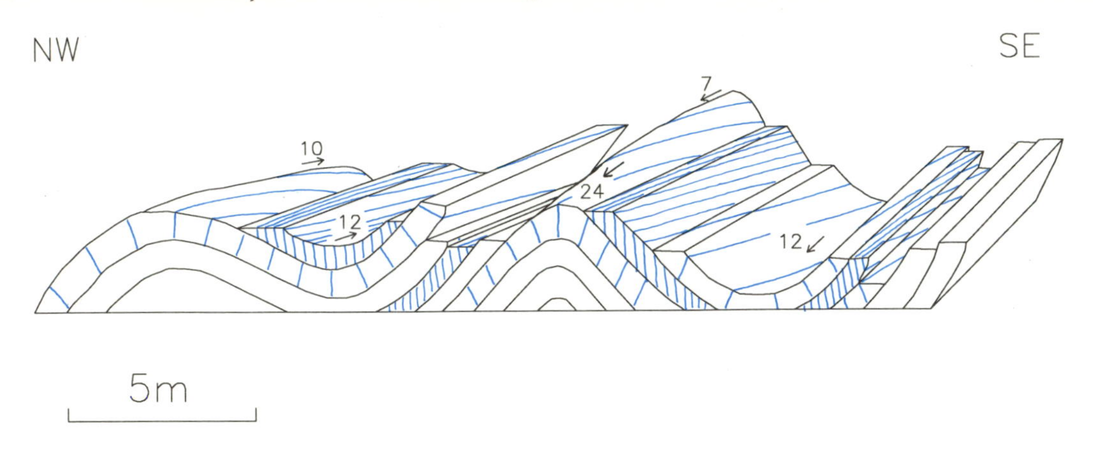


**Figure 16.7** Relationship between cleavage and folding in the southern part of the Southern Uplands. Note that the cleavage strike is displaced clockwise from the fold axes. Arrows and figures refer to plunge of folds. (After Walton, E.K. and Oliver, G.J.H. (1991) in *Geology of Scotland* (ed. G.Y. Craig), figure 7.6.)

Archaean and Early Proterozoic, major shear zones are very common and it has been suggested that much of the continental lithosphere, although able to transmit a uniform stress field, was in general more 'deformable' than is the case today.

We shall consider three examples of Precambrian tectonics: the Early Proterozoic of the Eastern Churchill Province of Canada and of the Nagssugtoqidian mobile belt in West Greenland, and the Archaean Superior Province of North America (Figure 16.8).

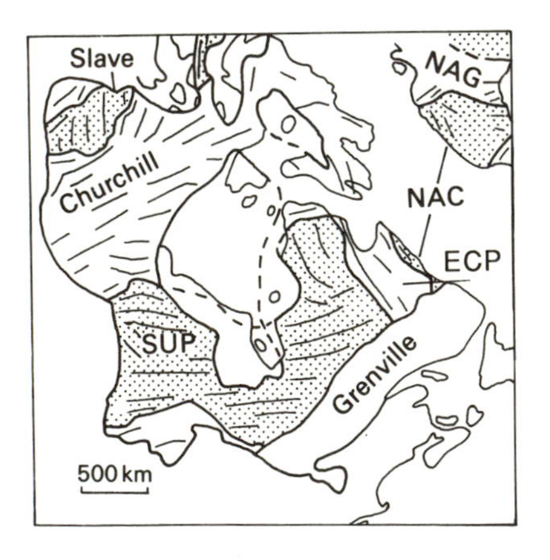


**Figure 16.8** Precambrian provinces of the Canadian shield. The map shows the Archaean cratons (dotted) and Proterozoic mobile belts of the Canadian shield. ECP, Eastern Churchill Province; SUP, Superior province; NAC, North Atlantic craton; NAG, Nagssugtoqidian belt.

## THE EASTERN CHURCHILL PROVINCE

This 400 km-wide belt is part of a continuous Early Proterozoic (c. 1800 Ma) orogenic zone surrounding the Archaean craton of the Superior Province (Figure 16.8). On its northeast side, it is bounded by the North Atlantic craton, which consists of the Archaean areas of eastern Labrador and southern Greenland. The belt consists of three parts: a western highly deformed sequence of volcanics and sediments of Early Proterozoic age (the New Quebec, or Labrador, zone), a Central zone consisting of reworked Archaean basement penetrated by granites, and an eastern part known as the Torngat belt which is dominated by a wide, steep ductile shear zone (Figure 16.9).

**Figure 16.9** A. Map of the Eastern Churchill Province. B. Simplified structural sections across the New Quebec orogen showing the westwards overthrusting. (After Wardle, R.J., Ryan, B. and Nunn, G.A.G. (1990) Labrador segment of the Trans-Hudson orogen: crustal development through convergence and collision, in *The Early Proterozoic Trans-Hudson Orogen* (eds J.F. Lewry and M.R. Stauffer), *Geological Association of Canada Special Paper*, **37**, figure 2.)

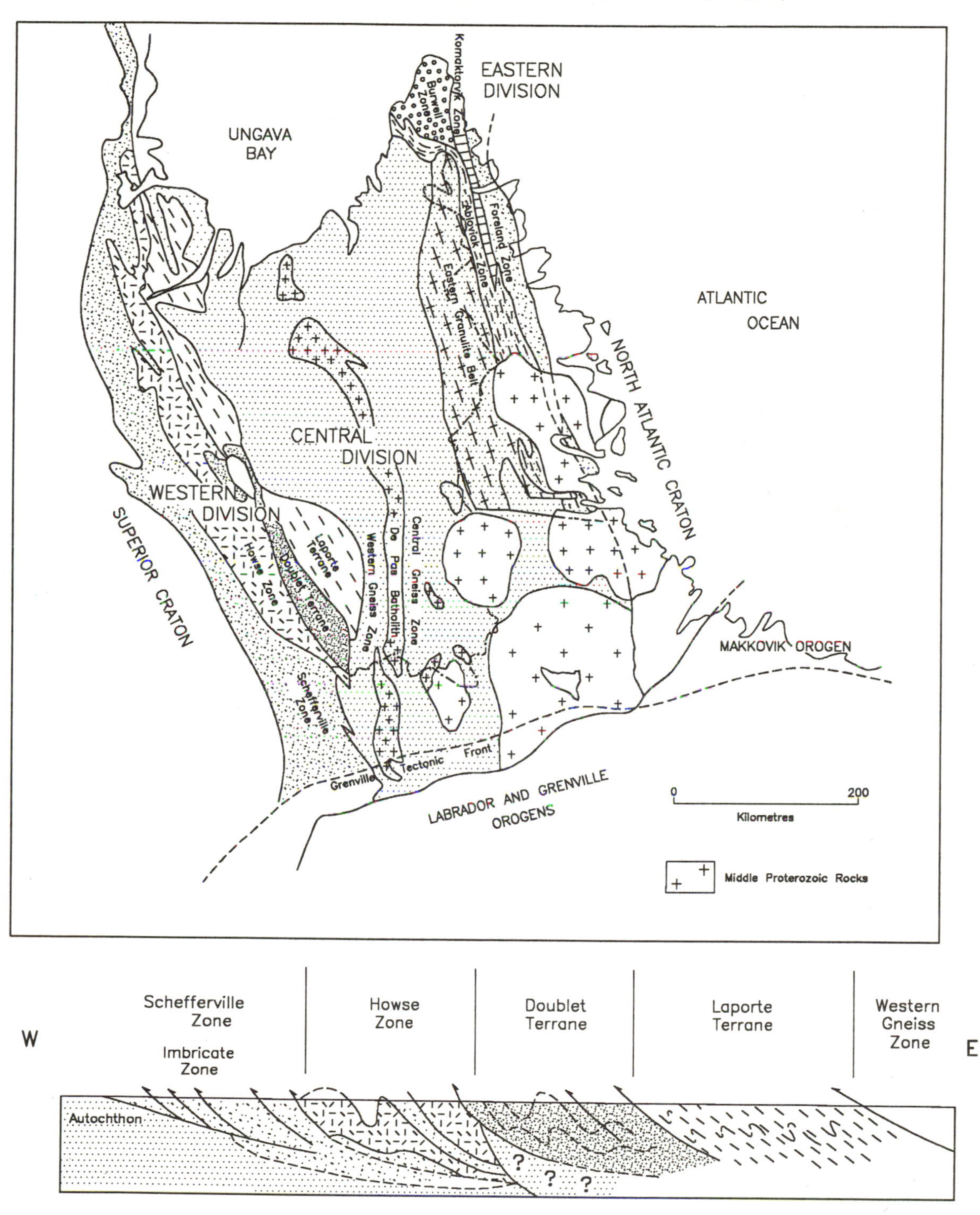

UNGAVA BAY
EASTERN DIVISION
Komaktorvik Zone
Burwell Zone
Abloviak Zone
Foreland Zone
Eastern Granulite Belt
ATLANTIC OCEAN
NORTH ATLANTIC CRATON
CENTRAL DIVISION
WESTERN DIVISION
SUPERIOR CRATON
Howse Zone
Doublet Terrane
Laporte Terrane
Western Gneiss Zone
De Pas Batholith
Central Gneiss Zone
Schefferville Zone
MAKKOVIK OROGEN
Grenville Tectonic Front
LABRADOR AND GRENVILLE OROGENS
0
200
Kilometres
Middle Proterozoic Rocks
W
E
Schefferville Zone
Imbricate Zone
Howse Zone
Doublet Terrane
Laporte Terrane
Western Gneiss Zone
Autochthon
?
?
?
50 km

The New Quebec zone is about 100 km wide and contains a thick sequence of 2.1–1.8 Ga-old metasediments and metavolcanics which have been overthrust westwards on to the Superior craton (Figure 16.9). The deposits thicken towards the centre of the zone, where they reach about 10 km in thickness. The folding and associated thrusting, dated at about 1.84 Ga, are strongly asymmetric, directed towards the southwest. Later dextral strike-slip shear zones occur locally.

The earliest deposits are continental clastic sediments resting on Archaean basement of the Superior craton. These are overlain by shelf quartzites, carbonates and iron formation, which are in turn succeeded by greywackes and submarine plateau basalts. This sequence is interpreted to represent the passive margin of the Superior continent.

The Central zone is about 100 km wide and consists of deformed and metamorphosed Archaean rocks intruded by granitic plutons thought to represent an Early Proterozoic magmatic arc produced by eastwards subduction beneath Archaean continental crust. The Central zone is therefore interpreted as the upper plate of a subduction zone and the New Quebec zone as a collisional orogenic belt.

The Torngat belt is between 75 and 200 km wide and separates the reworked Archaean rocks to the west from the North Atlantic Archaean craton to the east. This belt consists of a mixture of reworked Archaean rocks, Early Proterozoic granitic and dioritic plutons, mafic dykes and a band of highly deformed and metamorphosed metasediment, the Tasiuyak gneiss. The metasediments are thought to be derived from an Early Proterozoic accretionary prism, and the Torngat belt is therefore interpreted as a collisional suture zone between the Archaean block of the Central zone and the North Atlantic craton. The western edge of the craton is intruded by a suite of Early Proterozoic calc-alkaline plutons thought to indicate easterly subduction beneath the craton.

The earliest structures in the Torngat belt are west-vergent ductile thrusts, but these are succeeded by later movements on the steep sinistral Abloviak shear zone. The earlier ductile movements are accompanied by high-grade metamorphism dated at 1.87 Ga and are attributed to oblique collision between the North Atlantic craton and the Archaean block to the west. Later stages of the collision transformed the motion to sinistral transpression, steepening the previously formed structures and causing the sinistral strike-slip movements on the Abloviak shear zone.

Thus the plate tectonic interpretation of the Eastern Churchill Province views the origin of the belt in terms of two separate collisions, separated by about 30 Ma, with the Laurentian hinterland, the North Atlantic craton approaching from the southeast, and the Superior craton approaching from the southwest.

## THE NAGSSUGTOQIDIAN BELT OF WEST GREENLAND

The Early Proterozoic Nagssugtoqidian belt of Greenland (Figures 16.8 and 16.10) is the northwards continuation into West Greenland of the Eastern Churchill Province and consists of a 300 km-wide zone made up mainly of reworked Archaean gneisses but including some Early Protrozoic metasediments and volcanics (Figure 16.10A). The belt is bounded to the south by the Archaean North Atlantic craton and to the north by the Rinkian belt, also of Early Proterozoic age, with a small Archaean block at the boundary between the two belts.

**Figure 16.10** Main structural features of the Nagssugtoqidian mobile belt of West Greenland. A. Map of the Nagssugtoqidian mobile belt in West Greenland showing structural trends and main rock units. (After Escher *et al.*, 1976, figure 71.) B. Map of the Nordre Strømfjord shear zone in the central part of the Nagssugtoqidian belt (see A). Note how the sinistral shear zone is defined by the various lithological units and structural trends becoming aligned in a NE–SW belt in a zone of intense deformation. Black, amphibolites; ruled, metasediments; crosses, granitic and charnockitic intrusions. The blank areas are granitic to tonalitic gneisses. (After Olesen, N.Ø., Korstgård, J.A. and Sørensen, K. (1979) *Grønlands Geologiske Undersøgelse*, **89**, 19–22, figure 1.)

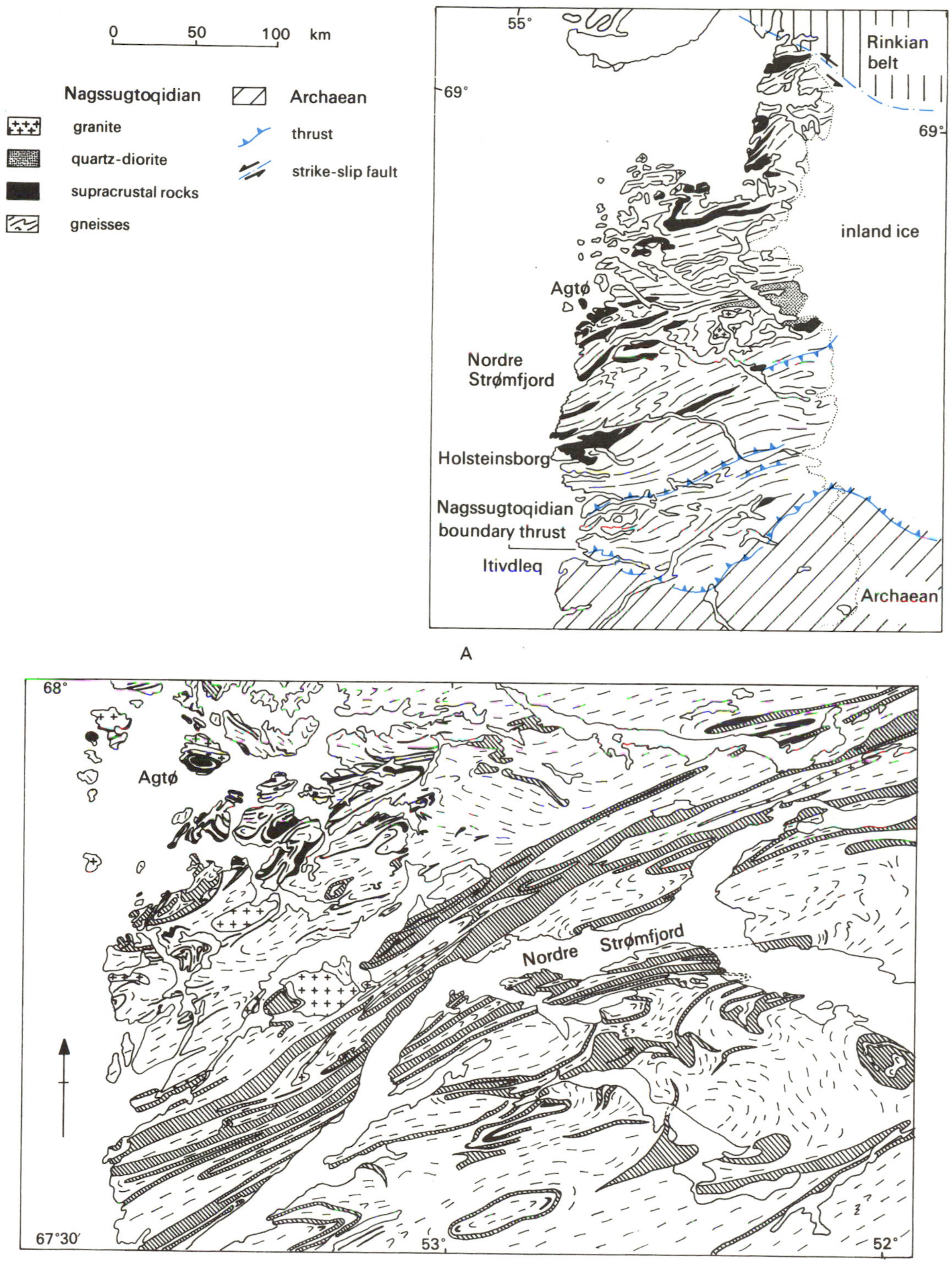
0
50
100
km
Nagssugtoqidian
granite
quartz-diorite
supracrustal rocks
gneisses
Archaean
thrust
strike-slip fault
55°
69°
Rinkian belt
69°
inland ice
Agtø
Nordre Strømfjord
Holsteinsborg
Nagssugtoqidian boundary thrust
Itivdleq
Archaean
A
68°
Agtø
Nordre Strømfjord
67°30′
53°
52°
B

The southern margin of the Nagssugtoqidian belt is a broad ductile shear zone (the Ikertoq shear zone) in which granulite-facies Archaean rocks like those south of the margin have been transformed into intensely deformed amphibolite-facies gneisses. An early phase of strike-slip displacement on the shear zone was followed by a later phase of more intense over-thrust displacement towards the south. This shear zone, which is about 50 km wide, grades northwards into a zone of rather less intense deformation, which is in turn followed northwards by another, narrower, shear zone at Nordre Strømfjord. The latter belt is steep, about 15 km wide and exhibits sinistral transcurrent displacement.

Figure 16.10B shows clearly how the structures and rock units north and south of the shear zone bend into parallelism with the shear zone, with individual rock bodies becoming greatly extended and thinned in the process. This shear zone was formed at granulite to amphibolite facies and is a classic example of a deep-seated ductile shear zone.

These belts of intense ductile deformation exhibit very high strains. Most of the fold axes are subparallel to a well-developed lineation that defines the $X$ strain axis and is subparallel to the direction of simple shear. Folds are typically similar in form, and boudinage of competent layers is common. Original discordances have been virtually eliminated and a marked parallelism of all previous structures, whatever their orientation, is characteristic. These features are all typical of high degrees of homogeneous strain produced by simple shear (see sections 8.8 and 10.6).

The sense of movement on the marginal ductile shear zone is consistent with the northwestwards convergence between the North Atlantic craton and the Laurentian hinterland described above in the case of the Eastern Churchill Province.

## 16.3 THE ARCHAEAN SUPERIOR PROVINCE

There has been much debate among geologists as to how far back in time the plate tectonic model can be extended and how it should be modified to explain the earlier history of the Earth. Some geologists believe that lithospheric subduction did not commence until later Precambrian times, and others that greater heat loss in the earlier Precambrian was accomplished by the presence of many small oceanic basins, with more rapid subduction, producing a higher rate of circulation of oceanic crust and smaller oceanic plates. It is often impossible to subdivide the Archaean crust into mobile belts and cratons, since evidence of mobility is so widespread. However, the Superior Province of Canada is a well-known and intensively studied piece of Archaean crust where the plate tectonic model has been applied, at least to the later part of the Archaean record (2.7–2.6 Ga).

The Superior Province (Figure 16.11) is a fragment of a larger Archaean continent assembled in the late Archaean between 2.7 and 2.6 Ga. It is composed of several subprovinces, all but one of which were formed in mid- to late-Archaean times (3.0–2.7 Ga). The exception is the Minnesota River Valley subprovince, in the southwest, which contains 3.6 Ga-old rocks and is thought to represent an older continent that may have accreted to the remainder of the Superior Province in late Archaean times. The subprovinces are of four types: volcano-plutonic (granite–greenstone) terrains, high-grade gneiss terrains, metasedimentary belts and plutonic terrains.

### VOLCANO-PLUTONIC TERRAINS

These are composed mainly of volcanic and sedimentary sequences intruded by abundant granitic plutons and exemplify the **granite–greenstone** type of Archaean terrain, where belts or irregular outcrops of volcanic and sedimentary rock (**greenstone belts**) alternate with large areas of granite or granitic gneiss. The larger greenstone areas often form basins with volcanic and sedimentary successions several tens of kilometres in thickness. The largest of these terrains is the Abitibi subprovince, which is a very rich mining area, having produced large quantities of gold, together with copper, zinc, silver and iron, and has been intensively studied. Geophysical evidence shows that the granitic plutons are tabular sheets 3–5 km thick and that the typical granite–greenstone type of crust generally extends to less

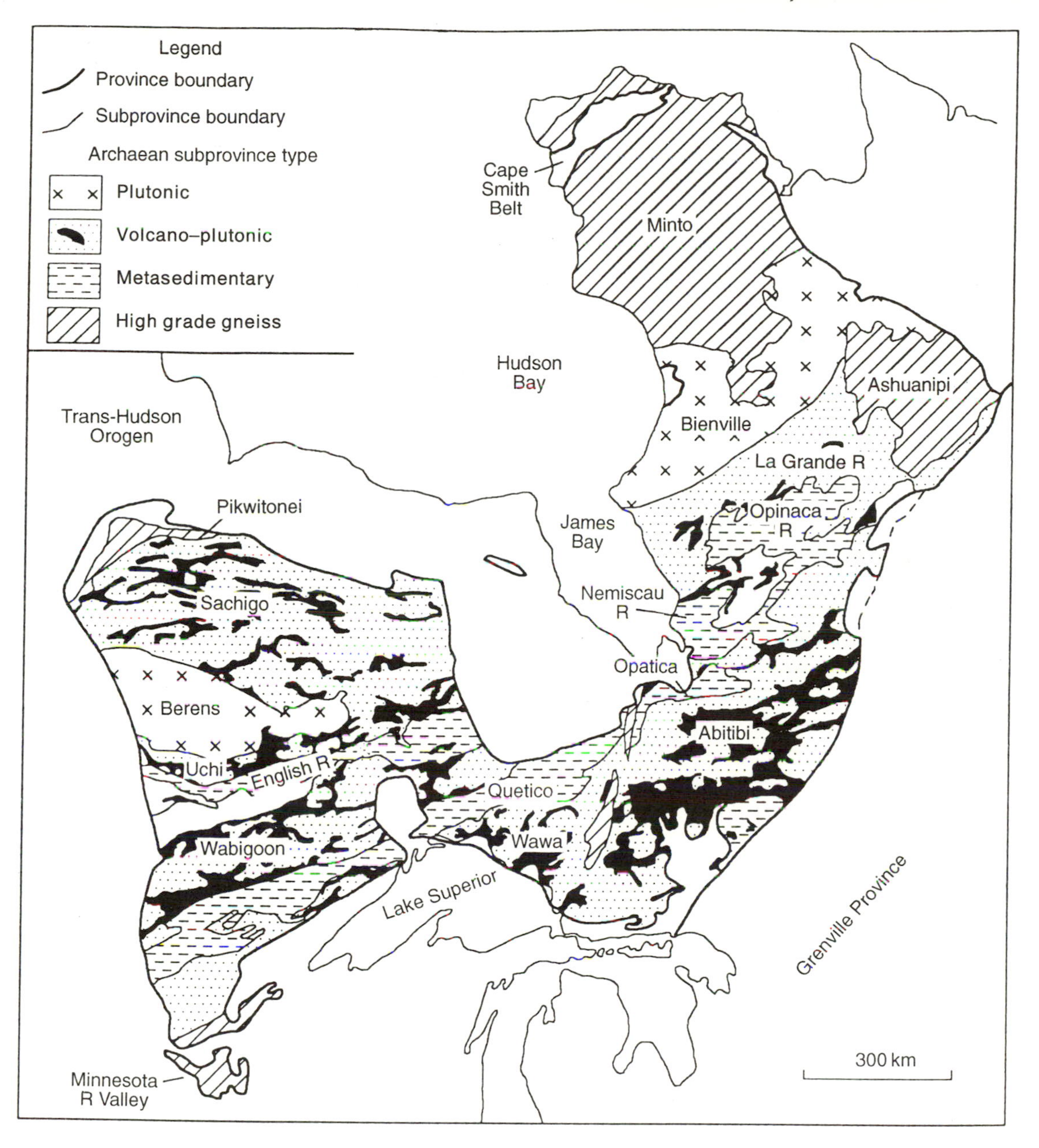


**Figure 16.11** Map of the Superior Province showing the distribution of the main types of terrain (see text) and the locations of the subprovinces. (After Card, 1990, figure 1.)

than 5 km depth, where it is replaced by the high-grade gneiss type.

The supracrustal assemblages within the volcano-plutonic terrains may be divided into the following types: (1) typical continental shelf sequences dominated by quartzites and carbonates; (2) mafic to ultramafic lavas interpreted as submarine lava plain sequences, and including minor banded ironstones, cherts and mudstones; (3) bimodal mafic/felsic volcanic sequences with abundant volcaniclastics interpreted as is land arc assemblages; and (4) clastic sequences composed of greywackes and conglomerates interpreted as representing clastic aprons around the volcanic island arcs. These terrains therefore represent environments that evolve from ocean-floor to island arc to continental through time. The earlier calc-alkaline granitic plutons of these terrains are considered to form the cores of the volcanic island arcs resulting from the subduction of oceanic crust. Later plutons are ascribed to the remelting of thickened granitic crust.

The structure of these terrains is dominated by major domal antiforms cored by granite and surrounded by narrow, tight, upright synforms containing the supracrustal rocks. At least two phases of deformation are recognized: an earlier phase producing major recumbent folds and thrusts with a strong foliation, and a later phase forming the more obvious synforms and antiforms that dominate the outcrop pattern and typically follow a strong regional ENE–WSW trend (Figure 16.11). Later structures consist of conjugate regional shear zone sets indicating north–south compression at a late stage in the development of the craton. The earlier deformations took place under generally low-grade metamorphic conditions, typically greenschist facies.

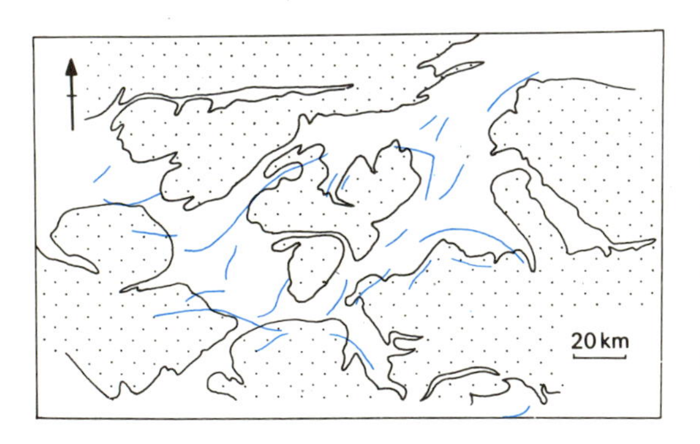


**Figure 16.12** Granite–greenstone relationships in the Kenora area. Note the shapes of the granitic plutons (dotted) and the variable trend of the fold axes (coloured lines), which 'wrap around' the pluton margins. (After Goodwin (1972) in *Variations in Tectonic Styles in Canada* (eds R.A. Price and R.J.W. Douglas), *Special Papers of the Geological Association of Canada*, **11**, 528–623, figure 6.)

In some parts of the Superior Province, the earliest deformation seen in the greenstone belts is variable in orientation and often follows the arcuate margins of the granites, suggesting that it may be related to the granite emplacement (Figure 16.12, see also section 12.4). Where the granite–greenstone terrain is comparatively unaffected by later deformation, the granites are approximately equidimensional in shape, and the early fold axes in the greenstone areas have a variable trend, often curving around the granite contacts. There are thus alternative explanations of the earlier structure: (1) that it is essentially subduction-related, as in the Southern Uplands example described above (section 16.1); (2) that it is related to the emplacement of the granitic plutons by either gravity-driven diapiric emplacement or a ballooning mechanism.

The volcano-plutonic terrains range in age from around 3.0 Ga in the north to 2.7 Ga in the south.

### HIGH-GRADE GNEISS TERRAINS

These consist of upper amphibolite to granulite-facies gneisses of both metasedimentary and meta-igneous derivation, together with various plutonic rocks. They are characterized by strong, polyphase, ductile deformation. Terrains of this type occur in the north (the Minto subprovince) and in the southwest (the Minnesota River Valley subprovince). These terrains are thought to represent uplifted lower-crustal rocks, broadly similar to those of the volcano-plutonic terrains but with a higher proportion of plutonic material.

### METASEDIMENTARY BELTS

Belts of this type include the English River and Quetico belts, which join eastwards to form the Opatica belt. They contain high-grade, mainly gneissose, metagreywackes and metasiltstones with minor mafic gneisses thought to be of volcanic origin. They are highly deformed and now dip

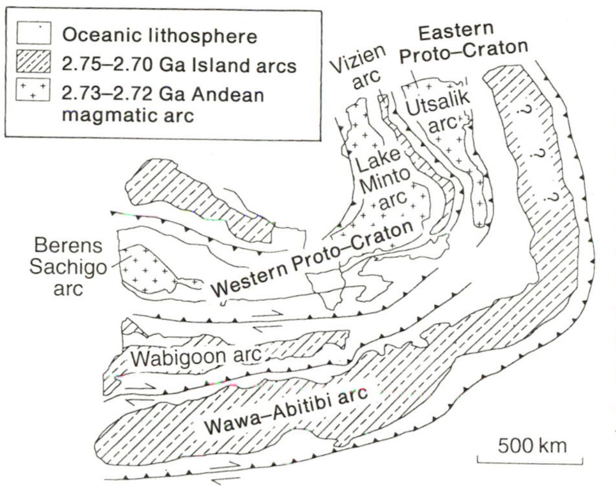


**Figure 16.13** Plate tectonic interpretation of the Superior Province at ~2.7 Ga. Map indicating the positions of the various arc terranes separated by subduction zones. Note that younger Andean-type magmatic arcs are postulated on continental crust of the earlier-formed Sachigo terrane. (After Percival, J.A., Stern, R.A., Skulski, T., Card, K.D., Mortensen, J.K. and Begin, N.J. (1994) Minto block, Superior Province: missing link in deciphering assembly of the craton at 2.7 Ga. *Geology*, **22**, 839–42.)

steeply northwards. These belts are interpreted as accretionary prisms marking collisional suture zones along which the terranes on either side have accreted.

## PLUTONIC TERRAINS

Terrains of this type are composed dominantly of tonalitic to granodioritic batholiths intruding gneissose basement of high-grade gneiss type, and include the Berens and Bienville subprovinces. They are interpreted as magmatic arc terrains developed on previously formed continental crust.

## PLATE TECTONIC INTERPRETATION

It has been suggested, on the basis of the distribution and ages of the terrains described above, that the Superior Province is composed of at least five magmatic arc terranes accreted together in the late Archaean, between 3.0 and 2.7 Ga (Figure 16.13). Individual terranes consist of typical ocean-floor/island arc sequences and contain calc-alkaline plutonic suites similar to, but more sodic than, those of subduction-related arcs in more modern times. Earlier accreted crust, such as the Sachigo subprovince in the north, were covered by platform sequences and later hosted Andean-type magmatic arcs. It is possible, but unproved, that the later, uniform, province-wide structures may have been formed in response to late-Archaean collision between the accreted arc complexes of the main part of the Superior Province and a Minnesota River Valley continent in the south.

## FURTHER READING

Card, K.D. (1990) A review of the Superior Province of the Canadian Shield, a product of Archaean accretion. *Precambrian Research*, **48**, 99–156.

Dewey, J.F. and Shackleton, R.M. (1984) A model for the evolution of the Grampian tract in the early Caledonides and Appalachians. *Nature*, **312**, 115–21.

Elliott, D. and Johnson, M.R.W. (1980) Structural evolution in the northern part of the Moine thrust belt, NW Scotland. *Transactions of the Royal Society of Edinburgh, Earth Sciences*, **71**, 69–96.

Escher, A., Sorenson, K. and Zeck, H.P. (1976) Nagssugtoqidian mobile belt in West Greenland, in *Geology of Greenland* (eds A. Escher and W.S. Watt), Geological Survey of Greenland.

Leggett, J.K., McKerrow, W.S. and Eales, M.H. (1979) The Southern Uplands of Scotland: a Lower Palaeozoic accretionary prism. *Journal of the Geological Society of London*, **136**, 755–70.

McClay, K.R. and Coward, M.P. (1981) The Moine thrust zone: an overview, in *Thrust and Nappe Tectonics* (eds K.R. McClay and N.J. Price), *Special Publication of the Geological Society of London*, **9**, 241–60.

Roberts, J.L. and Treagus, J.E. (1977) Polyphase generation of nappe structures in the Dalradian rocks of the southwest Highlands of Scotland. *Scottish Journal of Geology*, **13**, 237–54.

Thomas, P.R. (1979) New evidence for a Central Highland root zone, in *The Caledonides of the British Isles – Reviewed* (eds A.L. Harris, C.H. Holland and B.E. Leake), *Special Publication of the Geological Society of London*, **8**, 205–11.

Wardle, R.J., Ryan, B. and Nunn, G.A.G. (1990) Labrador segment of the Trans-Hudson orogen: crustal development through convergence and collision, in *The Early Proterozoic Trans-Hudson Orogen* (eds J.F. Lewry and M.R. Stauffer), *Geological Association of Canada Special Paper*, **37**, fig. 2.

Windley, B.F. (1993) Uniformitarianism today: plate tectonics is the key to the past. *Journal of the Geological Society of London*, **150**, 7–19. [Discusses orogeny and the possible application of plate tectonics throughout geological times from the Archaean to the present.]

# APPENDIX: STEREOGRAPHIC PROJECTION

Stereographic projection is a graphical method of portraying three-dimensional geometrical data in two dimensions, and of solving three-dimensional geometrical problems. In geology, the method is used mainly for solving problems involving the orientations of lines and planes in crystallography and structural geology. Such problems involve the angular relationship between lines or planes rather than their spatial relationships. Planes plot as lines (usually curved) and lines as points on such a projection.

The orientation of a plane (e.g. bedding, foliation or crystal face) is represented by imagining the plane passing through the centre of a sphere of radius *R*, the **projection sphere**. The plane intersects the sphere in a circle with radius *R* called a **great circle**. The two-dimensional projection of the plane is produced on a horizontal plane through the centre O of the projection sphere (Figure A.1A). In structural geology, the lower hemisphere only is used for projection, whereas in crystallography, the upper hemisphere is used. In the following account, the lower-hemisphere projection is described. Upper-hemisphere projections will be mirror-images across a vertical plane of symmetry. For lower-hemisphere projection, each point on the lower half of the great circle is connected to the point T where the vertical line through O cuts the top of the projection sphere, producing a set of lines TA, TB, etc. that intersect the projection plane in an arc known as the **cyclographic trace** of the great circle. The cyclographic trace of the horizontal plane (i.e. the projection plane) is called the **primitive circle.** The cyclographic trace of a vertical plane is a straight line through O. If the primitive circle is given geographic coordinates, then the orientation of any specified plane may be represented. Thus Figure A.1B shows the cyclographic traces of a set of great circles with a N–S strike dipping at 10° intervals from 10° W, through the vertical, to 10° E.

In practice, stereographic projection is carried out by means of a protractor termed a **stereographic net** or **Wulff net** (Figure A.2A) which gives the cyclographic traces of the complete set of

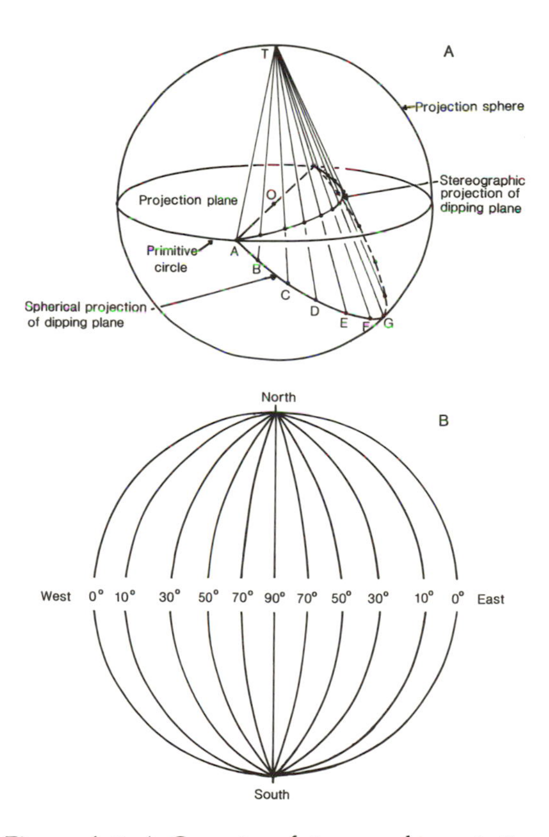


**Figure A.1** A. Geometry of stereographic projection. B. Stereographic projection of planes dipping due east or due west at the various indicated angles. (A and B after Hobbs *et al.*, 1986.)

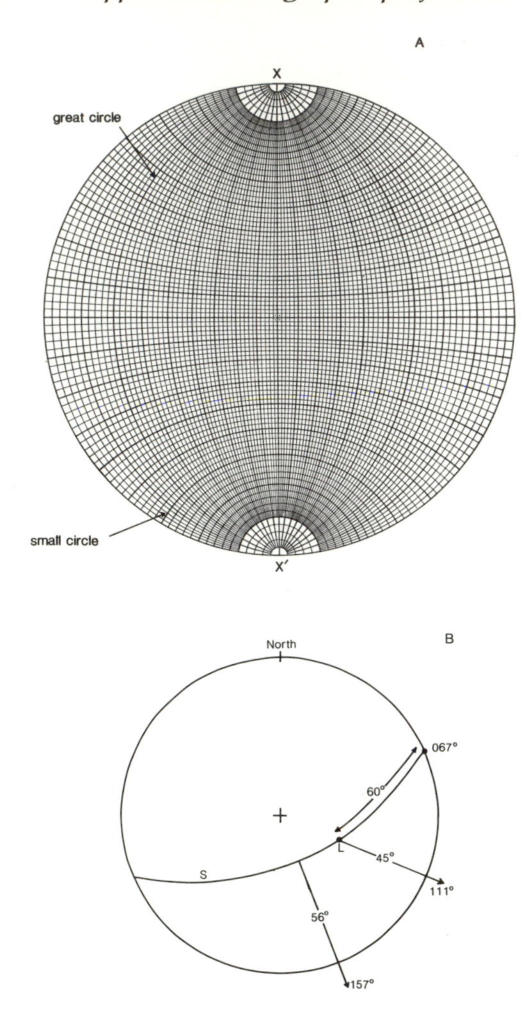


**Figure A.2** A. Wulff or equi-angular net. B. Stereographic projection of a plane with a strike of 067° dipping towards 157°. A lineation, L, pitches 60° NE in this plane. The plunge of the lineation is 45° towards 111°. (After Hobbs *et al.*, 1986.)

great circles at 2° intervals about a common axis XX′. The net also gives the cyclographic traces of a family of circular cones about the same axis XX′, again at 2° intervals. These traces, termed **small circles**, are used to graduate angular distances along the great circles. Equal angles on the surface of the sphere project as equal distances on the net. Thus the Wulff net is often termed the equal-angle net in contrast to the Schmidt, or equal-area, net described below.

The procedure for plotting geometric data is as follows. A piece of tracing paper is laid over the stereographic net and fixed with a pin through the centre O. The primitive circle is marked, and given geographical coordinates. The plot produced on the tracing paper is termed a **stereogram**. To plot the position of a plane with strike bearing (azimuth) of 067°, the position of 067° is marked on the primitive circle, by counting clockwise from N, and the stereogram rotated over the net until 067° corresponds with the axis XX′ of the net. If the plane has a dip of 56° to the SE, the appropriate great circle trace is found by counting 56° inwards from the perimeter (primitive circle) towards the centre (Figure A.2B). Alternatively, if the dip azimuth is given (56° to 157°), the position corresponding to 157° is marked, and the net rotated until the point situated 90° clockwise of X corresponds with 157°. The appropriate great circle trace is found as before. It should be noted that planes dipping to the east plot in the right side of the stereogram and vice versa.

To plot the position of a line with a plunge of 45° on a bearing of 111°, say, a similar procedure is followed. The bearing 111° is found as before. Since a plunge is measured in a vertical plane, one of the two vertical planes (represented by straight lines on the net) must be rotated to correspond with the position of 111° on the stereogram. The plunge angle is counted inwards from the perimeter to a point situated 45° along the line towards the centre (Figure A.2B).

Angles between any two lines may be found by placing their projected points on a common great circle trace. Angles between any two planes may be found by plotting the poles (normals) to the two planes and obtaining the angle between the poles. The **pole** to a plane is the line (point) situated 90° from the centre of the great circle trace, measured along the vertical trace.

Many geometrical problems in structural geology can be solved using various combinations of these simple procedures. For example, the plunge of a fold axis may be obtained by plotting the intersection of great circles representing two planar fold limbs, or representing various positions on the fold surface.

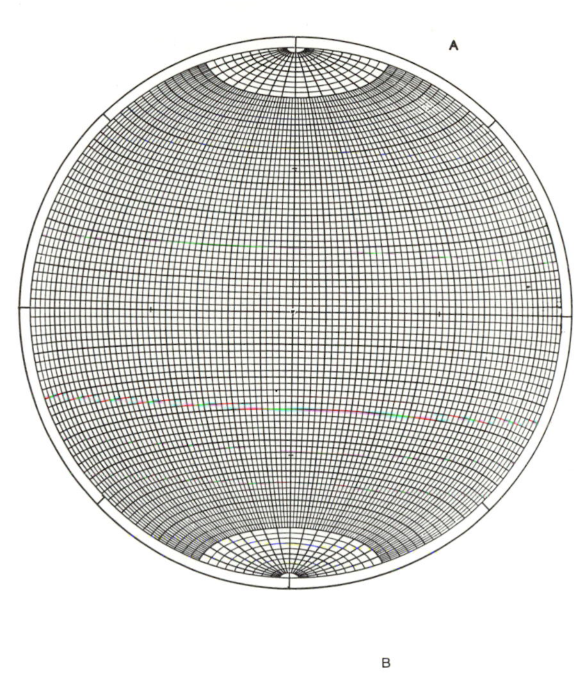


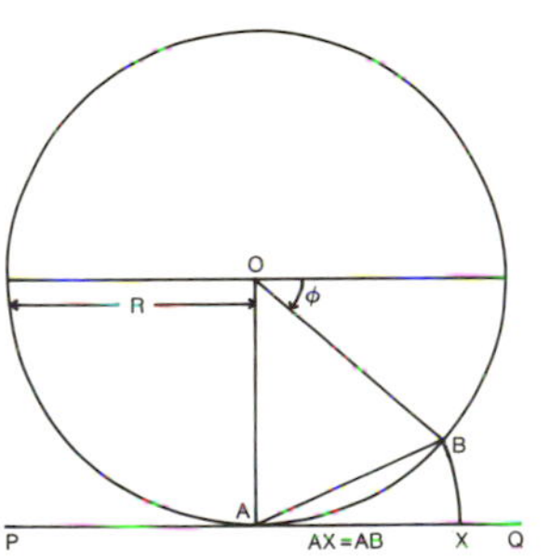


**Figure A.3** A. Schmidt net. B. The principle of the equal area projection. O is the centre of the projection sphere of radius *R*. OB is the trace of a plane dipping at angle $\phi$. X is a point on the equal area projection of this plane, PQ. (After Hobbs *et al.*, 1986.)

Poles to a cylindroidal fold surface plot on a great circle trace, the pole of which represents the fold axis.

In the statistical analysis of the distribution of structural elements, it is more important to be able to compare the density of lines (points) in different areas of the projection sphere than to compare angles between lines. For the former purpose, a different type of projection is used, in which equal areas on the projection sphere plot as equal areas on the stereogram. The net used for this purpose is termed an **equal-area net** or **Schmidt net** (Figure A.3A). The method of projection is illustrated in Figure A.3B. Equal angles on the projection sphere are not projected as equal angles on the stereogram, and the cyclographic traces of the great circles are fourth-order quadric curves. The Schmidt net may be used to plot planes and lines, and measure their angular relationship in exactly the same way as the Wulff net, and is often employed by structural geologists in preference to the latter.

For details of procedures in plotting and analysing density distributions of structural or fabric data see Turner and Weiss (1963) and Phillips (1971).

## FURTHER READING

Hobbs, B.E., Means, W.D. and Williams, P.F. (1986) *An Outline of Structural Geology*, 2nd edn, Wiley, New York.

Phillips, F.C. 1971: *The Use of Stereographic Projection in Structural Geology*, Edward Arnold, London.

Turner, F.C. and Weiss, L.E. (1963) *Structural Analysis of Metamorphic Tectonites*, McGraw-Hill, New York.

# INDEX

Pages where terms are defined or explained are shown in **bold**. Illustrations of the usage of structural terms in the wider geotectonic context will be found in the higher page numbers, from p. 141 onwards.